PRAISE FOR JANE BROX'S FARM BOOKS

"Arresting. With a poet's sensibility and an essayist's search for meaning, Brox gives us keen and sensuous observations of the land ... with compassion, honesty, and restraint."

—*Boston Globe*

"Brox's double-charged language turns family drama into something bigger [in writing] so arresting that even the most urban reader feels the author's sense-memory as his own."

—*Chicago Tribune*

"A clear-eyed and beautifully written celebration of a place where family, community, history and nature all still matter deeply. It's been a long time since I've read anything that's caught and held my attention and admiration quite the way this wonderful memoir of life in the Merrimack Valley has."

—HOWARD FRANK MOSHER

"A loving, precisely written evocation of a New England place and its people ... reminiscent of Thoreau in its exactness and breadth of implications."

—from the judges' citation for the PEN/New England Award

ALSO BY JANE BROX

Here and Nowhere Else: Late Seasons of a Farm and Its Family

Five Thousand Days Like This One: An American Family History

Clearing Land: Legacies of the American Farm

Brilliant: The Evolution of Artificial Light

Silence: A Social History of One of the Least Understood Elements of Our Lives

IN THE MERRIMACK VALLEY

Here And Nowhere Else

Five Thousand Days Like This One

Clearing Land

IN THE MERRIMACK VALLEY

A FARM TRILOGY

Jane Brox

FOREWORD BY SUZANNE BERNE

AFTERWORD BY THE AUTHOR

BOSTON

GODINE NONPAREIL

Published in 2024 by
GODINE
Boston, Massachusetts

Cover: 1924 Soil Survey of Middlesex County, Massachusetts, published by the United States Department of Agriculture, Bureau of Chemistry and Soils in cooperation with the Massachusetts Department of Agriculture, Division of Reclamation, Soil Surveys, and Fairs.

Here and Nowhere Else and *Five Thousand Days Like This One* were first published by Beacon Press. *Clearing Land* was first published by North Point Press.

LIBRARY OF CONGRESS CATALOGING-IN-PUBLICATION DATA
Names: Brox, Jane, 1956- author. | Brox, Jane, 1956- Here and nowhere else. | Brox, Jane, 1956- Five thousand days like this one. | Brox, Jane, 1956- Clearing land. | Berne, Suzanne, author of foreword.
Title: In the Merrimack Valley : a farm trilogy / Jane Brox ; foreword by Suzanne Berne ; afterword by the author.
Description: Boston : Godine , 2024.
Identifiers: LCCN 2024000072 (print) | LCCN 2024000073 (ebook) | ISBN 9781567928181 (paperback) | ISBN 9781567928198 (ebook)
Subjects: LCSH: Brox, Jane, 1956—Family. | Brox family. | Family Farms–Merrimack River Valley (N.H. and Mass.) | Farm life–Merrimack River Valley (N.H. and Mass.) | Merrimack River Valley (N.H. and Mass.)–Biography.
Classification: LCC CT274.B779 B765 2024 (print) | LCC CT274.B779 (ebook) | DDC 974.4/5 [B]–dc23/eng/20230321
LC record available at https://lccn.loc.gov/2024000072
LC ebook record available at https://lccn.loc.gov/2024000073

First Printing 2024
Printed in Canada

AUTHOR'S NOTE

I finished writing *Clearing Land*, the third book in this trilogy, in 2003. Between then and now much has changed, not only in fact but also in my heart and mind. If I were writing these books now, I'd write with a different understanding of our family and of the story I have unfolded here. I would also, in some instances, use different words and phrases to describe historical figures and events. I haven't altered the narrative, nor have I changed my original language, since these books are not only the story of our farm and the valley, but also a record of the time in which they were written.

CONTENTS

CLEARING LAND

Foreword

I FIRST MET Jane Brox at a small dinner party in Cambridge almost thirty years ago and I remember being startled when she said she lived on a farm. People in the Boston area sometimes visited farms, usually to go apple picking or to take their children on hayrides, but no one I knew lived on one, certainly not people I sat next to at Cambridge dinner parties. We got to talking and I was even more surprised to discover that Jane worked on the farm, that it had belonged to her family for generations, and that she had recently published a book about it.

Part of my childhood, too, was spent on a farm, a horse farm in Virginia that dates back to the eighteenth century, although my parents weren't farmers, or even from Virginia—they were both Midwesterners—and the stables were rented out, so other people took care of the horses; you could say we were more like tenants. Yet when my father gave up what had been a romantic idea of country life and moved us to the city, I was heartbroken. I had been used to so much space and light, to air that smelled of cropped grass and creek mud, to the sounds of horses shifting in their stalls and cattle lowing as night came on, and then to seeing fireflies spark the darkness. Now everywhere I turned there were houses, people, cars. It took me a long time to get my bearings, and I'm not sure I ever forgave my father. Some of this I recounted to Jane at that long-ago dinner party, probably thinking to establish my farm credentials, and she listened politely, probably thinking what I knew about farming wouldn't fill my wine glass. But I sensed kindred feeling. When I read *Here and Nowhere Else* a few weeks later, I glimpsed what that feeling might be: the apprehension that, in one way or another, every loved place is bound to be lost.

In those pages Jane describes coming back to live on her family's farm in Dracut, Massachusetts, in the Merrimack Valley, after years spent on her own. Her parents are elderly and her father unwell; a troubled older brother is struggling to manage the farm and its orchards; she'll have a separate house to herself on the property, where she can write. It makes sense to return. And yet as deeply as she belongs to where she was raised, land that her grandparents, Lebanese immigrants, bought in 1902, land that has literally fed the family ever since, it's a complicated arrangement, as any "going home again" must be.

Her ambivalence, her desire to distance herself from her family's way of life even as she tries to maintain it, results in an extraordinarily sensitive account of how a small working farm etches itself into the people who live there. Location takes on a different meaning when your surroundings decide who you are and what you do. Take the view from Jane's windows: "apples to the north, peaches to the south, a field of corn to the west, blueberries to the east." This is a world's whole compass, both sustaining and confining, the horizon clearly demarcated in every direction, enclosed even by sounds: "the steady progression of the mower, the jangling harrow, and the high-pitched fan of the orchard sprayer."

Work and chores define every season and every day, as they have for a hundred years, a cycle of planting and harvesting that feels too relational to be routine. Particularly when it comes to apples, which have names like a street full of neighbors—Baldwins, Cortlands, Astrakhans, Gravensteins—and the orchards themselves, planted and tended with paternal care by Jane's father. Inside the barn, the history of farmwork is tangible, held in "long-stored harnesses reeking of must, and all the laid-aside hoes and scythes, their hafts salted and polished by labor." It's impossible to read a description like this without being struck by the sensibility that registered it, one that also seems "salted and polished" by the long discipline of paying close attention to whatever, at the moment, is worth looking at.

In his five-volume *Modern Painters*, John Ruskin argues that without developing a habit of conscious attentiveness, you risk going through most of life blind: "objects pass perpetually before the eyes without conveying any impression to the brain at all; and so pass actually unseen, not merely unnoticed, but in the full clear sense of the word *unseen*." It is, therefore, in Ruskin's full clear sense of *seen* that Jane describes her mother making jelly from trellised grape vines by the farmhouse: "It is high summer and the air doesn't move even with the windows thrown open. Her forehead is damp, her hair curls with the steam. The metal spoon is stained purple and lies on the stove top." Or describes her father cutting up a Blue Hubbard squash, using a cleaver to get through the thick rind, his aim "as deliberate as a stonecutter's, and the two halves cradle away to reveal a thick wall of orange flesh, its surface mottled with white seeds."

The restraint and precision in these instances are not just beautiful; they grant significance to what we might otherwise never notice, they change *how* we see. I'm thinking of what happens whenever I walk out of a museum after spending an hour or two looking at paintings and sculpture. For a few minutes, everything I encounter on the street comes into vibrant focus and seems charged with meaning and energy. This is, in fact, what the world looks like—something I always forget until I'm shown again.

Here and Nowhere Else and the two books that follow—*Five Thousand Days Like This One* and *Clearing Land*—offer a combined master class in full, clear seeing, not only of the fast-passing proximate world, but also of worlds gone by. *What am I from*? is a question that breathes naturally out of these pages as Jane trains her gaze into the farm's recent history and then farther and farther back, into centuries before her father's father signed the deed to it, when Indigenous people grew corn there before being driven away by European colonists, and earlier, all the way back to the glacial drift that left behind the stony, alluvial soil of New England's fields.

Her mother's parents also settled in the Merrimack Valley, arriving from Italy in search of more than what they left at home; but though they lived just six miles from the Brox farm, their lives in the city of Lawrence might have been conducted on another pole. Her maternal grandfather, a weaver, was among the wave upon wave of immigrants who helped produce most of the nation's cloth in the nineteenth and early twentieth centuries in Lawrence's textile mills. With him as a touchstone, Jane researches and imagines her way into mill life along the river: the brutal pace of the factory floor, the choking air, the narrow tenements, the strikes, the hopefulness and holiday moments, the mass displacement when the mills began to close.

This is Jane's own legacy: the immense agricultural, industrial, and social shifts that have shaped the Merrimack Valley, both sides of her family, and the farm itself, rapidly being encircled by suburban housing developments as neighboring farmers sell up and leave. Her ambitions for this legacy are quietly grand. At the beginning of *Clearing Land*, she writes:

> As if understanding can alleviate loss, I am trying to place our own time within the story of cultivation. I hope constraint amplifies, that by giving shape to the stories, to the persistent half-lives of

> the vanished who roam even on stinging winter days, I can see more clearly where we belong in the accumulation of beliefs, ideas, violence, necessities, and desires that have determined this country.

Thoreau also recognized that constraint can amplify, and Jane puts herself firmly in his tradition by tuning into the profound complexities of one small American place to hear historical reverberations otherwise drowned out by the thundering cascade of the past. *What am* I *from?* becomes *What are* we *from?*

It's the intimacy of Jane's findings that present the most passionate argument for asking that question. I'm thinking of a moment when she is sifting through her father's farm accounts after his death and comes across a list of cows he owned fifty years earlier. She's struck by "the plainness of the names, the absence of flourish," names like Mule Ears, Long Legs, Little Jersey. Followed by a kind of shock:

> You can't always explain how or why an aperture opens, how you fall into a larger seeing, but reading those names, in his hand, after he had gone, was the first time I believe I truly understood how far he'd traveled in his life ... all of a sudden I saw the deep hard work of it, the aspiration to be gotten bit by bit ... He had tried growing a fig tree, walnuts, champagne grapes. The same land plowed for so long out of necessity may have been working its way toward desire. Still, if you were to

> ask him why he farmed, he'd simply say he liked
> to see things grow.

What these books bring home is that we are from the work and care of other people, those who came before us, the vanished as well as the closely remembered. People who, amid the often exhausting necessities of their lives, had aspirations, desires, their dreams of figs and champagne, who hoped for more than they probably achieved; but the lucky ones appreciated whatever they managed to grow out of their allotment on this earth—and what they grew they gave. Any larger seeing of history is an account of giving as much as losing or taking. To have traveled far within a small landscape, to have worked it hard and deeply, is to be part of the tremendous ongoing human project of cultivation, whether you are a farmer, a factory worker, or a writer.

Lost worlds are never entirely lost, that's the other discovery, so long as there is someone of great compassion and acuity to show you a larger way of seeing them. Out of the fields and hills, the orchards, rivers and glacial till of Jane Brox's Merrimack Valley, my own childhood fields and hills come nearer, along with my own father's dreams when he brought us to live in what was, for him, a new country, and then left it to follow other dreams, each time hoping, as we all do, for something more.

Suzanne Berne
2023

HERE AND NOWHERE ELSE

Late Seasons of a Farm and Its Family

For my family

Their care keeps us awake.
Their starry fear
shines in the tree of our speech.

—JOHANNES BOBROWSKI

Prologue

WE LIVE THIRTY miles inland along the old road to the coast, a road laid down on an early wagon track, which followed the Indian trace—a long day on sure feet giving way to oxcarts that took half the week to return from the sea with their burdens of salt hay. Now the coast is a scant hour's drive along Broadway, North Lowell Street, Main, and also River Road and Water Street, since the way sometimes skirts the muscular currents of the Merrimack, which salts at Newburyport, and pours into the Atlantic.

By the end of its journey the river is almost two hundred miles from the cold rose of its source in the White Mountains. In many places it flows through a yielding channel older than the ice ages. Where it courses over stubborn ledge, where the rock wears away at an incremental pace, are the waterfalls that were once the gathering places and fishing grounds of the Algonquin tribes.

Merrimack is their word. River of sturgeon, swift water, strong place.

To the south of our fields and woods the river flows broad and braided and eastward. It is the strong line of our landscape. The low-lying hills slope towards its channel, and every vein of water—the icy melt and the murk, the mineral-rich source and the field-drained runoff, waters that taste like metal on the tongue, waters redolent of balsam, and some of smoky tea—every vein drains into the Merrimack. Even the cut of the road depends on the river, since it nearly ghosts the water's course though we sleep beyond earshot of a steady current, lulled instead by the fine-tuned motors of the night freight trucks that approach and then pass.

1
House

"BUILT BY A frugal man in a frugal time," my father likes to say of my house, and even now, when the plumbing or heating gives out, he'll mutter that it would have been cheaper simply to raze the place and build a new one. So many surfaces made level with shims, and supports that have had to be supported. Newspapers and *The Saturday Evening Post* pulled from gaps in the walls, and the back stoop propped up by loose bricks. When a robber kicked in the side door, and I came home to the September air swirling through the rooms, I found two rusted nails in the hingehole where one screw should have been. Such jerry-rigged things turn up as unexpectedly as Indian flints, and like those flints, I imagine I'll never stop finding them.

The frugal man was Ben Prentiss, and the time, about 1920, just after he had married Clara. He bought two acres of land from my grandfather on what was then the edge of the farm, though by the time my father bought the house from Ben's oldest daughter, he had also bought up land to the east, and the two acres stood in the middle of our farm. Ben sited his house on a small rise near the road, which gave him a good view of all the surrounding acreage. It's the natural site for any house, even now, though it's a house no one would build these days. Half of it is a ramble of unheated storage rooms leaning down to the shed, and the hallways take up almost as much space as the six rooms I live in, rooms small enough to keep to their singular purpose. In the bedroom there's only enough space for the bed and dresser; the dining room, four chairs and a table; the study, my desk.

I easily take up the place myself, and I have trouble imagining how Ben, Clara, and their two daughters lived here, with the one closet, and even a slight cough carrying through all the rooms. The shallow well must have gone dry summer after summer. In my first years here—I had moved to this house when I returned to the farm to live—there were months when I couldn't water the garden and take a shower in the same day. The new driven well took no time at all to drill—an oversized rig backed in the driveway and bore through three hundred feet of ledge in a scant six hours. The source puts out so much water my brother Sam uses it to irrigate the tomatoes. Drawn out of the deep bedrock, clear, hard, cold, it courses through the skewed elbows of Ben's copper pipes—old pipes, still, and if I am away for a while the water comes up rust.

Of the grace Clara likely added to the house—the wallpaper streaming with flowers, the frilled curtains, the china cups—not a vestige remains, though much of what she planted in the yard still comes back every spring. She must have loved blues and purples—there are two islands of bearded irises among the meadow grasses, and violets spread under the west window. Lilacs bloom come May and soften the frame of the house—so upright, white, and unshuttered.

She had died, the two daughters had grown and moved on before I was born, so I have always called this house "Ben's house." As roaming children, in our boredom we would stray here. Most of what I remember now is a dark kitchen steeped in smoke and cooking oil. One of my cousins remembers shards of colored glass strewn on the floor of the shed. My sister, chickadees pecking at the sills where Ben left peanut butter for them. We all remember his pale, lank frame, and a fine nose with steel-rimmed glasses pinched on them. There were few words, though I don't think he was ever unwelcoming.

Most of the evidence of living was in the kitchen by then—the chair camped at the woodstove, and a table beside it with the day's paper, a flowered cup and plate, and a steel boning knife. The stove itself could be kept low since it only had to throw off enough heat to warm the one man rocking there. The quiet rush of fire contained in metal. No flame to draw his gaze. My mother would invite him for Sunday dinners, and sometimes he'd come, but only after long coaxing.

Dust in a slant of light. The grass springing back in his footfalls. Things that must have been stored in the shed while Clara had been living he had moved into the house—a good supply of cordwood, bundles of newspapers, empty bottles and cans. After her death he saved everything, right down to wood scraps and string. When he himself died, what his oldest daughter didn't take down to Virginia my father and uncles sorted and cleared away, and they ended up burning much of it on a windless spring morning in 1980.

What in these rooms would Ben and Clara remember as theirs? The plank of bird's-eye maple that runs half the length of the dining room floor—surely they'd know that—and the rough pine shelves in the kitchen, smoothed now from years of paint. But here is another life, with its own dishes, and no curtains, and a Turkish rug in the front room. The floorboards are sanded and glossy with urethane; the walls, painted a creamy white. On the fine shelves that must have held their keepsakes, my books are a square and solid weight. I think the place would be strange to them, they who had lived here so long their window glass rippled like quiet water, and their view softened with the distortion.

It was a view that must have had its familiarities from the start—both Ben and Clara grew up nearby, and always lived bounded by the Merrimack to the south and the low hills to the north, mill cities to the east and west. At times I try to imagine this house containing a whole long life. It feels small when I think about it, confining, and sometimes comforting—as it would, I suppose, to anyone who has lived in more places than she ever could have imagined. I can't gather everything under one roof anymore. Not enough space, enough time, no way to close the distances between all the things I love.

Sometimes there's no sound where I think a sound should be. The air warms unexpectedly, a dense fog slips into the orchard behind the house, and my ears strain for the three foghorns—I used to count to eleven between them—warning down the bay. And then I think how I should be hearing the drawn-out cries of oceangoing birds, and smell salt on the thickened air, and my chest tightens. How far away they now feel, the old friends who are strangers to the slope of these hills, and the names on the nearby graves.

By the time I moved back to the farm, the windows in this house were shaky in their frames. The ropes on the lead weights had frayed, the glazing compound had gone dry. That first winter I listened to the glass knock against the wood on fierce nights, and to the reedy wind. I could feel the drafts in every room. As soon as the warm spring days came, I replaced all the old windows.

Now that the glass is new, the far hills don't shimmer through it, and the skies don't have their distortions—and they won't—not for another half-century at least. The glass is so clear, it sometimes feels as if there's nothing at all between me and what I look out on, which is our apples to the north, peaches to the south, a field of corn to the west, blueberries to the east. Or I read the compass again: my parents to the south, my aunt to the east, my brother to the west.

When the irrigation pond turned the color of coffee, my father knew it had to do with the trench Gil Johnson had dug just to the west of our land. Runoff from the trench flowed into the brook, which in turn carried silt to the pond. My father knew this without ever seeing the trench or the runoff beyond his property. He was just used to the habits of the land. They were in his bones—the drainage patterns and the cold spots where frost lingered, and where the marginal soil was only good for a side crop of gourds. He knew it was nearly impossible to grow lima beans and that he'd sometimes lose his peach crop. It was the accumulation of his lifetime and of his father's life, too.

All through the growing season, that accumulation was right in front of him. He couldn't move to the right or to the left of it. It told him when to irrigate, and when to harvest, and when to call it a loss. Even after the workday had ended it was there at the dinner table. "You should eat some tomatoes," he'd say, "they'll be gone in a few weeks, there's been so much rain." Or we'd be trying a new variety of corn: "Sweeter than Seneca Star," he'd say, to no one in particular. After dinner he'd tune in the weather one last time and doze in the easy chair until bedtime.

His longest days were during the months of August and September, and in an exhausted hour he'd wish he'd done something different with his life. His brothers had gone on to other occupations—electrician, teacher, contractor, mechanic—and their lives seemed buoyant in the world, designed orbits. And he: nothing other than his father's son.

But such regrets slowed as the season slowed. November came and the moon rose over bare maples. The boughs of the apple trees were sprung and still. Winter rye was up in the cornfields. Only cabbage and cauliflower remained standing—they could be picked even after the first frosts, and then the ground was too hard to turn over. My father had more energy in the longer evenings and began a book that would take him all winter to finish. He liked James Michener and would read *Alaska* one year and *The Caribbean* the next. He'd garner a handful of facts out of all the pages, and the more he read the less accurate was his recollection of those facts. What he brought to the dinner table was something half-recalled and misremembered. All the same, he relished such talk, as if he was just singing for the song, since the words weren't tied to silt or runoff or coffee-colored water, since for once his words flew away as he said them.

MY MOTHER'S EAR is tuned to the distance, to a hawk's cry—asthmatic, from out of the blue. Or the sound of someone driving a wedge into wood: *chink, chink, chink* for so long, then his arm must give a little since his aim falls off its mark and the sound off its pitch. Of course, there's the engine noise of a tractor making its way across one of the fields. A change in the engine marks a change in the land or the end of a furrow. And every day at noon the truck door yawns and snaps shut, then the door to the mudroom, and my father is there at the kitchen entrance. He scrubs his hands at the sink and turns on the TV to get the same weather he heard at breakfast. Then he turns down the volume on the set and lets the news run on while they have their lunch.

They eat at the kitchen now, sitting kitty-corner to each other at the small round table. They look at the salt, the pepper, the bread, the way in winter they'll look at the fire when they talk to each other. They talk about us children, of course, of money and errands to run, the health of their friends. There's silence, and sometimes one of the old stories.

In other years we ate lunch in the dining room or on the long porch table, since there were four kids and Pete, my father's right-hand man. My father and Pete had their own talk: codling moth, cultivator, rye spreader—they were going over the morning's work or mapping out the work still to be done in the day. There was some borer in the next piece of corn, the tomatoes needed water, the orchard should be mowed. My mother served lunch, and the conversation went on around her.

SHE HAD NEVER in her life made a pie, so those first Tuesdays of her marriage my mother would walk up the road to watch them bake in the farmhouse kitchen. She'd stand to the left of my grandmother and my aunt with her eyes fastened on their hands as they cut the lard into the flour. The steel tines of the pastry cutter were long off their gleam, and the wooden grip was every bit as smooth as the scythe's. Straight into the flour they'd cut, and then they'd give the bowl a quarter turn and cut again. Again, until all the fat had disappeared into the flour. "It should look like soft bread crumbs," they told her as they flecked the mixture with water. "And the water must be very cold, but don't use too much or the dough will be tough." "Don't handle the dough any more than you have to," they'd say, as they scraped the mixture up to form two rough cakes, which they let rest under a towel as they made the filling.

Since she was married in late spring, she learned first to make rhubarb pies. They didn't measure anything for the filling, but simply tossed in the sugar and spices and cornstarch. "When you roll out the crust don't use too much flour on the board." "Make sure it's not too thick, but don't roll it too thin." "And do it right the first time because it won't be flaky if you have to roll it out again."

In her own home, she added too little water at first, and the ball of dough crumbled under the weight of the rolling pin. She had to patch crusts. She had to finger them into the pans. She had to roll them out again. Filling seeped through the bottom crusts and spilled over into the oven. Late in the day all she could smell was sugar and fruit burning off the oven floor. She stared down the hours to dinner, stared down all the Wednesdays and Sundays to come. She stared down latticework crusts on the blueberry pies, the star of steam vents on the apple, the fluted edges of the squash pies, and the fork-tined edges of the mince. Easier to pass through the eye of a needle than to learn this.

A house settled. Lines no longer plumb to the design. I remember my mother working without falter. I see her scooping the flour from a canister and leveling the dry measure with a butter knife. She cuts in the shortening with quick, efficient strokes, and the water she uses is as cold as the brook that spangles my wrists when I plunge in my hands. To roll out the dough she rocks the pin on top of it at first and then works out from the center, giving each stroke even weight. She rolls away, then to her right, towards her, and to her left until she is north once more. She lifts and turns the dough, then follows the compass again, rolling in all the directions of the wind.

She'd make rhubarb pies first thing every spring, followed by sour cherry—until the birds got to the trees. Blueberry and peach ran down the summer and overlapped the first apple pies made with Gravensteins. She made one or two pies with McIntosh, and then worked with Cortlands. But as soon as the Northern Spies were ripe she used only them—so deep is their flavor—until the last punky ones were pulled from the apple cellar sometime in March. She'd make squash pies in the fall and winter, too, of Blue Hubbard and butternut.

She's making a Blue Hubbard pie today, and after forty years my father still doesn't know that she needs only two cups of cooked squash for the filling. He has brought her the heel of the largest seed squash of the season, which must weigh six or seven pounds, and she has to struggle to cut away its skin and cube the orange flesh. To the steamed, mashed squash she'll add milk, eggs, sugar, cinnamon, nutmeg, and also ginger. It will be the one pie she makes this week now that it's just the two of them, and she'll work slowly, since she's bothered by a little arthritis in her wrists.

❦

IT IS EARLY March. Old snow in the woods, and mornings, anyway, frozen ground. Just visible in the matted grass is a set of wheel ruts leading down to the orchard. My father has begun to prune the young apple trees—a stir at the edge of a long sleep, nothing more. Any spring snow will put an end even to this light work. But the good weather makes him want to start the season, and for now he thinks about pruning two rows every good morning. He can work from the ground, and the young trees go quickly. Each needs only a few cuts, where two branches cross each other, or where a secondary limb competes with the leader. And the branches are slender. A snip, or a few draws of the saw is all. He keeps a pruning hook in the bed of the pickup for the rare branch that's out of arm's reach.

It's the older trees that are a job of work. They were planted so long ago, some of them, not even he knows who first pruned them. Now they've grown almost too large to pick with ease, and to shape them means using ladders and crawling up their slant limbs. Now and again a chainsaw. And a long time spent deciding on which cuts to make. He has to prune around years of wind damage as well as the work others have done—some trees are as skewed as homes with seven previous owners.

But the best are a good part trunk. He likes their gray, bull-necked strength. And their turned grace—branches spare as antlers contain light and air even in the center of their crowns. They'd show the smallest neglect. He's seen whole hillsides let go—apple trees shaggy as hermits, branches turned in on themselves—a tangle of nests—or flung wide or strayed to the ground and flailing in the simplest wind.

Well, this morning's work has tired him more than he'd thought. The house feels overheated. Maybe a short doze . . . He hears my mother making lunch—no more than a pin falling—and now the morning's work seems far off in another country—like tomorrow's work . . . the work of sixty years . . . He sleeps the same as if he's seeded all the fields with hay. Just sweet hay.

❦

MY BROTHER SAM has lived here, only here, for more than half of any long life, and the land is still not his own—not his orchard, not his fields, not his to dream on, not his to lose. More than anyone, he's bounded by these stone walls, the pines, these furrows. And no one talks anymore as if there'll be another life for him. "Maybe Sam should have done something different," my mother says, "gone out on his own."

The corn snow draws back to the shadows, and the blue-eyed cold once more has lost its edge. My father finishes pruning the new orchard while fields away my brother burns piles of brush on top of sodden grass. Smudge casts off from his fires. It billows and drifts above the oaks, it smarts in his eyes and throat.

❦

THE SPOTS ON the tomato leaves have them both stumped. I see them on opposite sides of a trellis, both bent to the foliage, lifting their heads to speak to one another or nodding in agreement. Sam rises and shows the underside of a leaf to my father, who takes it and turns to his left to see in better light. Another nod. It's clear the things they know wash over the same territory.

I don't often see them like this. Usually they keep to themselves. My brother is grading tomatoes or picking the day's corn while my father is checking the color on the apples or cultivating a late field. They agree beforehand that come September my father will oversee the apple picking; my brother, the squash harvest, so if my father mutters that too many of the Macs have dropped, Sam only says, "If it was up to me I'd have started picking them earlier." My father might answer back by saying something about there being no market for green apples. My brother shrugs, "Fine." And then there's silence.

I know Sam's head is full of ideas. We have a small orchard, harvested the same way it's been harvested all along. There are apple boxes so old the pine is splintering away from the nails. Bottoms give way, and thirty or forty pounds of fruit spill onto the concrete floor of the apple cellar. He has worked it all out on paper—a system with bins, and the forklift to go with it. He'll have to enlarge the door of the apple cellar, and grade the approach . . . He used to rattle off ideas like this, but these days he simply says, "Nobody does it this way anymore."

I know my father is silently running down all the projects Sam has left undone. The shed half shingled, one of the tractors disassembled . . . He is biting his tongue. It's been a long time since he's talked to someone about the day's work—what to pick where, or how it's still too wet to plow. Or told someone a little of what he knows—say, how best to graft the scion to a root stock. How the cut has to be clean, and the cambium should match all the way around.

❦

ONE WINTER, YEARS ago, Sam went so far as to clean up one of the unused sheds and set up a carpentry shop: try squares and spirit-levels, chisels, gouges, planes, hammers, and saws; cherry, oak, and maple spaced and stacked to finish seasoning. Bright nails gleamed, and the air was sweet and thick with sawdust and the scent of fresh cuts—a promising fragrance above the must of old canvas and leather.

At the dinner table he would talk as if he was born to it. He'd talk about applied geometry and dovetailed joints, and how he was going to build a cabinet of cherrywood; the clean, broad planks cut easily into a roughed-out shape in his mind. Or he envisioned an oak desk worked from dimension timbers—ponderous oak, with the grain dreaming through it, which would deepen under his own hand as he rubbed oil into its surfaces.

For a while we believed he'd found his place at last, but such craftsmanship lives in the patience and work of real time—the same as this farm—and if his troubles have canceled out anything, they've canceled out an understanding of such time. He's used one drug or another for years, and by now his habit has settled into a rimey life of its own. He goes unwashed and unshaven, and his clothes are ragged. He is sullen and alone. I can imagine his night sweats, with his pillow and sheets soaked through, his eyes shining and feral.

It was no surprise when he began to spend less and less time in the space he'd cleared and set up, saying he was waiting for a certain tool he'd ordered, or for the wood to season a bit longer. And as always, he said, "If you're going to do it, you might as well do it right."And then: "You never think I can do anything right . . . You think I'm just a screw-up, go ahead and say it."

And sometimes I do. I say it when the apples are unsprayed or the corn is late in getting planted. I say it when the crew is hanging around the packing shed waiting for him while he lies in bed dreaming of building an addition to the farm stand or a new greenhouse in which to grow early tomatoes.

2
Back

"I THOUGHT YOU were living off somewhere else," the customers at the farm stand will sometimes say. Or they ask, "What's brought you back?" I mention something about how the place is so much work now, or I talk about my parents being old.

When I ask it of myself, there's a winter's night I remember. It was before I returned here to live, and we were all home for Christmas. "I've never seen Sam looking so bad," my sister, down from New Hampshire, said. It was true, too, for as much as I could know. I was only living outside Boston, but I saw Sam about as rarely as my younger brother did, who was in from the West.

Sam lived then, as he still does, in a place stalled mid-renovation. We walked into a comfortless house, with insulation half-packed against the exposed two-by-fours and a plywood floor with grime scuffed in. He'd drawn the shades against the daylight and had a few lamps on. I don't remember any color. A small engine taken apart in a corner of the living room. Empty takeout containers on the table. Coke cans. Candy wrappers. Cold.

My sister, my younger brother, and I sat there in our winter coats and tried to talk to him over the TV about the way his life was going. His eyes were sparks, and he himself, thin. His talk was all farm. We heard again how there was too much work here, never enough help. His hands were tied. Dad would never change anything. If we tried to turn the conversation towards his own life, towards his unpaid bills, the cocaine and mood swings, he'd ride over us and say, "You don't know. You're not around. You try working here for fifteen years." And then he'd get up and pace and spin off into some of his dreams. "I could do so much with this place—we should be planting the first corn under plastic. We could go into greenhouse tomatoes. It's early money—Dearborne makes a killing every spring. If I had some help, someone to run the stand . . ." We were sunk low in his couch and looking straight at him. He stopped pacing and, for a minute, there was silence. When he started up again he looked at each of us hard, and in turn, as he said, "You left. You left. You left."

My first season back I was stunned by the business, how it now takes four, sometimes five people to run the stand, and still, on Sunday mornings we can't pack the corn fast enough. We sell all we can possibly pick—at least one hundred bushels, and I hardly look up from my work as I answer the customers' questions: "The corn? We've had it for about a week now." "Yes, thirteen to a dozen—just like last year." "Bring the water to a boil, and cook it four to six minutes." "Four to six minutes—that's it."

When I was a child, and my father still sold most of what he grew to the neighborhood markets in Lawrence and Lowell, the farm stand was a three-sided shack, just one step up from a table on the lawn. It had built-in counters on each side. The left one was at a slant, with the corn piled on it; the right one, level and waist high to the customers, held tomato baskets full of beans, peppers, squash, and tomatoes. The strongbox was tucked in the back. The scale, with its round face and white enamel pan, hung from a crossbeam. Against the back wall, bushels of corn were stacked, and a grape box, hung as a shelf, held half a dozen jars of honey, which I'd arrange in a diamond pattern, then rearrange in straight rows, and again in staggered rows. The honey always stood in shadow, and seemed a solid chestnut color. Sometimes a scrutinizing customer would take a jar and hold it up in sunlight to check its clarity, and it turned to translucent amber.

I couldn't have been more than ten when I started to take my turn minding the stand. The midday shift went to the youngest because the road was quietest then. The summer air: quiet, and hot, and bright. I remember times when not one customer would come, though mostly you could count on a trickle of people—many older ones who would come for a few ears of corn, and a pound of tomatoes:

"That tomato's not soft, is it?"

"No, sir."

"Are you sure you counted right?"

"Yes, sir."

"An extra one for me? That's a girl. How's your father? All right?"

"All right."

After they drove off, I could hear the car engine clear to the bend in the road. Then, the quiet again. I'd tilt my chair against the back wall, and draw the same scene over and over again on the back of a No. 12 bag: the view from the stand. In the foreground, a telephone pole, the mailboxes, a clump of grasses spiked with chicory. Across the road were the stone garage, the red shed, fields, and woods.

When I went off to college, and afterwards lived away, I remembered the stand as that same white, dusty place, even as I heard how they were building the new stand, which was closed in and at least a dozen times the size of the old one. They bought a cash register and put in a phone. They paved the parking lot. Sam hung a carved sign by the side of the road, which had steadily grown busier over the years as housing developments sprang up, and the corner markets in the cities closed down. He dragged the old stand behind the greenhouses and used it as a place to store empty crates.

The feel, for instance, of a ripe ear of corn was something I'd never forgotten, or how to spot signs of borer, and the grade on the apples and tomatoes. Such things kept me in good stead when I returned, but it had been fifteen years, and there were many things I had to learn for the first time. I knew nothing about the wholesale orders or the preferences of the customers. I heard things like "Well, they always let me pick out my own corn" and "They used to save me the large zucchinis for my bread" and "Last year . . ." I was a stranger, too, among the others who work here now, among Sallie who seems to have been here forever, and David who'd tagged along beside my father since he was twelve, who'd taken over for his brother, who'd taken over for another brother before him. There were small, long-running jokes between them—just the way David said "beans" or "harrow" or "water" could make Sallie roll her eyes or laugh.

I don't even know when I could first discern David's forthright stride from across two fields, or when I could sense, without looking, Sallie's presence at the side door, and I'd know she was resting her brow against her knuckles as she leaned against the jamb. We'd count the days down by the varieties of corn—Sprite giving way to Seneca Brave, then to the slow-growing August varieties—Sweet Sal, Sweet Sue, Calico Bell. I'd lift a bushel onto the table and David would appear out of nowhere to help me lift it the rest of the way. Sallie, beside me, packed the dozens and half-dozens into the brown bags, and our small talk bloomed during some of the brief respites, then fell away. "Once you know this work," she'd tell me, "there's nothing else." I'd always agree with her, even if I haven't always believed it for myself.

And I no longer believe it for Sam. The talk between us is still that iron wheel—orphaned and unwieldy, and I'm always afraid it will run loose. I wonder at how I could have thought it would be weightless. I have a harder and harder time remembering the boy who built a treehouse in all his free time, his scrapwood steps nailed to the trunk of the specimen maple. I don't even know if it's a memory or a dream—him building his raft or his radio or leaning into his stride as he skates along the circumference of the pond. And what of five years back—all that I felt—where has it gone?—that winter's night, when it was me he turned to and said *You left*.

When I close the front doors of the stand at six there is a sudden, pronounced quiet. The bushel baskets, the bins, the cleared counters emanate a new coolness, as if I'd descended into a cellar. It's the first moment I've had to myself for at least ten hours, and I take a long breath. My own house is quieter still, and when I enter, it feels as if I had left it ages ago. I'm surprised by the scent of cut flowers—snapdragons, zinnias, bachelor's buttons—and by the sight of them in the vase near the window. I don't remember leaving the coffee cup on the counter after breakfast. Or the morning paper spread across the table. A lace curtain stirs. I can hear the clear, minor notes of a sparrow.

Corn husks rough up my hands just as much as sandpaper would, and the day's dirt lodges in the loosened skin. The first thing I do those evenings is soak and scrub my hands at the kitchen sink. It's an absentminded task. I lean against a high stool and gaze down the orchard to the darkening pines. The water is lukewarm, nearly the same temperature as the air, so I can't really tell where one leaves off and the other begins. The sound of water falling back into water. The landscape disappearing into the soft night. Some evenings I am so long there that my fingerpads start to wrinkle. I dry my hands, then rub lotion into them.

❦

FOR HALF A dozen years I lived on an island thirty miles into the Atlantic. On a winter's night, the town was no more than one small, uncountable cluster of lights, tight as the Pleiades. The dark outer reaches, swept by the lights of three lighthouses, were hardly inhabited, and beyond, even darker, and everywhere, was the northern ocean. Milk, butter, eggs, the obedient cantatas practiced by a choir—in rough weather such things could seem miraculous.

It was like nothing I had ever known. Weather reports contained words that had hardly ever crossed my family's lips—buoy bell, moontide, small craft warnings, and moderate seas. Near gales—so different from our inland winds. There were plants and trees I had never seen: scrub oak, bayberry, sweetfern, and heath. Even the white pines—long familiar to me—grew there in unfamiliar ways. By rights I think of them as slender and straight, sistering the masts they were once felled for. There they are hardly taller than the houses, and their jagged crowns lean into the land as if the work of all their years was to endure winds coming off the entire compass.

When I think of the island now, I like to remember the shortening dusks of September and October when

I'd walk to the end of the road to watch the full moon rise over the Atlantic. I'd shelter myself in a dune, with my back to moorland that had turned burgundy and rolled away like a sea of its own. The surf sounded innocent in the distance. An orange glint over the gray-green water, then the orange moon itself rose pale—nearly transparent—on the horizon. Warmer days I'd stay and watch it hang low there before making my way home in the dusk. On the road back the winterberries blazed out from their bark. The white trim of the summer houses—shuttered where they faced the sea—was luminous in the failing light.

The trim would shine, too, in my headlights on my early drive to work. I'd leave home a good hour before sun-up and drive the seven miles to town past sleeping houses lost in a land darker than the sky. Until their white paint came up in my lights, only their angled outlines distinguished them from the low woods and fields. It was close enough to morning so that the deer had begun to move, and clusters of them would appear in the middle of the road. As my car approached, they'd raise their heads, turn, and stare at me. I'd stop and shut down my lights, which would bring them to, and they'd look forward, then break into a run. A rustle of underbrush, a glimpse of their white flags going away, then they were gone.

At that hour there couldn't have been more than a handful of people awake—I'd count them to myself—the third shift at the power plant, the night officer at the police station, and the other bakers like myself. For the first several hours I worked alone making batches of bread and shaping the morning pastries. It was steady work at a long maple table, and small sounds told its rhythm: as I divided the risen dough into equal pieces, the dough cutter struck the wood regularly and with a dull thud. White, oatmeal, wheat. The cut dough sprang back into crazed shapes that gave under my touch as I'd knead one after the other into uniform smooth oblongs and set them in the black tin bread pans. Then I'd turn to the pastries, which required more painstaking work. It took the same kind of patience my mother had when she rolled her grape leaves. You had to find an efficient rhythm, but you couldn't rush. One after another of the almond pastries twisted into the same figure eights, cheese into spirals, cherries into circles, then set to rise in staggered rows on parchment.

The pastries would be in the oven and the bread well into its second rise when the first streaks of winter dawn appeared. If I had time, I'd stand out on the back stoop to finish my coffee while the sun came up over the bay. The long stretch and wake of the town had begun. As I looked down the slope to the water, I could see upstairs lights, and beyond, a dozen scallop boats making their steady way to the far reaches of the harbor. The drawn dredges like tucked wings, the fishermen stock still, one hand on their wheels, staring ahead. The small wake each boat made unfurled into others, and then attenuated into loose waves that lapped against the shore. I could see the early ferry easing out into open ocean. As it passed the jetties, the seals—indistinguishable from the granite at that hour—stirred, and slipped off the rocks into the water.

The life of the restaurant began to wake too. Steve, who cooked breakfast, stumbled in at the last possible minute, and within minutes after his arrival the smell of bacon frying overcame the aroma of the bread in the ovens. Onions and potatoes sizzled on the grill. His whisk slapped against the metal bowl as he made his pancake batter. Our talk. What we did yesterday, the boss, the customers.

And more conversations as the waitresses arrived to set up the dining room. From back in the kitchen, we couldn't hear more than murmurs when they clutched around the coffee machine. Then they'd scatter to set up the tables, and their voices would grow louder as one would call across the dining room: "You *have* to tell him . . ." They'd meet again at the coffee machine to finish the story in voices low and intimate as if between sisters.

Once the customers arrived, the place was awash in voices. A dozen conversations among tables and between tables. Births, deaths, real estate, a story circulates. A death or a scandal could be all the talk in such a closed world the way one topic could have consumed days on the farm years ago. Something as small as honey. "Pop was always wishing for honey," my father has told me, "but we didn't have the money for it. Then a wild swarm landed in the Baldwin tree. What an event, when Pop captured that swarm. He took the gray mare and went to see Mrs. McLaughlin, who told him how to make a box for them. It was all we talked about for weeks. It kept him hoping for months. But the bees starved that winter, and we never did get any honey."

By the time I finished my shift it was broad day, and the life of the island was fully awake. With all the daily bustle it seemed so much less an island than when I first awoke. Always the night accented its tossed place on the sea. I clearly remember the evenings I went to gather a friend from choir practice. I could hear the music from streets away—interthreaded voices cast into the immense night, and thinning into the greater dark beyond us. The sung words I heard were the same simple old ones: *so that I may be a child of peace; and let me come to thee from my sorrows.* What made them complex was the steeped unity of the voices singing them. Our own comfort, they seemed to say, or none at all.

A comfort it was hard for me to conjure when the wind came out of the east—the same direction where that moon had so calmly risen. There were stories all over the island of people going mad from such wind. They would be found, the stories go, their fire gone out, sitting and staring into cold ash. I'd keep a fire going all the time on such nights, and sit right next to the flame. Sometimes I'd sleep the night beside it. Even so, there were times when the wind sounded so fierce—the brays, the creaks, my whole house sounding like something in the way—the fire couldn't calm me. In the end, I opened the door to find my peace. Outside, only the sound of pine boughs sweeping in that wind. The mist hid the cold gleam of the stars. Bare branches glistened in the glow of my houselights. By the time I was calm, my hair and face were soaked by a fine rain.

My father was always saying that island wasn't fit for a billy goat. He has lived all his years among the water-worn hills of this cultivated valley, a land shaped by hands and tools and machinery. Most of what we see here are the thought-out lines of our own making—wheel ruts, staked rows, stone walls, and the pliant trace of a furrow accommodating the rises and curves of the earth. Here, every inch has been built on, given up on, reclaimed and built on again. In the pathless woods you can read the history of European settlement—cellar holes within shouting distance of each other, dry wells, metal dumps, and wild asparagus. He remembers when the road was one lane and unpaved, and when it took a team of oxen two days to reach the coast. He remembers the first airplane to fly overhead. Sometimes the memory of these things is more vivid than the words he still lives by—harrow, plow, seed, soil, frost.

While I lived on the island I returned to the farm maybe two or three times in a year, and with each visit my parents seemed to grow much older. Slow to get up from their chairs. Lost thoughts. I'd greet my mother at the far end of my ferry ride, and I'd see how her hair had grown so white, and my heart would sink when I noticed—we had been the same height for so long—how I was now inches above her.

Even with such moments, the farm felt well in the past to me. And the distance seemed to widen more every time my father or mother—they had both lived with their own parents until the day they married—would suggest I come back, or my mother would ask, "Why are you living there?"

I was never able to say. I went to the island to live after I finished college—because I had friends there—only meaning to stay for a little while, the summer, maybe into the fall—until I decided what I would do with my life. I hardly imagined I'd stay for years.

So small and exposed, seven miles wide, fifteen miles long. I don't know if I'll ever live in a place again where there's terror and beauty in such measure. I missed it for the longest time after I left, but it has been more than ten years now, and when I return for visits, even my longing for it seems remote, though sometimes I wish I could just take a walk along those dirt roads that score the moorland there. I used to walk the same way again and again—a hundred times—and then one ordinary day an arrowhead would turn up at my feet—its honed edge or its worked shape would catch my eye. The rains, they say, washed them onto the roads. Anyway, they were a wonder to find. When I moved outside of Boston, I ranged them on the varnished sill of my bedroom windows. Against the polish of that wood, and the mild sounds of traffic, they seemed only foreign and rough, and no more than curiosities.

I couldn't have moved to a more different world. My city neighborhood was a sturdy one, built up sometime between the world wars and ambling down just a bit since. Streetlights and houselights burned bright enough to obscure the night sky—a complacent sky, milky almost, the color of a pigeon's throat. Away from the salt-laden air, it was hard to get enough of my own breath. And a sheltered grace seemed to weigh down on me—those beautiful shade trees, straight and wide-crowned, arcing over the streets and the small yards. Even so, it was a world

I longed to fall in with. If I were asked finally why I left the island I'd have answered it was just too small, and far. Now I had a commute to work with everyone else, though among so many strangers, I couldn't comprehend the years and effort it would take to feel as if there were familiar faces around me.

I'd see my family more often—the farm was only thirty miles away. When help was short I'd lend a hand on the farm stand or with the greenhouses, and what had touched me only for a moment when there was a sea between us—my mother's stoop, my brother's rough face—gradually became my cares. *If I had some help* . . . My father owned Ben's house by then, and when the old tenant died I moved in. Sometimes it's seemed a last resort. And more than once, especially of late, a thieving choice.

My mother and I are preparing a Sunday dinner. From my place at the stove I hear her setting the dining room table, and after she finishes placing the silverware beside the plates she moves around the dining room patting things, putting the slightest skewed things straight. Sam comes in, and she asks, "Why don't you wash up for dinner—comb your hair." He smirks and walks past her. The days he does give in, she seems so happy, happy with any little inch he will give, the smallest thing parceled out.

I hear Sam in the living room talking to my father about the plowing: "We're ahead this year—I'll have the plowing done by June." He sounds assured as always, and none of us says anything, though we know it's not true. Yes, the times he works long into the dusk tug at me for more faith. But my faith even in those good days has started to falter.

Sam's talk dominates the meal. Sometimes my father says a thing or two, but I know he is holding back what he really thinks for the sake of peace. I've grown as silent as my mother, and excuse myself after I finish the dishes by saying, "I have things to do."

I drive nearly an hour to the coast and walk for miles along the waterline. An expanse of sky, the slow lap of a calm sea. Far out, broken touches of white surf. Of course I try to imagine what my life would be like if I hadn't returned. The possibilities turning over in my mind, turning weightless and slow as a tethered astronaut. What kind of work would have filled the years? Would I have a family of my own? Where to from here? Could I live here and not be part of the work? From my dining room table or my study window, almost all of the farm falls within my view. Just a quick glimpse and I know it's Sam's blue truck trundling down the edge of the berry field. There are sounds I always hear: the steady progression of the mower, the jangling harrow, and the high-pitched fan of the orchard sprayer. And silence too often means the work's not getting done.

When occasional terns stray inland on the last reaches of an ocean wind, my father simply sees the same white birds that he has seen for eighty years. Birds he knows from here and nowhere else. They eddy over the fields, higher than the hawks, glinting and white as salt itself. If I see them when I'm especially tired, I'm likely to slip into a dream of that island where terns are countless among other birds we never see here—shearwaters, long-tailed ducks, cormorants, and grebes—birds whose names are part of a vocabulary that can't help me here. I should be concentrating on the words at hand: harrow, plow, seed, soil, frost.

3
In Season

LAST NOVEMBER A night wind tore the plastic sheathing from the ribs of the greenhouse, and the exposed frame became one more stark contour in a land of bare trees and empty fields. The inside froze and filled with snow. Now, after a month of melt and spring mud, I can walk through what remains. The ribs still arch strong above me. Below, watering cans and plastic trays are strewn across the benches and floor. A stack of four-inch pots lists next to a heap of potting mix. There's a stray nozzle, baskets and hoes that belong in the shed, and an uncoiled hose. It looks as if a night wind had also torn through the inside, or as if we'd fled—from what?—overturning everything on the way out. But it's only that we have never taken the time to clean up.

I have registered so many winter months since, so much orderly time, that it's hard for me to comprehend the summer rush now. The mess seems only senseless, and I want to set it right, to check the hose for leaks, to rinse the pots, and wash down the benches.

It's more work than I think, taking me a long day to sort through the pots, discarding much and rinsing every salvaged thing with a weak solution of bleach. It takes even longer for my father and brother to raise a new roof over the old ribs. The poly is light but cumbersome to draw up and over the frame, and it's subject to the smallest breeze until it's secured to the wood and metal. Once that roof is on, though, the greenhouse feels almost new. If I stand in its center, the shapes and colors of the world outside—seen through translucent plastic—seem to be muted versions of themselves. It's warmer inside than out, and I can no longer smell the spring thaw—soft air, sap, the pliant earth—for the peat.

The day begins with a flat of tomato seedlings that need to be transplanted into pots of their own. I break open a bale of potting mix—closer to chaff than to soil it's so light—and fill each pot nearly to its rim. It takes a long time to saturate the pots with water, so I pass over the pots again and again with a running hose until they finally take on water, becoming dark and nearly faithful to earth. I make a hole in the center of each pot, then pull a seedling from the flat. Each one has already sprouted a set of true leaves beneath the growing tip, yet the entire seedling is hardly longer than half my little finger. I never feel my bulk so much as now, bent over these small starts with a wisp of a seedling grasped between my forefinger and my thumb. It would take no effort at all to bruise the cells in those stems, and I have to transplant five hundred of them.

By the end of the morning the greenhouse once again seems as it should, with filled pots spread across the benches, and the air warm and humid. It is a day without wind in the spring of the year. The roof above me is hardly measurable it's so thin—a frame for the ephemeral, made half of work, half of dream, sited, squared-off, saying *here*. Here is your house built nearly of air.

A SWEET AFTERTHOUGHT of a crop, these five rows of Knight peas that no one has had time to trellis amid the toll of June work. Now each row is a tangle of vines, leaves, tendrils, pods, and blossoms. To spot-pick the rows, Sallie and I have to pull back the whole green confusion. We can't tell which peas are ready for harvest by sight alone, since the pods are often blown full before the peas have ripened. We feel along the pod for the certain curve of a mature seed, or discern an incremental increase in weight. Slow-going, two-handed hours, and all through the first morning we check ourselves by shelling a pea now and again—they never taste sweeter than amidst those tangled rows under a long morning sun. There's no keeping them, really, since once off the vine their sugars turn to laden starch.

Seeded in cold ground and thriving only in a cool season, peas are among the first of the crops to ripen. It is late June, and there may be raspberries, too, in an early year, along with lettuce, say, and cucumbers. A small start, but enough of one to open the doors of the stand. Ours is broad and beamy and, at first, as orderly as whitewash: a scrubbed, granite-cool interior, overwintered, empty, and full of shelf space. Is there a scent? A rub of resin, maybe the fragrance of milled pine, but nothing that speaks of time or history the way the barn does with its long-stored leather harnesses reeking of must, and all the laid-aside hoes and scythes, their hafts salted and polished by labor, their blades rusted, honed, and rusted once more. It's daunting, the thought of filling all the unstoried space—these first tentative side crops hardly take up a corner, and when familiar voices break out of a seven-month silence, I believe I hear an echo:

"I've been watching the corn."

"Been waiting for you to open."

"Have a good winter?"

"Yes, yes, and you?"

Those five rows of peas are never again what they were during the first days of picking. Sallie and I spread blight with our boots and hands as we work over the rows, and we scar the undeveloped pods as we search among vines whose leaves have begun to yellow and curl. That single planting of peas is spent and harrowed down even before the first corn has ripened, and in some years the land is replanted with a late crop of carrots or beans.

People are always saying that they haven't had their fill of peas, that we really ought to grow more of them, but in truth they soon forget about them once all the other crops begin to ripen: the early corn is also sweet, and there are tangled heaps of green beans and translucent wax beans that some still call butter beans; summer squashes warm from the field; beets and carrots pulled from cool, mineral earth; the spines of cucumbers and all the lettuces—Romaine and Boston and Red Leaf. Green onions, bunched radishes, and amber, hive-shaped jars of honey. Astringent thyme and parsley, basil, tarragon, wild marjoram, the mints. There's hardly any shelf space left in the last days of July. Even so, there are some who run ahead to a later season, some who have been dreaming of tomatoes for months and are impatient with the pink star at the blossom end of a stubborn green fruit. They ask after peaches as well, and pears and butternuts.

Now there is much talk happening at once, and conversations begin in midstream as if I was already familiar with the story:

"Those berries are for my son, you know he still can't leave the house."

"We'd be up to camp by now but she has her treatments."

"That's it, Kentucky Wonders—does anyone grow Kentucky Wonders? At the Center Street house my mother set them against the white fence . . ."

I can't follow the voices talking themselves into August, when we can no longer keep up with all the ripe tomatoes and another scent has insinuated itself. What is it? Something deep-seated—wild slip-skinned grapes, I think, or apples bruised by a picker's thumb. I hear flock-notes, and dusk is earlier by an hour. Half the corn has been harvested. It never ceases now, the sound of the disk harrow going over the old fields. A boy of sixteen spends his days perched on a tractor, sitting straight-backed as the one just man, breaking up the straw-colored stalks of corn, and breaking them up again.

❦

IN MID-JULY, GREEN washes the landscape. Along the brook bed, brambles and swamp maples. Lone pines, deciduous woods, the soft contours of the cornfields—not one yet plowed under. Summer runs down from the day they start to pick corn. By 6:00 a.m. half a dozen men are in the fields, dressed in oilers if the dew is heavy. They work between the rows, snapping the ears off and tossing them onto the flatbed, which hauls along beside. My brother drives the flatbed and knocks down the spent stalks as he goes. The corn is so high it hides everything—the pickers, the truck, my brother—and at some distances all I can see are ears of corn—ten or twelve at a time—arcing over the stalks as the men throw them onto the truck.

When I was growing up, for two solid months we'd have corn for lunch and dinner. My mother put the water on to boil while my father headed for one of the nearly ripe fields. He was checking to see if the corn was ready to pick, and he'd scour that field for a dozen ears ripe enough for us to eat. If it was a new variety he'd ask us what we thought. Wasn't it sweeter than Harmony? Didn't it have more flavor than Seneca Star? In truth we liked the squat, bright, early varieties the same as some slender, tapered ear with kernels tiny as milkteeth.

It didn't really matter, though, what we thought. To my father, the measure of all corn is Gold Cup, hardy and sweet, though each year he grows less and less of it because people clamor for bicolored corn now. Butter and Sugar the customers call it, or Honey and Milk.

THE VARIETIES OF peaches that grow in this part of the country have hardiness bred into them. Reliance. Elberta. Not names to dream on, though to grow peaches here, near the northern limit of their range, is something of a dream. My father was well over sixty when he planted the peach orchard in a longtime meadow. That was the last piece of land to be hayed—cut and cured and stored. I barely remember it, the rectangular bales spaced like footsteps across the land.

The peach trees—fifty or so of them—were no more than whips to start with, and it hardly seemed they'd survive a January. In truth, the trees don't always winter well—some years there is barely a crop and, once in a while, no crop at all. The peach orchard blooms early and pink. It will never look as crabbed and thickened with the years as an apple orchard—the trees aren't long-lived and they are pruned in their own way—without a central leader—so the limbs curve out and up to embrace an open center. Each crown, a ballroom of its own.

I have never seen relic peach trees crowding a cellar-hole the way apple trees often do. Apples are old in these parts, a cash crop talked about endlessly. How many times have I heard about the Ben Davis, no good for eating, that would cling to the branches until spring. About the long search for a McIntosh that would redden on the tree. Or the original Red Delicious fenced and locked at Stark nurseries. About cider years. Snow apples. Astrakhans.

Somewhere to the south there must be like stories about peaches. People traveling up from Georgia or the Carolinas will say our peaches just don't have the flavor of theirs. They'll name varieties I've never heard of that ripen right down the summer. Here the peach season doesn't last long—a few weeks in August and maybe the first of September when we have a fruit that isn't meant for storage, that doesn't smell sharp as the cold side of the mountain, and whose flavor isn't deepened by the frost. Reliance and Elberta are hardy names but, like all peaches, their fragrance blooms at the back of your throat as it passes.

SALLIE STILL MOVES easily although she is well over seventy. Both her hands work the bushes, snapping the beans from the stalks and filling basket after basket. Now and again she might stand to stretch her back or rest her eyes on a stark, distant object—a lone gap-crowned pine or the white gate at the edge of the woods. Otherwise, on any good day, she keeps a level pace.

She won't work the bean field when it's wet—handling the crop then could spread rust—and she might quit early in too much heat. Sometimes the heat or constant rains drive two plantings of beans together. Then she can't keep up with them—no one can—and at last the earlier planting, mealy and tough, will have to be abandoned for the sake of the later one.

Sallie has worked outside so many years you can't see her sister in her anymore. Sometimes she'll drive the tractor if need be, or help when the tomato crop starts to get away. But mostly she keeps with the beans, accumulating bushels in some shade at the edge of the rows where my father collects them at the end of his day. And she'll mark tomorrow's place with a stack of empty baskets overturned against the chance of rain.

EVERYONE HAS BEEN wondering who'll pick the blueberries this year. Betty died over the winter, of a stroke, and the berries had been her job for nearly a dozen years. The ten weeks of a good berry season—July into September—was just enough for her. She had come here in her retirement, as Sallie had, for some seasonal money, maybe a shape to the day.

During the peak weeks in August my father would send the newest, youngest help down to the berry field to help her out. "Pick where Betty tells you to," he'd instruct the boy as he fitted a coffee can with a makeshift neck strap, and the boy would lope through the apple orchard to the berry bushes happy to think that, at least for the day, there was no stooping, no dust, no cornhusks. He'd come back to the stand hours later dazed and sullen. "She kept after me like my own mother about twigs and leaves in my bucket," he'd say, and describe how she nearly stood over him to make sure he didn't pick anything green, and picked every bush clean. And he'd walk away muttering that he'd rather pick corn.

Well, it had come to be her berry field. She, herself, was meticulous and proud about the way she picked the berries. It seemed she barely touched them as they tumbled into the can. She would already be at work when I woke at six—she liked to start just as soon as it was light enough to see, and many days she'd have picked her sixty quarts by noon. I could see her white hat inching along above the bushes. Her filled buckets were covered with towels and set under the bumper of her car where it was cool—there every morning of every day in the slightest hint of dawn just as the first heat settled on the night-freshened grass.

Most likely the berry-picking will be shared among several people now. The payroll can run to eight or nine at the heart of the summer. Some have been here for as long as Betty, others skate along the surface for a few weeks, maybe for the apple crop, or for the September work. September can be difficult because there's still so much to do, and the high school and college students have gone back to their classes. Over the years there have been many who've started to work here at thirteen or fourteen and worked summers all during high school, and if they went to college, during college—moving irrigation pipe, picking corn, hoeing. My father has run through whole families of boys. When one grows old enough to move on to a job beyond the farm, his younger brother takes over. They still come around to pick up some corn at the stand and to say hello now and again. Richards. Davidson. Stark. By now some are middle-aged, and have grown children of their own.

The boys work the summers alongside two or three older men who begin the year with short days in March—pruning, tending the greenhouses—and stay on through the last trickle of work in November—cleaning up and grading apples for wholesale. Years ago, the full-time men came from town, but these days most residents are skilled or professional, and work elsewhere. Many who work here now were born in Puerto Rico or the Dominican Republic, and live in Lawrence. The radio in the packing shed is tuned to the Spanish station. Men who've worked in fields all their lives—outside of San Juan, and then in Texas, New Jersey, Florida. Roberto, who has been here seven years, is always remembering picking berries in New Jersey, or tomatoes, or zucchini. And, here or elsewhere, it's not likely his work will be different in the future.

Every April, as soon as things start to stir—Sam out working on the harrow, the shed doors open—sedans full of men pull in. One does all the talking, saying they know the work, they have a car, they can be here at seven every day. They'll work on Sundays. Sometimes a priest will come by with a man who knows no English at all, and so speaks for him. "He's a good man, and dependable. He's in church every week."

"We really have all the help we need."

"He has six children . . ."

We never had this much help when I was a child—the stand was much smaller, and we didn't grow such a wide variety of crops. I only really remember Pete. He had come down from Nova Scotia for work, and he came to the farm after losing three fingers in a cannery accident. He hoed, harrowed, planted—the same as my father—and as no one had before or since, he'd eat lunch with us. Every summer of my childhood he sat across from me on the porch table, his head dipping to half meet the plate with every bite. He ate steadily until he had cleaned his plate twice, hardly talking except to thank my mother, or to say "no thank you" when she offered him eggplant.

Pete passed away years ago, in his old age, during the clear, windless days of a January cold snap. Coming back from his woodpile he lost his footing on some ice, knocked his head against the frozen ground, and lost consciousness. An armload of split maple scattered around him. No one passing had seen him lying there in the yard behind his four-room house by the Merrimack. The short day slipped into night, and he died of exposure.

❦

As a child I'd wake on hot and clear July days to my father calling my brother from the bottom of the stairs. At 6:00: "Sam, time to get up." At 6:20: "Are you going to sleep all day?" At 6:30 my mother would begin to call—just his name—and after he eventually got out of bed he'd have to catch up with the crew in the cornfield.

Twenty years later, it's me who is calling him to work. I can't get into his house since the path to the kitchen door is grown over, and he took out the front door years back and walled in and shingled over the gaping space.

The mailbox, once hit by a skidding car, has never been set right. He rarely answers his phone. My friend tries to tell me there are people like him all over New England. I don't think he ever goes anywhere. His truck is always in the yard, and at night his windows are dark. All I know for sure is that, somewhere far inside, the TV's on.

It's well past seven in the morning. I drive up the road to his house in a fury, and punch on the car horn—short beeps at first. When that doesn't work I lean into it with all my strength. Sometimes his confused face appears in the bedroom window, and I drive off, knowing he'll shamble down to work in fifteen minutes or so. Other times, the more I lean on the horn, the more deeply he pretends to sleep.

❦

HIS OLD FRIENDS ask after him. "I haven't seen Sam in a long time," they'll say. "Tell him to give me a call."

"I'll tell him you were by the stand," I say, "but, you know, he doesn't get out much these days."

If I'm grading tomatoes as I talk to them, I'll gesture at the work I have left to do, as if to suggest we're all strapped for time. They nod. They know the real reasons they don't hear from him, but let my brief reply stand in for everything that's wrong. Otherwise, where would I begin? Sam's troubles have become lived-in, as lived-in as the way my mother worries how he'll look each day: when will she see him, and what will his mood be like? Holidays, weekends, and workdays the same, winter, spring—she's waiting for his good days.

Lived-in, too, are the arguments between my father and brother. I have seen it so many times: my father behind the steering wheel of the truck, my brother on the outside braced against the driver's side door. My father's voice strains to keep in the anger: "You don't know how to order seed. Christ, this is enough tomato seed for five years."

"We need to plant more tomatoes this year."

"And who the hell's going to trellis and sucker them all?"

"If you'd hire more help . . ."

"If you'd get up in the morning . . ."

PERSEVERANCE IS BUILT into the work. It's one of the reasons I love the long rows of tomatoes to be trellised and the flats of seedlings to be planted. I love the patience it takes, the repetition, the deliberate pace, the way under the June sky, everything has to be handled just so; how it's always the same earth under my feet and that the tasks hardly differ from year to year, except maybe now there's an improved variety of corn, or the disks on the harrow are made of a more durable metal. But perseverance can't make up for everything, say, for the insufficient time I have to learn what I need to know about the late land and the cold spots. My parents are already old, and I came back late.

When Sam said he couldn't handle it on his own, did I really think he was talking about the greenhouse and the transplants, or pruning the tomatoes—those things that can be learned just by looking straight at them? I spend so much time on such things, on learning a knot sturdy enough to hold the late vines and simple enough to fall away after a frost. I make the days long and many and filled with work, and in August, if I finally break down at the end of a day, just break down, everyone can tell me I'm working too hard.

But it's not the work, the way it's not anything that perseverance can help. What can help if I'm simply afraid of my brother's exhausted face, afraid of what makes him so late and so sullen? What, when I can't remember a word between us that hasn't been misunderstood? Yes, there are times when he'll sit down and join in the talk about the spring planting or the apple crop. At such times that talk comes easier to him than it ever will to me. He's the one who learned early on about the cold spots and the late land and where the carrots should be planted. On those good days you'd swear he could take apart an engine and put it back again.

But even then, the distance between us is as small as my glance away when he looks at me, and as infinite. It's his eyes I'm most afraid of. When they have that cornered look and are glassy they're as incomprehensible to me as the spaces between the stars, spaces compelling me to stand fast and stare or else become a lesser creature.

IN A DRY August, after the first piece of corn is plowed under, the field turns to a bald patch of rocks, rough clumps of soil, and wheat-colored stubble. The first piece of corn is almost always in the east field, and when that corn goes down, I can see clear across to the purple loosestrife that mats the swale around our irrigation pond. Loosestrife lines the brook bank, too. *Millweed*, my mother says, and tells me the seed had come here snagged on raw wool from Europe. It worked itself down the Merrimack from the textile mills in Lowell, and took hold on the river's wide-curved banks, then along its tributaries, then along every leaky little stream and wash and pond in our parts.

Dry years it thrives the same as wet years: narrow as candles, the spires of countless petals grow so close together they seem one—a fray of magenta in all the low places. Everything else in the fields and woods is paled by drought. A white haze hangs like mull over the land. A man hitching down a weedy row, heat bugs searing, the pickup trailing dust. You can't see the pond for those weeds—they're tall as a grown child sometimes.

OLDER PEOPLE STILL say "drouth." They remember the rasp of sere grasses and one bird's thin cry. They remember the plow scoring parched soils and a whole season of crop loss. The corn stunted and dry-tipped; the lettuce, bitter. Bitter the stem ends of the cucumbers. A season begun under such stress will never fully recover, and so the late tomatoes were tough-skinned, and the peaches so small they'd nestle in a young girl's palm. What could they do to feel a little less like clay? When they needed rain, the joke goes, they'd cut hay.

I say "drought," and I've always heard an engine in the distance along with the tick of the irrigation nozzles as the water turns in slow enormous arcs over the crops. Water measured in hours and inches, and falling with precise intensity on the corn and beans and tomatoes. Water that smells of iron and musk, and always enough as long as there's enough labor.

The irrigation pipe is moved by hand and laid down between the rows. When one field is watered for several hours, the pipe is moved to another, and after everything has had its inch, it's time to begin again. "That storm saved me a thousand dollars," I've heard my father say in the aftermath of a good rain.

This summer has been so dry that the west pond has been pumped down to the lowest level anyone can remember. It's all we seem to talk about: "Have you seen the pond?"

"No more than a few feet deep in the middle now."

"I know. Christ."

Field grass has taken hold on the banks, and three drums from a raft broken up in my childhood are at last exposed. There's an old lawn mower, too, and the remnants of some apple crates. Frogs squat along the edge, but the water has sunk below its inflow and outflow, so now it's too warm for the fish, and the heron is usually absent.

ALWAYS THAT FRAIL, white-haired woman, when she comes to the stand, rubs a leaf between her thumb and index finger and lingers over the fragrance of mint. Mrs. Mansur buys three bunches of flat-leaf parsley for her tabbouleh, and all the Italians, a sprig or two of basil for their sauce. Occasionally someone buys thyme or tarragon but, for herbs, there isn't much more demand than that. At most, they're a grace note, not a tried thing to most of the customers—even the old ones. And unlike the fruits and vegetables we grow, they don't often work their way into the talk among neighbors here, which goes on all day long:

"... Ned's not doing too well."

"That's too bad."

"Bill and I went to see him yesterday. He was pretty confused. Wanted to come home with us, but I don't know if he'll ever make it home again."

"You never know. Look at the Judge."

"You're right. You're right."

"George says the relatives are already coming out of the woodwork."

"Always the way. What have you got there—tomatoes? They look good, I should pick some up—I was just coming in for corn—the kids want a cookout ..."

Corn, tomatoes, beans—everyone buys them without a second thought, but in the corner where the herbs are displayed a grown daughter and her mother stop, puzzled:

"Ma, what's this?"

"That's the thyme, I think."

Or a woman wonders aloud: "Tarragon. What would you do with tarragon?"

"Some people put it in their salad dressing," I answer, "or you can try cooking it with chicken—it has a little bit of a licorice flavor."

She crinkles her nose. "Chicken ... Really?" her voice having grown softer as she drifts off toward the summer squashes.

I imagine she's weary of cooking—like my mother—preparing the same few things the same way through all the years of her marriage. An arrow-straight task to get through the three meals she prays will yield no surprises and no complaints, and won't make too much of a mess to clean up. Tarragon. She'd wonder how much to use or when to add it in. And for what? For her husband to ask, "What have you put in this?" and then eat around it. And the kids. There they'd be, leaning their chins into their palms as they scraped all the green bits off to the sides of their plates.

We've only grown herbs for a few years, and I've always been the one to plant and take care of them. The seedlings are so small I don't even need a trowel—it's easy enough to push aside a bit of the fine tilth, place in the seedling, then tamp the soil back in. I set them along a white string, which keeps the rows straight, and then water them in. Most herbs don't attract pests, so they don't require much care except to keep them weeded and watered, and of course to keep on top of the harvest, since once they blossom, their strength goes into the flowers, and the flavor of the leaves is lost, or becomes bitter.

And though they are the least in demand of anything we sell, I give the herbs one of the best spots on the stand. Both my father and Sam shake their heads, and mumble under their breath, "You should be putting the cucumbers there." "All that space—and there's no money in it." I take far too much time with them, too, picking them myself every morning. Parsley, basil, chives, mint, tarragon, thyme. Each treated just a little differently. I shear off the parsley near its base so the new growth will come in evenly, and snip the basil just above its lower leaves. The perennial herbs are slower to grow, so I trim each plant sprig by sprig and bunch them in arbitrary sizes as I go—if there's plenty the bunches are big, if not . . . Each bunch sits in its own pint canning jar half full of water—a dozen or so jars in the coolest of corners, which only gets some western exposure. On a clear afternoon the water in the jars glints and there are skeins of sunlight on the leaves.

I know we are a long way from the old herbals I love to read, which tell me parsley sprang from the blood of a Greek hero, that thyme will keep up courage, and tarragon will cure the sting of a venomous reptile. Tinctures, tisanes, auguries. Vinegars, juleps, conserves, something for the cranial nerves. A pink bloom will decoct to a balm for a tired heart, and a deeply lobed leaf will re-create the eyes. In those books, something as small as a sprig of rosemary has the power to strengthen memory, and flourishes only in the gardens of the righteous.

What we have in these years is pure fragrance. Late in the day, a middle-aged man stops at the door of the stand. It's been humid lately, and sweat stains the back of his workshirt. His boots are unlaced to finally let in some air. He rubs his brow as he tries to remember what it is he's supposed to pick up for supper. I hear a long drawn-out breath, then he turns single-mindedly towards the wax beans. As he passes the herbs, I see him brush against a bunch of basil, enough so its scent quivers into the air. I can smell it too. He has stopped still—I don't think he's sure of what it is, though the more the scent diminishes, the more he seems to strain to know it, as I myself have strained to hear a note sustained after the bow lifts from the strings.

❦

GRAPES HAVE ALWAYS grown wild here. The vines tumble along the stone walls and the old mail road, whose faint wheeltracks I can still follow through the back woods. Sometimes they climb the trunks of the white pines and curl along their lower branches, then cascade back down in a freefall, but mostly they grow back in on themselves, twining among brambles, bark, and stone so densely I can't see my way through them except for where a few stray tendrils curl away from the thick of it.

The vines are too entwined to produce any quantity of fruit, and what fruit there is ripens unevenly, so that one grape tastes sweet, and the next, bitter. But the smell is so musky, and the late summer air so deep with it, it stops me still, and I breach the tangle of vines—so strong it's like working against a floodtide. All that redolence, and so little fruit—a few small grapes is all I ever find to eat—way in—in clusters of three or four.

Years ago, my father dug up some of the better grapevines from the mail road, and planted and trellised them alongside the farmhouse. He pruned them back, and he still prunes them every spring—you can see how severely in winter when, even with a season's regrowth, the vines sinew sparely along the wires and stakes. They remind me of flung arms, and I can see how they strain against their form—there are places where the vines perceptibly curve away from the wires, then are reined back in again. Some braid around each other, and only a few tendrils and tips stray away. In late summer the fruit hangs in heavy clusters, and they're the grapes my mother has always used to make her jelly.

She picks them when they are still a little underripe—there's more pectin in them then—so the jelly will set more firmly. I can easily conjure her at the kitchen table picking through the fruit, discarding any that are bruised or too soft, and removing all the stems. She boils them in a stockpot, and within minutes the grape scent thickens, the kitchen suddenly as heady with it as that low place on the mail road. It is high summer and the air doesn't move even with the windows thrown open. Her forehead is damp, her hair curls with the steam. The metal spoon is stained purple and lies on the stovetop, whose white enamel is splattered with purple, too.

As is the bleached cheesecloth she uses to strain the cooked fruit. The cloth catches all the skins and seed, and what passes through—juice the color of a dark gem—she returns to the pan along with sugar, and boils again. After awhile she begins to test the jelly for firmness by lifting the spoon from the pot and letting the mixture fall off its side. It falls first in two distinct streams. She dips the spoon and lets fall again, and again, until the streams sheet together and fall in unison from the middle of the spoon, which means the jelly is done. She skims the pot, then pours the mixture into six-ounce jars, which she seals with paraffin.

Next day she caps the jars. A few she gives away, but most she stores to have with their breakfast all winter long.

In June, long before the first scent of grapes infuses the air, when the leaves on the vines have just unfurled and are still a bright green, and the fruit just hard green promises of themselves, my mother picks grape leaves. There are only a few weeks in the year when the leaves are tender and insect damage hasn't yet set in. She likes to walk along the mail road and pick from the wild vines. There are countless ones to choose from in the mass and tangle, but she picks carefully so they'll be of a uniform size, and not too deeply lobed. She cuts their stems as she picks them so they won't pierce the other leaves.

These she stuffs with a mixture of lamb and rice. She places some filling near the base of the underside of the leaf, and tucks in the sides as she rolls it towards its tip. She rolls them firmly enough so they won't fall apart during their long cooking, and loosely enough to let the rice expand as it cooks. Each, a uniform cylinder, which she sets in rows in a large pot lined with lamb bones and torn leaves. A second row goes crosswise over the first, and then she places a heavy ceramic plate on top of them to weight them down. She pours in water to cover the plate, and simmers the pot on the stove.

When she brings them to the table they lie on the plate as symmetrically as they had in the pot. She ladles some broth into a separate bowl, and she sets out a bowl of plain yoghurt, too; it is the traditional way to serve them, though out of all of us, only my father dips the hot grape leaves into the cold yoghurt before he eats them.

"I guess they thought we were strange" is all he has ever said about the way the neighbors looked at his mother as she picked the grape leaves along the old mail road. She and my father's father had come from Lebanon, where stuffed grape leaves are an everyday food. But it's not something their neighbors here would have dreamed of making.

I often wonder what those neighbors thought of my grandparents who, having come for work in the mills, had bought a dying farm and settled among them. Maybe because my father has so rarely spoken of it, I imagine suspicion, and wariness. Maybe they waited for failure, though by the time my father was grown, I imagine he looked much like them as he walked through the late fields towards the woods, his rifle slung over his right shoulder, the dusk closing in around him as he called and then listened for his lost hound.

I REMEMBER WHEN my father drove the tractor as he sprayed the apples, everything but his eyes covered: work gloves, brimmed hat, shirtsleeves rolled down, a bandanna pulled over his nose and mouth. When he sprayed the orchard nearest the house I could hear the halting rhythm of the work—the tractor in low gear, an idle, the spray gun sounding off like pent-up air released. Low gear again, idle . . . Could I smell the spray from my bedroom window? There was something tangy and thick on the air those mornings—he sprayed at dawn, before the breeze came up, so the mist could drift straight down and settle in a fine bloom over the fruit and leaves. The whole orchard shimmered pale and silver-green when he was finished, and my mother spent the next week nagging us to stay out of the apple trees.

One year there was a hailstorm. Hail spat on the roof and rebounded off the granite steps. It sank into the Baldwins, Cortlands, and Northern Spies. The pocks healed, then spread as the apples ripened. Cider, seconds, drops. He kept to the spray schedule anyway, for the health of the wood.

❦

SOME OF THE apple boxes come from more than fifty miles away: Upland Farms, Peterborough, New Hampshire is stenciled into the side of several of the boxes, and so is Badger Farms, Wilton, and Moose Hill Orchards, Derry. Apple boxes are often bought secondhand, so the pine is weathered gray, the color of apple bark itself, and no longer smells of resin, but of old soil.

For most of the year the boxes are nested together and stored in a shed—more than a hundred stacks of them, and each stack six or seven feet high. They come into use gradually around mid-August, when the summer apples—Astrakhans, Gravensteins, and early Macs—begin to ripen. We have only a few trees of those varieties and the apples seem incidental among peaches, corn, and tomatoes. Nevertheless it's *their* tight-hearted scent you pick up on a diminishing wind.

September drives down, the Macs are ready to drop, and apple boxes are hauled by the truckload to the orchard. Then the place is littered with pickers—some on the ground, some on ladders, all slung with apple buckets, which are slighdy kidney-shaped so they'll fit snugly beneath the chest of a man. The canvas straps are worn low against their shoulders to prevent a sore neck, and I suppose it's those straps that make apple buckets look like collapsed marionettes when they are stowed away at the end of the year, waiting to be taken up by a human hand.

The galvanized buckets have a felt rim to keep the fruit from bruising, and they have a canvas bottom closed by a drawstring so the fruit can be lowered into boxes without being dumped or handled a second time. Eight or nine hours a day, one apple at a time twisted off its stem and placed into the bucket, the bucket filled and loosened into a box—this makes a gentle, clustery rumor—and then again . . .

Macs, Macouns, Cortlands, Baldwins, Northern Spies. So many apple boxes unnested, filled, and stacked beneath the trees—three, sometimes four high. Here and there one of those names reads out from the stacks—Upland, Badger, Moose Hill—orchards now lost to neighborhoods and industrial parks. Or maybe they have simply been abandoned, and the apple trees are rowed among saplings now, their crowns still shaped by the old prunings.

THE LONG HAZY days are given over to clear September light. It's on towards evening and I can see Sam's pickup pull into the near orchard. He smokes a leisurely cigarette in the cab of the truck before he steps down and walks toward the Cortland trees. He rubs his grizzled chin. I can hear him whistling. I imagine he's checking to see if those apples are ready to pick. He draws a high branch down and turns over an apple, another, and then releases the branch, which is so heavy it only sluggishly returns to its place. He goes on to another tree. No one could look more assured than he as he sets his gaze ahead and walks further into the orchard.

Why is it so hard for me to see? It's a moment that could keep me here. A moment when I can almost believe those are long strides he is taking straight into the future.

TURN-OF-THE-CENTURY PICTURES OF this farm show a place that is nothing like what I know now. It is almost another country—all pasture and sky, with the house and carriage house and barns in bony relief. Half of those buildings still stand, but the land is long since changed. You'd have to search deep in duff to find traces of the far fields, search beneath needlebed, seedbed, leafbed, because white pines have grown over those fields, pines that are tall and soft and have nothing to do with us. We keep only the near land clear, the choice land, rich and level. And an acre is less work than anything our predecessors dared dream. What took them exhausting days to turn over in a cold spring takes part of a morning now. "I can still plow," my father says, "it's no more work than driving a car."

When I was a child there were still a few cleared places that weren't cultivated. The land was marginal or stony and so kept as meadow—a fallow mix of grasses, wildflowers, and weeds—playing field more than anything. They were mowed several times in a summer, not for the hay, but to keep the pines from encroaching. My father was late in years when he began to plant apple trees in those meadows. Cleared places, they seemed, for his dreams and ambitions. He planted redder strains of Cortland and McIntosh, and varieties not found among our old standard trees: Macoun, Jonah Red, Spartan, Empire. Who were they for, those trees? He was late in years, as I say, and hunched down packing in earth around whips that were thin as fly rods.

Anyone who would plant an orchard must be undaunted by time, willing to wait long years with little chance of seeing the finest seasons. And since an orchard is land narrowed to one crop only, anyone who would plant an orchard must abide by the final decisions. The chosen rootstock, size, variety, the methods of pruning, are promises that can't be gone back on, promises requiring care to the end. So, who were they for? I had nearly forgotten the old faces. I had long swallowed the old names. And now those trees have come to trouble my own future. How much easier for me if those places had remained meadow, playing fields seeded with timothy and redtop. Easier for me if *all* the fields were hay. I can imagine the spare, expansive beauty of timothy and redtop seeded over our whole story, seeded over all the stories, the stones once again unturned, and the soil undisturbed. Nothing there'd be to mark history with, no narrow rings for drought years, no wide rings for plenty, no familiar crowns shaped by hand or shaped by the westerlies. Timothy and redtop give in to all the winds. They'd lie down easily beneath needlebed, seedbed, leafbed, the way this orchard never will. I have dreamt of the long sweep of a burnished inland sea, now glinting and now in shadow, always a bed for my weariness, a resting place for my eyes. I hear the way those grasses make a consoling sound in the wind, a harmony of reeds so unlike the winter branches clattering.

PEOPLE SAY THE Blue Hubbards look like seals resting on the curves of their stomachs. Round-shouldered, full-bellied seals. Blue Hubbards *are* a lumbering kind of winter squash—all middle tapering towards a blunt stem end, and the blossom end can perch like a small, inquiring head. Like seals in shape, yes, but a Hubbard is warty and frosty blue. Its heft is too great and the rind too thick for even a chef's knife to be of much use. You have to drop it on a resistant floor to split it open, or use a cleaver the way my father does when he cuts one up for seed. His aim is as deliberate as a stonecutter's, and the two halves cradle away to reveal a thick wall of orange flesh, its surface mottled with white seeds. He milks the seeds from the flesh and scatters them on newspapers to dry. *Milk* is his word for it, and it does look as if he's working a cow's udder, which he had done as a child and long into adulthood.

There are hybrid vegetables, improved varieties, disease-resistant strains. The kinds of corn my father will grow, or tomatoes, or peppers change from year to year, and he buys seed from Harris or Agway. Hubbards are one of the last vegetables he still cultivates from its own seed.

Elderly customers are always remembering the nutty, sweet taste of Blue Hubbard and what a smooth pie it makes. But the best Hubbards are upwards of thirty pounds—far too big, they say, now that they live alone. So we split the Hubbards and sell them in quarters or sixths. Some people will still buy whole ones to eat at Thanksgiving or Christmas, kept until then in a dry corner of the cellar—two or three snugged on a shelf as in other years, while outside a shutter whinges, and the blown leaves are covered with frost, then snow.

❦

ALL THOSE PEOPLE who would rush in at the end of the day to pick up a few things for dinner, they no longer come. There's little left to buy in these last days of October—field pumpkins, Buerre Bosc pears, four kinds of winter squash. The apple crop has dwindled to Baldwin, Northern Spy, and softening McIntosh. Earlier in the year, we harvested the cabbages with their outer leaves intact, and they resembled blown historical roses. Now the exposed parts of the plants are flecked by age and frost, so I strip the outer leaves before putting the cabbage out to sell. Shiny, tight, and pale, those heads seem strangely plain, though no more so than the swept, empty corners of the stand.

We face the same slowdown as always, with drawn-out time between customers who buy a bushel of squash for storage, or a bushel of mixed apples. A woman's brother in New Orleans wants the same Thanksgiving he's always had, and she spends a long while picking out four thick-necked butternuts to ship to him before the holidays. It's not a time to expect anything new. So late last week when I set the walnuts out for sale, every customer had something to say: "For Heaven's sake . . ." "Will you look at that . . ." "Your own walnuts now . . ." "I never knew they'd grow here . . ." Some would sift through the basket and test the heft of the nuts in their palms. Some would buy a generous pound.

The two walnut trees were a long-ago gift to my father—English walnuts, though a hardy strain from the mountains of Poland. For as long as anyone can remember not one nut had been successfully harvested from those trees. Some years the squirrels got to them; other years the nuts, dry and inedible, rattled in their shells. And no one can say for sure why there was a harvest this year, or if there'll be one again. All we know is that Sallie gathered five bushels from out of the grass and fallen leaves beneath the trees. The leathery husks of the walnuts, split and softened, left a dark stain on her hands as she gathered them and stained her hands again as she scrubbed the husks from the shells: a persistent dye that remained for days in the crevices of her palms.

Five full bushels gathered and scrubbed, then cured in the empty, tepid greenhouse where the men eat their lunch now that it is the warmest spot on the farm. And since there's little left to talk about, they might talk about walnuts while they crack them, as if they were cracking skim ice—with the heel of their palm, or a hammer, or the butt of a trowel. Boat-shaped half-shells litter the greenhouse floor. And the nutmeat, since it is still green, tastes sharp.

Five full bushels is nothing compared to the thousands of bushels of corn, tomatoes, and apples harvested in a year, but walnuts have starred the calendar now, and next year there'll be a handful of customers who say, "Shouldn't the walnuts be ready about now?" Or, "My wife tells me she bought some walnuts here last year." It will be about the time the year tightens. It will be dark before people are halfway home from work—the way it is right now, with the Boston traffic forming a chain of headlights that stretches all the way back to the eastern horizon. The many tired and abstracted, all following the same graceful curve of the road.

WHEN I WAS a child, the Thanksgiving turkeys arrived in late spring. Twelve hundred day-old poults huddled under brooders in the red shed. Some evenings after supper we'd trail after my father as he went to check them. The red shed was stuffy and smelled of hot sawdust; the poults, penned into half of it by high, tightly woven wire. When my father entered the pen, hundreds of poults scattered from where he stepped. He'd grab four, one for each of us. Heads tucked in, they were just palm-sized balls—all heartbeat and nerve. Warm, even to our warm touch. Nervous as they were, a young poult could fall asleep on me in no time. Then its beak might dig hard into my forearm. I remember times I'd jump back it hurt so much, and drop it, and it would tumble into the sawdust and scurry for the brood.

While we held our poults my father saw to the feed and water, and he scooped out the pecked-at chicks and settled them into a small pen of their own. Now and again he'd find a dead one.

It wasn't long before the poults began to grow long in the neck, their heads turned scarlet, their wattles sagged. White feathers sprouted at awkward angles from their backs and wings, and they were herded into the larger open-air coops. The evenings grew longer and we played outside after supper in the willow, or on the rope swing.

All November the phone would ring off the wall with people calling to order their turkeys. My mother would write down their names and the weight they wanted in a large notebook by the phone. Some would request hens or toms. A dozen or so, the largest turkey we had. Extra giblets if we had them. The turkeys would be slaughtered the week before Thanksgiving, and we'd sell them in the five days before the holiday out of what we always called the turkey salesroom.

It doesn't fall in line with the rest of the farm buildings. It's almost domestic: one story of white clapboard with green trim and double-hung windows, six lights over six. There is a small overhang, almost a porch. The door is front and center, and its center panel is glass. I found out a few years back that it had been built as a tea room for my aunt, called "The Red Wing." "You'd be surprised at the business," she told me. "By then there was regular traffic between Lawrence and Lowell. Your grandfather built it for me; brother Charles built the tables. We sold sandwiches and ice cream, and my pies." In remembering this she sat on her hands and swung her legs. She was silent for a moment, and then smiled: "Anything to keep me home."

In the days before Thanksgiving, my sister, my brothers, and I would spend much of the weekend and after-school hours at the salesroom. There was a woodstove in back, and between customers we'd camp around it drinking sweet milky coffee. The smell of wood heat and coffee held sway against the gamy tang of the fresh-killed turkeys just as much as the cold and the iron light of those short afternoons.

Customers would come and go and we'd try to help, though most of them insisted on seeing my father.

"Can I help you?" I'd ask.

"Is your father around?"

Those days marked the real end of the farm year, and everyone, it seemed, would want to close it down with a little talk. Friends of my father would come in with gifts—whisky, sometimes Edgeworth tobacco—and stay awhile. Stories swapped. A pipe lit and then drawn on. A bottle opened and poured into coffee cups. A final ease with nothing to do but sell down those turkeys, all ordered and waiting to be picked up. The last customers would come in at dusk on Wednesday.

My father would always save a few dozen turkeys and freeze them, which we'd eat on Sundays all through the next year. And he was always prepared for the inevitable call on Thanksgiving morning. I think he loved how the caller would be desperate. They had left their turkey on the porch—there wasn't room in their refrigerator—and the dog had eaten it. They had tried cooking their turkey overnight in a low oven just like the magazine had suggested, and now it was charred beyond recognition. They'd take any size, even a bruised one. Please, did we have any left?

There haven't been turkeys here for twenty years. The new slaughterhouse regulations made it impossible to continue without expanding considerably, and at a considerable cost. In the last years it had been hard anyway for my father to find men willing and able to help with the kill. The turkey coops are storage sheds now. The turkey salesroom—dry and tight—is where Sam stores the seed. Even after all this time my mother gets several phone calls every November—someone inquiring: "You still have the turkeys, don't you?"

❦

ONLY A COUPLE of men stay on to finish out the season, their day beginning in slant light and ending in shadow. There's no hurry now—if it rains or a bitter wind blows in, the work can be put off for a better day since they don't have to keep pace with ripening crops or a parching sun. Water jugs go untouched. It's coffee they drink with their lunch.

They are putting the farm away—cleaning the barn, draining pipes, grading out the last of the stored apples.

Machinery is neatly rowed in a shed: tractor, planter, sprayer, harrow, plow. There's frost nearly every morning, and in the orchard all but the last leaves have fallen. You can see how the pruned limbs work out of the squat trunks—the bare wires of a cared-for land.

Now all the fields are in rye. Winter rye is a cover crop only, planted in fields after the corn has been harrowed down. Its dense roots work to keep the long-broken soil from eroding. In mid-September the first planting of it shoots up a minty green, while in other fields squash vines wither to expose tan butternuts, and pumpkins absorb the late sun. By the end of October its deep, even green overlays burnished husks, cobs, and broken stalks. It's a nearly perfect green, without row or furrow, a green that brings up the well-worn roll of the land as it slopes towards the river beyond our view. To those who've worked here all spring and summer and into the fall, the farm now looks surprisingly plain, a landscape only, surveyed and wide under too big a sky. And your eye is drawn to all the slight movements—a hawk, say, and its shadow, or a rabbit scrabbling across matted grass. A deer at the edge of the woods raises its head—such a small stir, no more than an orphaned meteor—and I'm left wondering whether or not I saw it at all.

Wild grasses and orchard grasses turn sere, but winter rye does not. It remains green under accumulating snow, green as ever on south slopes in the January thaws, and everywhere green come the first March floods. Only on the late land—low and boggy, the last to be harvested, and so the last to be put in a cover crop—is the ryegrass sparse.

There you could count each blade if you wished, and in a dry season unanchored topsoil will blow across the thin snow when gusts come up.

Dusk comes soon enough. The two men hunch in the front seat of their sedan and let the engine warm up a minute before they head home. Lights from their small city star the east hills. One evening planet in the sky before them, the crescent moon growing whiter and harder in the sky behind. Straggling migrations of Canada geese settle in the rye to feed on what remains of the corn. Some evenings they are countless. Even so, they make scarcely a sound. It seems late for them to be this far north, but I suppose it's the plentiful grain that keeps them here, and frozen ground will drive them on.

THE BRIGHT COLORS are just behind us. No unfallen leaves remain except for clusters of wine-dark ones in the low centers of the oak, and a burnished singular few on the outer branches of the apple trees. It's late. A hunter's shots have scattered through the afternoon.

I am alone on the farm stand when dark comes on, alone as I hasten through the closing chores. I sweep and cash out, then take one last look around before the lights, the lock, the door. I see bare shelves and half-filled bins, a space nearly emptied of its season, its atmosphere filled with the scent of late apples and winter pears—the long slow draft of last things that sweeten only after their harvesting. It marks the end of all we've done, that fragrance, and it dissipates the moment I shut the door behind me, dissipates clear to the first stars of the evening. I sense nothing now but the cold coming on.

It won't be long before those flavors, so keenly deepened, turn at last. The time ahead will in no way resemble the hours already spent. And we, who know only the things of the day, the steady work of our hands, what will we do? It has proved less than a part of all we need, this sift of earth, this deliberate dream.

4
White Pine, Relics, Rust

WHITE PINE IS a given name. It has nothing to do with their blue-green needles or their branches, their cones or fissured bark. It isn't tied to their wintry scent. We call them white pines after the quality of their lumber. The dressed wood is the palest of all pines, close-grained, soft, and long-prized.

Always the stories tell how white pines reached over the Northeast—stands of them that neared two hundred feet in height, many so true they were felled for the masts of a foreign navy. The rest: squared off to exhaustion. What remains of those old forests are the stories only, and perhaps a mullion flecked with milk paint or a floorboard scoured to a deep patina. In early homes I've seen wood panels that must be almost two feet wide, and still I can't comprehend the vanished size of those trees. The white pines I know, the ones I've always believed to be characteristic of these parts, are no more than a century's growth. People call them old field pines because they took hold on abandoned farmland, thriving on thin soil and the bald heat of the days. Their wood is rarely as clear as that of the deliberate primeval growth—boxes, matches, pulp. Still, white pines are the most majestic tree I know.

Behind my house there are four of them that have been full-grown for as long as anyone can remember, and now they must be a hundred feet tall. They're so far above our accustomed sightline of berries, sheds, and fruit trees that my eye is drawn to them from wherever I am on the property. Even when I'm indoors they sometimes come to mind. During a strong night wind I might start thinking they'll go down, though they've withstood a handful of major hurricanes and all the squalls in between. In truth, they are shaped by such exposure: knotted, warty trunks that are too staunch to creak in rough weather, and crowns that lean in from the north as if they've kept the brunt of the wind in them.

In winter gales their boughs don't clatter like the birches. They don't work into your thoughts like the oaks, whose dry leaves rasp. White pine needles are sheathed and soft; their branches, supple as muslin. It's a steady, clear flame of a sound they make. I can hear it way above the roof—ash-throated and abiding.

In the woods where one of the larger trees has fallen, the surrounding ground is flecked with white pine seedlings thriving in that keyhole of light. They are nothing but long soft needles, those young trees, whorls of needles hiding trunks no larger than sticks. Elsewhere, the woods are too dark for the new pines to survive. Instead, hardwoods grow. It seems those other trees will take over later on, but for now white pines are prominent, and after the yellow and red leaves have fallen, those pines stand over the land.

It is then, in winter, that you see how their high-topped crowns are broad and irregular. Some branches shoot away from the rest—out and up—as they follow the lead of the sun. Lower limbs, long blocked from the light, have pruned themselves back to skeletal remains—broken, without bark, stained a moss green. The trees don't thicken and crook the way a solitary pine might, but are slender and straight.

I have never seen the owl, though I have heard its call, rounded and hollow. When the winter wind drowns out that call it's the white pines that sound in the woods. With all their length those trunks creak as they sway to the limits of their axis, and the creaking builds and decays, builds and decays just as if it were the old uneasy sea pitching beneath them. That creaking dwells in the woods, never clearing the boughs, which have their own sound—a persevering, swept note—distant, with nothing to stop it.

I KEEP MY bearings by following the stone walls that figure through these woods. There are breaches in some of the walls where stones have toppled, but many remain as they were built—adroit and lastingly precarious. Every once in a while I come to a place where the walls are seamed clean as fontanels, as if one stone had been fractured somehow. The stones themselves are rarely white or brown—mostly gray, of all shapes, and not often bigger than a man alone could hoist. Exposure has dulled and smoothed their surfaces, in dank places moss grows on them. They are mapped with lichen.

Not long ago these woods were pasture and fields and the walls were made from stones that had been cleared from the land. You'd think there must have been more stone than soil, since I can't begin to count the miles of walls that wind through these woods and the surrounding countryside. And the soil is unpromising still. In our fields, the frost and machinery work new stones to the surface all the time—coarse ones, the quartz still gleaming in them. As ever, they are cleared away. But now they are simply piled in a spare corner of the farm. They're glacial debris, really, rough granites deposited when the last of the ice retreated. The colonists called them fieldstones.

They say it was the European plows that brought so many stones to the surface—the land had never been so thoroughly worked before. And the time told here had never kept so close to the tick of the clock. It couldn't have taken long for the plowed-up stones to dovetail with the need for livestock fencing and a dream of fixed property lines.

Well, property lines vanish as quickly as freak snows, and now these walls seldom coincide with any boundaries on the surveyors' maps. They're relics after the fact, the light we see from collapsed stars. Some edge roads and fields as if still in place, but largely it's through these woods they maze—gapped and turned back on themselves like an old woman's furred-in thoughts.

❦

IT'S AFTER SNOWMELT and the early rains, and the brook has settled back into its old known bed. Just last week the flood current poured over the smaller sounds, but now that the flow has calmed, each disturbance—eddies, channels, near water falling over roots and tumbled granite—makes its own soft-throated sound, and I can't hear the end of them as I sit by this broad place in the stream where moss-covered stones scatter through the bed. A child could cross stone to stone, and here is where we did cross as children, tottering over the water, our arms outstretched for balance.

If I just sit in this one place and gaze at the sunlit water, and listen, and breathe in the spring green, I can almost believe it's as cool and removed as when I would come here thirty years ago, here just to be away. A child picking her way through the tangled fallen brown of these April woods. It smells faintly of skunk cabbage, whose purple hoods mottle the low places. The fiddleheads are curled tight in their papery cauls. The shaded remove of an already removed place.

Our farm lies at the eastern edge of what is now a moderate, sprawling town. Though this had long been an agricultural community, by the time I was school-aged it was edging towards suburbia, and there weren't many farmers' children left. School was on the other side of town, in the most built-up and densely populated area—a far journey there, and an even farther journey home—a crossing, really, back into my own world. The area around the school was a city to me. The cotton mill at the Beaver Brook dam and the blonde brick Catholic church rose above squat businesses. There was almost no breathing room between Turner's hardware—sacks of Blue Seal out front, pickup trucks pulled up alongside—and Brown's Appliances, where the white enamel stoves gleamed in a sturdy row behind the plate glass. Village Pizza, Dalton's Ice Cream, Rosie's Variety, which except for the Squirt and Marlboro signs seemed a shingled cottage among the storefronts. There were pots of geraniums on the stoop in spring. Becket's Cleaners, LeBlanc's Real Estate, Marie's Doughnuts—at that hour, a stretch of mostly empty stools, maybe an old man rereading his newspaper, a younger one bent to the contents of his briefcase, two women talking and sipping coffee.

After a few miles the businesses fell away, and our foreheads rested against the glass. The driver downshifted as we turned from the state road onto Coburn Road, the first of a network of back ways that made up the rest of the route home. They were the old two-rod roads, my father has told me, originally measuring two rods wide from stone wall to stone wall. They used to be the one way from farm to farm, from farm to town. At a few points along the route I could see them pay out across the territory for miles ahead, breaching the hills and scribing the broad valleys, but mostly the roads were so curved they only opened up right in front of us, suddenly revealing the old houses built into the bend in the road or at a convergence of slopes. They sit at the heart of their land, those houses—worlds of their own—surrounded by barns and outbuildings, maybe a windmill, a silo. Pollen clots the hand-dug pond. And their fields stretch away in every direction.

By the time our bus traveled those roads, many of the farmers, even if they kept a small dairy herd, had other jobs, too. The acreage had closed in on their homesteads, and the spaces between sprouted clusters of subdivisions, little mazes that stopped abruptly as they had begun, like those towns floating on the Kansas plains. The streets that reached back—horseshoes, or grids, or circles—weren't guided by fences, and had no destination other than to return to the old road. I could see down whole rows of houses waiting to settle. Clapboard Capes, half-bricked Garrisons, low ranches—more similar than the lives inside could possibly have been—were built where they had been marked no matter the lay of the land. They set squarely in the center of their lots, parallel to the road. And a planted tree—split-leaf or red-leaf maple—in every yard, equidistant between the house and the road. Basketball hoops on the garage peaks. Hopscotch squares chalked on the drives. The bus would slow at the first corner of the development, and all at once six, eight, ten children would gather their books, file out the open door, and scatter like struck marbles.

The busload lightened to a handful of children and the houses drew farther and farther apart. My cousins, my sister, my brothers, and I were the last stop in the afternoon.

Winters we'd step down into the approaching dusk. In the east a lone radio tower pulsed red just above the oaks. The evening paper had already come, and I'd collect it at the mailbox and go in.

The distance that bus traveled is almost nothing to me now, but it's also true, of course, that the world has closed in. To be a child here would hardly be the same. Even the far edges of these woods have been claimed by houses. I now see lights through the trees to the north where once the view had nothing human in it. The new people are more affluent than those who had come here when I was a child, and their houses are larger and more elaborate. The lots, larger; the landscaping, more elegant.

Yet among these subdivisions, there are still a good handful of dairies, orchards, produce farms, about the same size as ours. Every once in a while another may fold—the taxes are high, and the land values are high, as are start-up costs, so if there's no one in the family to take on the work, it's unlikely the farm can continue. There are flare-ups with each new development. Old residents can be cold-eyed about the new ones; new residents are never prepared for, say, the smell of Thayer's pigs or his piles of spare parts and tractor tires, the rusted rainwater pooling in his yard, the plastic on his windows all summer long.

Many of those families in the houses between farms are faithful to the locally grown corn and tomatoes, and they carry home their milk in glass bottles. They may stop by the stand three or four times a week. "I wish you were open year-round," one says to me. Another, "I tell my daughter this is our garden." And, another, spending a good while longer than really necessary choosing her apples, tells me it's nice we're here as I ring up her order. Then, "Don't ever sell."

It's not a sentiment I've ever heard expressed by those who live farther away. Many of the customers who come from Lowell or Lawrence speak Spanish or Khmer, and when they come to weigh and pay for their produce, their children, some of them no older than seven or eight, speak to me in English, "How much for this?" then turn to their parents and speak in their own language. And turn to me again with payment. If there are no children, we might negotiate wordlessly. A man holds up three cucumbers with one hand and I pick the correct change out of the coins he holds in his other, open palm. "Thanks." He nods or closes his eyes and then opens them again, and heads away.

As the last of the open land, this part of town is zoned for industry. A dreamscape for everyone who runs for selectman, since they can conjure jobs and an expanded tax base out of the deep woods. And there has been some development. What was once the old Coburn land quickly became a regional office of New England Telephone. As fluorescent surveyors' flags appeared, the talk geared up. I'd hear it at the stand. A man working his toothpick against an incisor saying: "I hear two hundred trucks a day will be heading towards the interstate."

And the man answering him shaking his head: "They say they'll be starting at 6:00 a.m."

"That early, really?"

"Don't think the road can handle it."

"I know, I know."

It's only been a few years since the phone company went in, but already it's hard for me to remember what the woods there looked like. We get a flurry of the linemen's trucks pulling in the stand at the end of the day. There's one who loves spinach—he keeps telling me to sauté it with garlic and olive oil. Another buys a Bosc pear and a McIntosh for his next day's lunch.

It's built up enough here now so that if you have a far point in mind you might pass by without noticing the butternuts full in the September sun, or—if it's mid-July—that the Seneca Brave is ripe and the talk of all our talk. There have been customers—strangers—surprised to find us here. I remember one woman who wouldn't believe our cleared land could produce enough to fill up the stand. She asked me, "Where's the farm?"

"Here."

"Where?"

"Here," I said again and swept my hands full circle.

"For all this? I don't believe it. They must buy some of this stuff. Are you sure they don't buy some of this?"

"We grow it here."

The rare times larger animals cross through anymore—they are being squeezed out to the north—they cause enough mayhem for a story in the local paper. It was front-page news when a black bear cub came out of these woods and squalled across the fairway of the nearby golf course. Sometimes the stories even make the Boston paper, as when late last summer a bull moose broke out of the pines and trampled across the backyards on Cedar Lane. By the time the first squad car appeared it had made its way to Hawthorn Road, where the officers called for backup. "It's the size of a horse" came the plea over the scanner, which brought countless curiosity seekers along with three more patrol cars. In the end they were as helpless as the first, except for keeping the people at a safe remove as they waited for the expert marksman, who was to shoot the moose full of tranquilizers. It took a while for the drug to take effect, and that huge mammal splashed through a built-in swimming pool—as he emerged the water sheeting off him sounded like a waterfall—and knocked over a gas grill before he turned, bewildered, towards the crescent of onlookers, and crumpled against a lawn chair.

Even so, some still come here, as I think the bagpiper must, because it does seem remote. I have heard him playing his pipes at the cemetery. Sometimes he's in full dress as he stands at the top of the hill. He has his back to the woods and faces out across the graves towards our fields as he plays. I don't recognize the tune—it seems a long sequence of variations. His drone pipes carry farthest on the cloudy days, a mournful bray, the music itself making the place feel lone and distant.

And by this brook, on a mild spring afternoon, there's enough woods left for much to remain hidden. I rarely see the deer, but I do see their sign—scat, ragged browse, and the path they've worn along the water's course. Sometimes I hear coyotes from my bed as I try to sleep. Their calls harp on the storyless night and, same as strangers, they set all the dogs barking.

❦

THIS LAND HAS a kind old roll to it. Subdued hills stretch on for as far as I can see, their slopes as gradual as the hips and shoulders of a figure in repose. Or I'm reminded of the fluid curve of starlings sweeping towards a lone tree. Nothing remains—not that the scanning eye can see—of the obtrusive, jagged peaks of earlier ages. Nothing of a time when lichen and moss were the only green; and sound, a stone tumbling down the scree. Erosion and long, slow accretions turned those ages into quiet country; and that country has been worn to these broad valleys and low rises by quickening streams veining the soil on their way to the sea. Geologists say these hills are what remains of a peneplain, which means *almost a plain*, and if you search back far enough you can find *patience* in the word.

The Pennacooks farmed it first, without a wheel or a plow, in fields they'd cleared with granite axes and fire. They didn't turn over the earth, but used spades made from the shoulderbones of deer, or kept to the lighter soils where a planting stick could make a hole for seed. Everything hoed and hilled and harvested by hand. They say you could smell the herring buried in the older lands, and that the fields looked nothing like the ordered rows I know—one crop planted alongside another so that the hard northern flint tasseled above beanstalks or a tangle of pumpkin vines. I imagine it was hard to tell where their care left off and the wilderness began. At the edge of their fields they thinned the grapes and wild berries. In the forests beyond, to improve the hunting, they set fires to keep down the saplings and brambles. Footfalls on duff sounded softer than the beat of a siskin's wing. In the long winters, cut pumpkins, dried beans, and braided seedcorn lay deep in their stores. Now a white stone washes onto my path, and I can't tell for sure if it's rough quartz or quartz worked into a spearpoint or knife.

The English settlers cleared those fields more thoroughly, and cleared their hunting grounds and fishing camps and settlements until the land resembled a woven pattern of holdings: small blocked-out acres of tillage and pasture divided by fences and hedges or a thin line of trees that bordered a stream. As before, corn, squash, and beans. Also cattle, poultry, and sheep. And teamed oxen—heads down, wide-eyed, straining against a yoke. You could hear their labored breath as they worked across the heavier soils. You could hear struck metal—curt, then reverberant—belling across the open fields as shoes and blades were hammered into shape.

One of the old plows hangs in the back of the barn. The handle is fashioned from clear-grained wood, and it has an iron coulter. I can see how the man behind guided his own dreaming, which yielded only to the contours of these hills. Dreams laid down on soil full of stones and early frost. Soil thinning down the hundreds of years of such dreams—children and children's children in a cloud of raised dust behind the oxen, or raising their scythes to mow a field, stepping into the stubble they themselves had made to sweep the blade again and further the swath in the grasses. The hay wilted, then dried, and as it dried its scent lingered to the day that one mower hung up his blade on the low branch of an oak tree and walked off to join the Union cause.

That story about the scythe is a story from states away, but it could have happened here, too. My father told us about it once when he came back from the woods after finding a horseshoe embedded in the trunk of a maple. He supposed someone had hung the shoe on the tree maybe fifty years ago and, as the tree grew, it grew around the shoe. "Just like that scythe in the oak," he said, to which the mower never returned. To this day it hangs on the branch, its grip worn by weather, the blade's arc gone into the growth rings.

I've seen the lists of the Civil War dead from the small towns near here. The same few surnames—Richardson, Varnum, Coburn, Clough—scroll down the page. And those who returned had seen firsthand the long growing seasons to the south and the richer topsoil to the west, against which their own toughed-out acres must have seemed like folly. Why wouldn't they dream of the tall-grass prairies, and of being the first to turn over such soil—stone-free, level, and deep in loam? I was thirty before I'd ever driven west, and I couldn't stop looking at the long horizon. The plantings disappeared straight into it, and there was always that sky. I remember the small towns floated like islands among the crops, and every made thing seemed outsized—the barns, the silos, those combines. When we say *farm*, I said to myself, we don't mean the same thing.

The war only hastened the long slow abandonment of the New England land, which the opening up of the West had made inevitable. There was no way to compete with crops being grown more cheaply and efficiently on better soils, or soils that simply had not yet been exhausted. The poorer upland farms were the first to go, though I still see one now and again—a handful of cattle wandering a rocky slope or picking out grasses among the pines, a wrackline of saved, rusted machinery alongside the house. One light selves the night, and every time I pass by I wonder *who* or *how*?

With their alluvial soils, the better valley farms hung on, but you could no longer shout the news from farm to farm. In some places, white pines reclaimed the abandoned land between them, and elsewhere mill cities—seemingly sprung whole from the heads of new dreamers—were built along the river. Within a dozen years the Merrimack had been dammed in a handful of places, and red brick factories lined stretches of its banks. In every city, high as the church spires, came a clock—four faces to the cardinal points so you could read the hands from any place in the city. The daughters of the farmers read them from their dormitories, which were often built of the same brick as the factories. On their way to work their footfalls drowned in the countless echoes of others, and once there, the air was white with cloth dust.

The pace of building outpaced the ability of farm families to supply the mills with labor, and eventually the workforce had to be drawn from other countries. Both my parents' families came here at the end of the last century to work in those mills, and sometimes I'm not sure which side of the family the stories come from—stories about strict time, the boss over them, Bread and Roses. How all the doors in the tenements opened at the same time every morning as the city set off for work. My mother's family stayed in Lawrence, and her father worked as a weaver all his life. My father's family bought this farm a dozen years after they came here.

"Jesus, I don't know how they got the money together," my father says, and I think he still remembers every trip to the bank and how much the corn on the School Road sold for in 1931. In his story about the red heifer, who had six or seven lactations and was still called the red heifer, he remembers how much she yielded and the exact price of her milk for years on end.

Most of the farms that survived survived on their dairy herds. Their connected buildings, which made for easy passage in winter, have created the classic picture we hold in our heads: a beauty of form and line, angled and neat against the lush summer hills. A moderate slope to the roofs; the doorways, plainly framed. Children made a song of those structures as they played their games in the south-facing yards, chanting *big house, little house, back-house, barn.* Front parlors, summer kitchens, the scent of lilac. Their marker trees shade our driveways, and the peonies they planted to make a tea for nightmares are arranged in vases on our tables.

I like to imagine the journey of that milk, how it was a scant seven miles from here to where it was finally set on a doorstoop in Lawrence. It began in the calm of predawn, and the cattle shifting in their stalls. The air in the barn is warm with the heat of the animals, and humid. It smells of hay and dung. The tin swallows its own resonance as the milk fills the pails. "I can't believe they trusted me with the milk," my aunt keeps saying these days. "I wasn't more than five or six—tiny—and they had me carry it all the way to the well." It would stay cool there until the wagon from the creamery came to collect it. As the wagon worked back out of the yard, the burdened wheels turned slowly. The sky just beginning to pale. In the east, the last slip of the moon.

At the creamery the milk was combined with the milk from other farms, poured into clear glass quarts, and loaded into another wagon to be distributed to the households in the city. The glass sweats. The road is cobblestone now, and the bottles chime softly as they nudge against one another. A child jostles among them. In his backward view, the buildings lining both sides of the streets rush away. The shod hooves clatter and set off sparks as they strike the cobbles. When the road crosses the Merrimack at the Great Stone Dam the water pouring over the boards is loud enough to take away his thoughts, and he can hear that roar until the wagon turns away from the river and into the maze of lanes and alleys where people live.

As he climbs the stairs to set the milk on the landings of the tenements, his boots scuff the sunk well in the middle of each step. The paint, if there had been paint at all, is worn to bare wood from all the comings and goings. Sometimes he hears a scurrying before his steps. He sees shadows—or is it a man?—moving among wooden barrels. So early, and already so much going on. Many voices through the doors—the murmuring of those who've just woken, and must get on with the day. And children calling names he's never heard—*tante* . . . He smells yeast proofing, and bread baking, and long-lingering spices. Alcohol. Urine. Naphtha and lanolin come in on a small wind. He is supposed to be quick at this—three landings in each tenement, and he has to leave the milk at the foot of each entrance and pick up the empty bottles left for him. As he runs back down the stairwells—the man at the reins is impatient—the empty glass against empty glass sounds just like ice falling from the pine branches in the January thaw.

The way the ice is falling now in the woods behind my house, from pines that have grown up just since the hurricane, on the same land that may have been Pennacook hunting lands, and then the cleared fields of an early settler. Even so, the lay of the land itself—those payed-out hills—is still so much the same. And the changes wrought—the ways of understanding, of forging tools, of telling time, the negotiated shapes we dream—what do they mean against such patience? That dairy herd is gone as well. There is a dump in one of the low places in the woods where my father has abandoned the things from those years. Milk cans, like nurse logs, are half sunk in the soil. Their exposed parts are already thinned to tracery. Some bottles are there, too, and on days like this—warm after a winter storm—I can't tell what is shattered and glinting—the fallen ice, or glass.

WHEN THE BARN was thick with animal warmth, snow melted as it fell on the roof. Now that nothing but tools and machinery are stored here, the barn is cold in winter. On windless days it is almost as cold as the open air, and snow accumulates on the roof, in some years lingering into March.

One day voices echo in the frost, the next they are lost in the sound of the thaw. This morning water blistered along the edge of the barn roof and began to drip—in measured time at first—and then it quickened as the day grew warmer. By noon it seemed that all the snow of the lean months was streaming off the roof. The dripping incised channels in the accumulated snow around the barn, and then worked channels into the ground itself—lines more true than those of the building, which has long since sighed and settled.

Lines as true as the ones in the carpenter's dream. One man built every barn still standing along this road. No one remembers his name, but you can tell his work by certain details. Above the door he always hung a row of six lights, which sometimes fogged with the pluming breath of the herds.

❦

THE MILK ROOM was always slab-cool and dank. A square fieldstone room in the southeast corner of the barn, it has one window—four lights painted shut. Years ago they stored the fresh milk there, but the last of the dairy herd was sold way before I was born. Pasture turned to cropland, and we store baskets of harvested crops in the milk room: beans, cukes, tomatoes, summer squash, green peppers.

Where some of the pasture has grown into woods, the stone walls are strung with rusted barbed wire, and every once in a while you'll see a tree growing up through the wire. There'll be a broad welt on the bark where the strands have healed into it. The disused mail road runs along some of the walls. Tangled slash and undergrowth have not yet encroached on it, and the wheel ruts are apparent. My brothers, sister, and I wore the right one to a narrow track of dirt the width of a foot. The left one is faint and grassy.

HEAPED IN THE corners of the barns and shed, or slung on driven nails there, are the scythes and harnesses, the horse-driven harrows and plows. A sower's apron, a grain scoop, a hay rake put away for the day that turned into years, then into seasonless decades. Crusted, creaky things, with gray dust sifting off them. Beyond the reach of the sun and the damp night air, unstirred by passing voices or the swallows nesting in the beams, they await the last say, like dear letters before the final parting—packeted, stored, and—since every end can still be believed—sometimes unfolded and read for remembrance.

When I poke around in those corners, I almost always find things I want to take home where I can scrub the must from them, and make the grain come up, or a blade brighten, or the cracked leather soften. Old tools become my doorstops and bookends and range on the kitchen sill. I've hung the draft horse's hammered shoe above my doorway, and packed soil into milk pails to set pansies in.

Many are the things I find that I can't identify; when I show them to my father his memory jars, and he'll ask, "Where did you find that?" or exclaim, "Why, that's an antique . . ." then tell me it was what they'd used to saw the pond ice, or cut the curd, or funnel sap. The day I showed him the fawn-footed handle of an ax, he turned it over, and tested the feel of it in his hands. He must have remembered his old strength as he tightened his grip. I swear his look was to warn me again that any tool will falter in an unskilled hand.

PEOPLE MARK TIME by the '38 hurricane. "That stand of pines has been there since before the hurricane," they say, or, "The stream didn't pass that way until after the hurricane." Big stories about the storm come from the coast: buses swept off roads, the late summer gentry caught watching from their windows—the way they always watched line storms—as seas broke over their houses.

We're thirty miles inland, and a good distance from the Merrimack, so the stories we hear aren't about floods and tides. Our place was all wind, sounded by things in its way—branches, clapboards, loose panes. Wind flattened the cowshed, they say, and pried the well cover from the well. It knocked over fences, both whitewashed and stone, and sheared laden branches off the apple trees. In the cemetery it uprooted a grave. Wind turned the woods into a foreign place—a haphazard horizon full of slash and sky; trees strewn every which way, exposed sapwood, and gapes to draw a stare, I imagine, like powerlines cut through the wilderness or ski trails on a summer mountain.

All my uncles love to tell how they worked long days hauling hurricane lumber to the mill. It took them years just to clean up their own little woodlot, they say. I've seen pictures of them, leaning on axes, hats jaunted back on their heads. They're posing by downed trees, and seem like victors in the photographer's eye.

❦

MONADNOCK. THE EXACT meaning of the Algonquin word is lost. Some people believe *manitou*—a guardian spirit—is stowed in the name. Others safely say that it means "mountain that stands alone." Monadnock rises only a few thousand feet above its surrounding countryside, but it *is* solitary, with long gradual slopes, and so seems imposing. And it is all the more imposing because its summit is bare rock, so that when the lower land is running with the thaw, bringing out new scents and bringing up the color of lichens, the snowed-in peak of Monadnock keeps its cold look.

Between the ice age and the present day there was a time when the summit of that mountain was covered in red spruce—a pure stand, dark green and resinous. Those trees must have played down the elements, or they took on the brunt of them. Anyway, they protected the soil, and the summit's amber duff—color of steeped tea—just kept getting deeper and softer.

The stories I've heard say the fires began soon after the region was settled. Fires to clear pasture. Fires that had spread too far up the mountain and burned until they burned themselves out. The one fire that was so hot it killed the red spruce and the soil they grew on. The charred trunks fell where they stood—a black tangle that became a haunt for wolves. It was early in the nineteenth century when the dead timber was deliberately set ablaze to drive the wolves from that place. What remained—ash and barren soil—easily blew away on the constant wind. So, the summit will always be bare, they say, since the weather is just too persistent for the new soil to establish itself, and the old berry-pickers' paths are just too well traveled.

At the summit, cold-chiseled into the rock, I find initials and dates that go back into the last century. As the surrounding land has become more widely settled, the climb has become more popular, since it isn't difficult, and from the top you can see far in every direction. To the northwest rises Mt. Mansfield in Vermont; east of north, Mt. Washington; north of east, in Maine, Mt. Agamenticus. To the south is a quieter breath, the low-lying hills of eastern Massachusetts. Those hills have no focus—except for the higher towers of Boston—since they're nearly even in height.

When I look out from the top of Monadnock I know the farm is off to the east and south, but I can't pick out where it must be. All contour is gone—the hills we plow and build into, the low place in the orchard where the frost lingers, the rise where, early on, the lights of a small city might climb towards Orion. It appears as if all the land has healed back into a plain, the working of all that water seeming to have disappeared, and along with it the barns and fields have disappeared, the back stoops, the millstack, the stone church. Detail lost to the color of a cold flame, as lost as the towns beneath the Quabbin Reservoir, which supplies much of that southern stretch of land with water.

The old residents of those drowned towns hold reunions now and again to recall the scent of the honeysuckle or the sweet taste of a sheepnosed apple—the last details, the way my aunt will recall, not the herd, but her cheek resting against the pied flank of the cow as she finished the milking.

❦

ONE OF OUR fields runs right up to the west wall of the Richardson Cemetery, and springtime burials can put a stop to work there. The sound of the harrow—like a jangling chain—and the entrenched engine noise of the tractor drown out voices, so Sam will stop the machinery as the funeral procession makes its turn into Richardson gate, and he'll wait out the burial service. The worked-over soil is to his left; unbroken land, to his right.

It's not the oldest of cemeteries, and many of the gravestones are from recent years: large, polished granites, headstones on shoulders of stone. On that small wooded hill they appear almost too large, the same as deep sea vessels anchored in a snug harbor.

Almost always you can see surviving family members attending those graves. By the first of December, balsam wreaths are strapped to the granite markers, and potted poinsettias—silk or plastic—are set atop the small garden beds in front of the stones. Every year one woman hangs miniature ornaments from the shrubs that flank her husband's grave. Nearly every grave is decorated for the Christmas season, and also for Easter, when cream-colored lilies plume against the polished side of the stones. At other times of the year the decorations are not as widespread, but they are consistent. On Mother's Day the cemetery will have its own pattern of floral arrangements, different from that of Father's Day. Every veteran's grave is flagged at the end of May, and when Halloween comes, pumpkins of all sizes are placed alongside the graves of those who died young. In the course of the year such configurations are nearly as predictable as the appointed turns of the constellations—when Orion rises in the east at sundown, when the Dipper begins it decline, when the Pleiades, the deep-seated Pleiades, is overhead at midnight.

To the north of Richardson Cemetery is a smaller, older graveyard—Bailey Cemetery. Its markers hold the names of the settlers—Coburn, Austin, Clough—and are mostly portal shaped, some of black slate, stained with lichen. They are half-sunk and skewed. At times passersby will stop to make parchment rubbings of their carved details. Or they'll puzzle over the epigraphs, and finally smile: *Beloved of Almon. She hath done what she could. I shall return to her, but she shall not return to me.*

No one remembers the last burial there, and I have never seen flowers set against the stones, though towards the end of May, veterans will see that the soldiers' places are marked just as they are in every other cemetery in town, and early on Memorial Day those veterans pay their tribute: seven guns fired three times, then a patient and clear trumpet. It is just after dawn. The road is still. The ceremony is witnessed by no one now that all the nearby children are grown. There must have been a dozen years when we'd crouch beside the cemetery wall to wait out the long, single line of taps so we could search the grass for shell casings before being called back to the house for breakfast.

THE WORKED-OUT TRACTOR stops an access road. The lame harrow has been pulled to the inside corner of a fence. And the old planter—at station among shattered grasses by the barn. It's likely they have been rusting for longer than they labored, having been hauled to marginal places after they broke a final disk, say, or were shadowed over by new machinery. After a good rain, the low ground around them puddles with copper-colored water. Rusted to fine things, they are, nearer to script or to vertebrae than the forged shafts, axles, and tines. Who, anymore, can imagine they were once bent to a human will—these that thin and crumble to so much ferrous duff?

And who isn't inclined to think of them as being closer to draft horses at pasture than to the machinery we now use? A new Case tractor parades across the years we have, hauling beamy harrows, planters, and cultivators. A giant bucket attaches to the front of the tractor and is used to move loam or to clear a way for drainage. We tell our time by those outsized wheels that make their slow, peremptory turns and lay down chevron tracks on the soil.

And yet, in the wake of that tractor, not much is changed from other years. Disks and tines disturb the earth at standard intervals. Hybrid seed funnels through the planter and drops out in traditional increments—so many inches between stalks of corn, so many between beans, between cabbage plants . . . East, west, and east again, measured lines follow the well-worn roll of the land.

Scours on stone made by glaciers, scours made by harrows. Where is the trained eye to distinguish one from the other? The distance between the past and ourselves is as thin as the saturated red enamel on the new tractor, though we can't bring ourselves to believe its durable surface will rust. And besides, the machines are too big to be simply left to one side. Imagine this tractor in disuse, and how it would block the easy line of the horizon. Sharp edges in a hazy season. Orchard grasses would never clear the hubs of the wheels, and the raised bucket stalled above the cab would remind us of a saurian jaw and the creatures of a vanished epoch.

5
Winter

In the mudroom, Hubbard seeds are scattered on newspapers to dry. Winter rye has come up in the cornfields. Apple branches are sprung and still except for a topmost bough here and there, still laden, that had been impossible to reach no matter how the ladder had been angled. The stand, swept clean and shut up.

No one ever buys more than one red cabbage—one lasts weeks. Six rows weren't picked clean before the hard frost and haven't been plowed under. The frozen heads appear from a distance: perfect, the color darkened to burgundy. This morning's windless snow is melting back, exposing the red cabbage, all that's visible.

❦

Even under a changeless sky my father tunes in the weather half a dozen times a day. The strong, unhurried voice of the forecaster seems to carry an honest weight across a great distance, since every sentence rises from and decays back into a sieve of static. He knows the litany by heart—daily extremes in temperature, the barometric reading, the extended forecast. Then the voice runs on through peripheral facts—the winds atop Mount Washington, temperatures in Hartford and Providence, Marblehead tides, Boston tides, Hyannis . . . A few are places he has never been.

Now the voice says rain will continue on into the afternoon. If this rain had fallen a few weeks back, he would have worked through the morning anyway, dressed in his yellow oilers that stand out against the drenched land like a blaze on bark. But it is November. There are no crops and little work, and the rain falls without consequence. Even so, by force of habit, he sizes it up. This was the best kind—teeming, soaking deep into the ground. It could rain a long time like this and not cause runoff or flash flooding. The damp air has seeped into the house and so, for the first time this year, he turns on the heat, and he can smell dust burning off the pipes and radiators. Though it is still midmorning he switches on the light above the sink. Water dripping and a fine water falling are the only sounds from outside, since the rain has quieted the mockingbird, which was nearly quiet, anyway, so late in the year.

"IS THAT YOU?" I asked as I pulled the old photo from a box of photos. It was a summer day in the picture, and the child who stared straight ahead was no more than four, chubby and dark haired, in an oversized dress and scuffed boots. She was alone and holding on to a fence. "Oh!" my mother said as she took the photo from me. That was all.

The stories my mother tells of her childhood are spare. I know her parents came here before they were twenty from a village in Campania. I know they sailed from Naples and continued to speak Italian for the rest of their lives, though my mother would answer in English. My grandfather insisted on it, the way he insisted on living on the third floor of the tenement for the light and air. "Those stairs kept me slim," my mother now says.

The photographs from her life before marriage are also spare. The one I pulled from the box is the only one I've ever seen of her as a child, though there are a few formal portraits of her in her twenties. The others were lost, she guessed, in the time after her father's death when they broke up her parents' apartment. Their keepsakes had been stored in the top drawer of the dresser, which they had sold along with the rest of the furniture. And that apartment—it was razed thirty years ago, though I can still recall it. Well, a few shards of it. I remember a low coal fire and a bed by the stove. And the way my grandmother gestured good-bye—with her palm facing away from me, and opening and closing her hand, so that I thought she was beckoning me back.

To me, my mother's past is made up of such discrete moments, almost as neat as the attic she keeps, which she is always cleaning. If she adds to its store, she is sure to take away by sending my father's old clothes or our old toys to the Goodwill. Of course, there are things that will always remain, though they may never be used again—a silver service, the wedding dress, a box of baby slippers and bracelets, the pictures of us children.

She was always carrying a camera as we were growing up. In some of the shots we're in the same poses year in and year out: the four of us in descending order in front of the fireplace or crowded on the stairwell on Christmas morning. Each of us stands alone and in white under the variegated maple on the morning of our First Holy Communion. I've found a whole box of photos of my sister in her ballet outfit. She is standing center stage in no recognizable pose. Or she is tiptoeing in a circle with other children. Although I know she threw a tantrum before every class, there are enough photos so you'd think that dance became her life.

My mother is always remembering us in fine detail, even now as she is growing forgetful about her daily life and has to write everything down: a list by the telephone—*dentist*, *sisters*, *milk*; a scrap of paper marked gas stuck through the keychain; and days on the calendar written over with *Maggie*, *trash*, *lunch*. It's not unusual for her to forget what we talked about yesterday. Or if she remembers, she is no longer sure she remembers correctly: "We're going to Lawrence . . . at three . . . tomorrow. Right?"

Sometimes we joke about it, sometimes I grow impatient. A little afraid. How can this be when she even remembers my father's past clearly? She has heard his stories so many times that they have begun to lap over her own early years. She might just as well have been there, too, to see his uncle waste away with a blood disease, though he died before she even knew my father. She nods as my father tells how he came out to the farm to spend his last days in the summer air and all you could see of him was his wool blankets as he sat out back huddled against the Baldwin tree.

ON WARMER DAYS my father speeds the pickup across the hardened field to the last standing rows of cole crops. The cabbage is punky and smells high, but there might be a small cauliflower bound tight in its leaves and only touched by frost. My mother thaws the cauliflower by soaking it in a bowl of water until it comes up to temperature, then she pares away the damaged parts. It is not a necessity, this. He just likes to rummage around in the ragged end of the season and salvage what he can. She goes along with it—or waits it out—until the deep snow.

She has lived more than half her life in this sprawling, sturdy house among fields to which she brought little more than her parents' picnic basket woven of ash. I have heard most of the stories about the city of Lawrence from my father who never lived there. He is the one who tells us again that the main street was cobblestone, just like Paris or Rome, and the shod horses sent sparks flying. He tells how the Italians buried their fig trees to protect them from frost, and he describes the trench they'd dig alongside the trees and how they'd bind the crowns and cut away the far roots before tipping them into the ground, just the way her father did in his garden in that city where the snow slushed under hooves and boots, and turned brown as the canals.

Here snow weighs on the bull pine and the turned branches of the pear, it settles on the lead-white sills and in the empty open greenhouse. It ghosts the sere vestige of a garden. The darkened outshot of the house is closed off, and snow accumulates in perfect arcs against the frosted windows of those rooms. In the lit rooms where my parents live snow flecks the panes—a sound softer than the click of needles shaping a sleeve of nubbed wool, sound of a mouse too small to leave tracks on solid ground.

THE TWO MUDDY wheel ruts have been paved into even, cambered hardtop. There's traffic at all hours, especially around dinnertime. From the farmhouse kitchen my aunt sees headlights going west and taillights going east, more lights than stars, more outside than in. She shuts one lamp off behind her and turns one lamp on ahead of her as she moves along the downstairs rooms. A house lived down to a single light isn't something she ever would have imagined. For so many years there were two shifts at the table for dinner, and she couldn't get a word in edgewise. Tuesdays she'd bake at least eleven pies, and Wednesdays the bread.

What has remained the same? The carriage house has always been white, and the weather vane always a mare. The house itself has been painted brown, and then painted gray; the upstairs fireplaces sealed, the silo blown down in a hurricane. One brother after another married, except for Cal, who always had to have two eggs over easy for breakfast and a hot meal for lunch and dinner.

There were the three of them for the longest time—she and her sister Alice, and Cal. She did all the cooking except at holidays, when Alice would take all that time making the cookies, hundreds of them shaped the same. Cal has been dead these three years, and Alice three more. Their clothes hang in the closets. Visitors have slept in their beds. My father left that house well over forty years ago, when there were still seven at home, and now he stops in every day to make sure she has taken her pills. He knocks, and enters before she can answer. The rooms are quiet enough to stun anyone at first, and he starts with the smallest talk: "How cold did it get here last night?" "Has anyone called?" "Are your groceries holding out?" He goes over her mail and sets aside the bills to be paid. If it's winter he might suggest she visit the brothers in Florida. "There's nothing to worry about. I'll put you on the plane at Logan and Charlie will meet you at the airport in Tampa." She never agrees. He checks her pill box and the expiration date on her milk.

There are sparks you can blow on for a little heat. Sometimes old lost things take flame. Today he has stayed on longer than usual, and they are talking of the time before the silo blew down in the hurricane. The streetlights have flickered on, and the traffic's picked up going west.

❦

SHADOWS SHARP AS substance. The long road white with salt. Through the bare branches I can see every buried field, all the neighboring houses, and further—seven miles at least—to the millstacks and stone church of Lawrence. The work is all inside now. My father's desk is covered with paper, and we bend over the columns of wages, sales, net worth, and assets—the mute, unalterable facts of the finished year. If only the coming one were as solid and clear. He is getting too old for the work. Some days, lifting his head from the ciphers, even he admits it: "I'm not going to do this next year. I'll just prune the fruit trees and seed the fields with hay." Other days he'll say he wants to plant new peach trees on the northeast slope of the orchard come spring.

We watch the clear drop poised at the end of the branch and wait for it to fall into the thawing stream. If we begin again, we'll have no choice but to see it through to the end. The seedlings warming in the greenhouse, in damp peaty growing mix describe a certain and exacting path. We only work alongside, putting one foot in front of the other, and by high summer we can't see beyond the near woods. We can't see the far fields.

Sometimes I think we have too much time in winter. Too much time to mull over numbers. Time to consider our place. Time to exaggerate our fears the way reasonable people do during a sleepless night—there's a presence behind them, a presence in the shadows, and no way to return to the journey of their dreams.

Yes, he can think about what to do all winter, but all he knows how to do is to continue. Already the seed order trickles in, and I find stray packets underneath the papers on the desk—the small crops—Nantes carrots, say, or Detroit Reds. And here are the Lady Bell peppers. I lift the packet up and my father says, "That seed cost me forty-five dollars." One quarter ounce. Germination rate: ninety-four percent. It will produce fifteen hundred plants. The nursery will set them in cell packs, which I'll be able to carry in my two hands. We'll transplant them into pots which will take up half the greenhouse, and when the ground warms up enough to plant them, they'll range across a quarter of the back field. People from town will buy them, people will come out from Lawrence, and some from even farther. Those plants will yield green peppers and red, sweet peppers that will have to be picked every other day from July to the frost.

MY FIRST PUBLISHED poem was written so long ago now I've lost track of it, though I remember it was a lyric about the summer fields. My father and Pete were mentioned in it, and when I showed it to my father, after a short silence, he smiled and said—half to himself—"Who'd ever have thought poor Pete would end up in a poem?" Since that time he's occasionally asked, "How many poems did you write today?" or "How much do they pay you for these?" And I mumble across the gulf of lives lived differently: "Oh, I really didn't finish anything today . . ." or "Well, you know, it's not like plowing a field . . ."

I can see him laboring to fit my answers in with his own idea of work, of a steel blade cutting through the thick April rye, and his wake of turned-up earth. One works by going back and forth as the sun arcs across the spring sky, and there's sheer physical exhaustion at the end of the day. Against which all I can muster is: "You get a different kind of tired."

My mother keeps most of her thoughts to herself, though once she said, "It's always about the farm, isn't it?" I bring her most of the work I've published, and she looks at my name on the back of a journal and shuffles through the pages—she'll read them later, never in front of me—then sets it on the coffee table along with *Woman's Day*, various catalogs, and the Boston paper. Or she'll hand it to my father, who'll thumb through until he comes to my work. This last time he looked at the magazine and said: "How about that. You see—there's more to life than planting tomatoes."

It's strange for me to hear that, with its hint, I think, of regret about his own path. He is forever telling me farming is a hard way to make a living, forever warning me away. But I also know he is ingrained in this world, and for him there is no other. Every time I suggest we fold our hand here, he won't hear it.

So much of what he remembers is full of savor, as much as the winter apples in their cellar or the wild grapes in summer. Seventy-five years collapse when he says, "I remember helping Pop plant the potatoes—I couldn't keep up with him—I wasn't more than ten. He'd dig the holes and I would run along behind him and drop the seed potatoes in—right there, where we planted the Hubbards last year." And just yesterday he brought up the hurricane again, this time a story I have never heard, about the orchard he had planted not many years before the storm, and how the wind half-uprooted his trees so that they listed eastward. "I collected the downed wires—they were everywhere—and used them to stake the trees back upright . . ."

A story just told, and, soon, that one too will go untold. What, then, will be made of these times? Will they be remembered? There are years of sleep between the sparked stories, and every time I try to trace it—just when I think what's recollected shines—it shifts, and takes a shape I have never seen.

So much to tell, and even with all the quiet of winters here, there doesn't seem to be enough time. The connections between things takes so long to become apparent. I always seem to want to say more than I'm able to set down. I put down the words, then take them back. Go down a different path. Work, senses, anger, the rust, the relics, the seeds. To give everything its true weight, and not to weigh the anger more than its measure.

Cold and still this morning, and a fine, dry snow has started to fall, deepening the winter cover on the field I see when I look up from my work, the same field where only a few months ago Sallie and I untied a thousand trellised tomatoes, and the spent plants collapsed one after the other at our feet. Now I see white falling into white, and beyond, the sagging side of a disused barn, its paint gone gray. Few sounds come in from outside other than the small worryings of the nuthatches and chickadees, and—muffled by the storm glass—the long hush of traffic. Indoors, too, is cauled in quiet. Key clicks no louder than the songless birds, and soft lead scratching out words.

I know how it will be when the wind comes up at dusk. The snow drifts just as if it were fine sand, and off the crest of those drifts, a handful of snow swirls up and sweeps across the open fields. It keeps an upright shape for a moment, then scatters and is tossed up into the starhung night.

6
Leaving

A DRY WEIGHT. A small measure. But in the right hands, even a winter seed speaks for itself. In the still of every January, my father gathers up the seeds from a Hubbard cut up months ago. He takes the corners of the newspaper they are lying on and makes a funnel to slip them into a mason jar. There's always talk that seems to go with it: "That first Hubbard, the flesh was the width of my hand. I saw it in the Boston market—fifty years ago. I said I'd buy the whole squash—it was forty, forty-five pounds, but that I wanted the seed, too." Another time I remember him saying, "One year I sent some seed to Agway—they gave it a field trial, they said, but it didn't yield enough for them. Imagine that . . ." His face brightens as he speaks, just as when he's talking with his sister or brothers. He tightens the lid and labels the jar with masking tape. The best squash won't have a lot of seed, and they usually fill the jar halfway.

One year he spent much of November in the hospital—in Boston—and my mother couldn't negotiate the traffic, so almost every day I'd take her there for a visit.

"How are things at home?" he'd ask.

"Fine," my mother would always say.

"How's Sam doing?"

"He was in for lunch yesterday. How are you feeling?"

"Oh, lazy. They keep poking me. They wake me all night to check something or to take some blood . . . Don't ever get old."

My mother would sit at the foot of the bed, her ankles crossed, her purse clutched in her lap. They'd talk about the weather. My sister. The news. I'd look to see if the numbers on his chart had changed. He'd tell me what bills he was expecting, which ones to pay, and which ones to wait on. I'd ask about his creatinine level and his iron level as if we'd talked about such things, too, for our entire lives.

The one time I went to see him on my own, he was half in and half out of sleep, and his voice was groggy when he asked, "How's Mom holding up?"

"She's OK," I said.

"Make sure you check the oil in the car for her, and the pressure on that front tire."

I remember voices in the hallway. Someone wheeled a cart by, and the glass vials chinked against one another.

"Did Sam sell the Hubbard yet?"

"It's still in the shed, but he put a tarp over it. It'll be OK for a while, unless it gets really cold."

"But it's going to get cold. He's got to get rid of it before Thanksgiving. There won't be any market after that."

We heard a nurse in the next room saying, *You have to keep the patch on, Mr. Browning.*

My father said, "I still need some seed for next year."

"Do you want me to put some squash aside for you?"

"A few. You can put them in my cellar and I'll break them up when I get home."

I couldn't begin to imagine him lifting the Hubbards or taking the cleaver to them, but I said nothing. I just nodded as he told me to make sure they were heavy for their size—the paler ones. No bruises.

He closed his eyes for a while and then said, "You should be going before the traffic gets too heavy." The color in the eastern sky was already deepening. Thousands of lit windows defined the skyline. I gazed out on the quiet river and Beacon Hill as I put on my coat. I heard him say, "I remember when the State House was the biggest thing on the horizon." What I saw was a small dome distinguished only by the mild luster of its gold leaf, and nearly drowned among the office buildings. A memory in the same place, the world surrounding it changed. As I turned to leave he said, "If anything happens just pull in the tent around you."

There was frost on the tarp, but the squash beneath it had stayed warm enough. I'd say there were at least four ton nested together in one big pile. I walked around the circumference and saw many that would be fine for seed, but they were all mixed in with ones of lesser quality, and I had to lift and shift parts of the pile. The squash were awkward and heavy and not easy for me to handle. Some I could barely manage. And there were beautiful large ones that, when I picked them up, felt lighter than they should. Heavy for its size, he had said . . .

My labored breath plumed into the cold air. The road was quiet, the rye paled with rime. Sometimes I hear what I want. Behind me, a scattering of oak leaves scraped across the pavement, and for a moment they were the footfalls of someone coming towards me.

THIS WEEK'S TALK is about cemetery plots. My father has bought the last four in the cemetery, halfway up the hill, behind the white birch. From there you can look straight across the late land to the sheds, the stand, the farmhouse. "At least I think I bought them," he tells me. "Carroll, when he arranged for it, had his bottle with him. But I wrote him out a check. He said he'd mail me the deed."

I make a joke about how I want my ashes scattered into the ocean. But he can imagine nothing other than lying in the lee of his farm. My mother will lie there too, even if it's a Protestant cemetery. And Sam. Though after my father is gone, it isn't likely to look out on farmland for long.

Sometimes my father talks of the future as if there were still a clear path. He says to me: "Someone could make a good living here if they're willing to work hard." Maybe it's really his way of speaking about our own lost chances. Though I don't think it's that we haven't worked hard enough. This failure is smaller than that. Smaller than the economy or the weather. Too much silence. Questions that echoed, and went unanswered. And anger thrown into the day's work the way Sam boots a shovel into clay soil. Moments of anger, misunderstandings, and silence accreting over the years until they've become as large as the pines that now shade the eastern edge of the orchard. Their shadows have assumed and deepened the slighter shadows of the apple trees, and hold the frost well into the fall mornings.

A failure that's come on so gradually we took it as usual, and then inevitable. Just like the old horse-drawn coach we left out in the weather. I have no trouble picturing it, the way it listed to its right side. All the times I tried to count the wooden spokes, or gazed at the way my face distorted in the beveled edge of the oblong windows. It stood by the stand for years, a conversation piece for the customers, something children would rush up to just to sway on its handles, and jump on its velvet cushions. A man would come by off and on down the years wanting to buy it and refurbish it. "Not for sale," my father would tell him, and no more than that.

Of course it moldered away, though I seem to remember vague talk about restoring it. My sister had a friend . . . But really by the time such talk surfaced, the wood was already badly weathered, and the metalwork was caked with rust. At last only the glass was fully intact. It would have taken a special eye at that point to save it, and more work than anyone could imagine. It crumbled through one winter, and collapsed in a gray pile in the snow. Gone its grace, though we all remember it, and the way we loved to climb in and out of it.

I WISH IT was August and the corn was running together, and there wasn't enough help, and Sam and I had argued and argued. But it's January. The work and arguments have settled down, and what I say comes out of a long-thought-out silence. I tell my father: "I just can't work with Sam anymore."

He must have known this decision was coming—his voice is small, too, as he says, "He's my son." Then, "I guess it wasn't meant to be." Not much more talk than that—my mother greets my decision with silence, and believes, I think, I'll still relent. Sam—Sam just shrugs his shoulders and walks off.

All the voices: *I figured you'd be opening soon—I saw the corn tasseling out . . . Tomatoes—finally . . . The peaches taste sweeter this year . . .* They are all alive in my mind right next to the old questions. *What's to come? Has my being back here changed anything? Maybe it's made things harder for everyone.* Alive in the long evenings when I wonder how it would be if Sam never dreamed tigers were stalking him. I try to imagine how he would fare in the world, and what kind of job he'd hold down.

Finally, when it grows late, and I just can't worry anymore, I come back to my red-eyed justifications: *The land is tired, a long stretch of fallow years would be the best thing . . .* Balm for a hundred tiny cuts, yes, but it can't take care of my anger. I remember pounding on the truck horn to wake him, and I clench my teeth just as then, and mutter to myself as I always have: "I *won't* be my brother's keeper."

It's Orion's evening season, and the air is so clear I can see Betelgeuse for the red giant it is. Even with the bitter cold, I can't help but take in the sky for a while before I step into my parents' quiet low-lit house. If I have dinner with them now, we finish more quickly than before. With no more easy talk about the farm, I think of how frail they seem, I hear the sound of silverware scraping the china. The refrigerator kicks on. My father hunts up a long story about hauling potatoes out of Maine, or of the parade in Lawrence on Armistice Day 1918. My mother asks if I've heard from my sister.

"No, no," I say, "not since last week."

We fall into silence again before I bring up the one thing so much on my mind:

"Have you thought anymore about next season? Why don't you just let the fields lie? You could rent out the orchard easily enough if you wanted to. Or you could just keep the trees pruned."

"What will Sam do?" my mother asks.

"Sam can figure things out for himself." They say nothing, and I continue: "You could plant enough for the family—a block of corn, a dozen tomato seedlings. Beans. Carrots. Those Knight peas you like so much."

It's hard, even for me, to imagine such a pared-down future. The snowed-in fields are immense, and any small attempt would seem no bigger than candlelight against a starless night. But I prattle on, anyhow, about how he wouldn't have to plant beets again, until my father's temper grows short and he says, "That's enough—I haven't made any decision yet."

The cold months wear on, and I keep at the subject—a winter finch at her seed. I've brought it up again today, and after my father's temper flares and subsides, I promise myself that I'll hold my tongue from now on. "Drive me up to the apple cellar—please," he asks. "Your mother wants some Baldwins." We're quiet in the truck, and when I come to a slow stop by the cellar door, he takes a long time getting out. He is eighty-three. By now both knees are bad, and he is stone blind in one eye. No sound other than his walking stick striking the ground. Then he says, "I worry where people will get their corn."

LATE SNOW. LATE spring. But finally I hear a call longer than the two notes of winter, and the earth gives beneath my feet. Mist rises off the last stretches of snow—the day is shifting mist—and pours down the gradual slope of the orchard to catch in the branches of the pear; it curls into the harboring pines and thins across souring ice on the pond. Mist obscures the barn and yard and carriage house, then shifts towards the rye, and our buildings suddenly appear clearer than I remember.

To our north the deepest snowpack in years has begun to melt. Already the streams are breaching their familiar channels, and once the rains begin water will erode new beds across the fields, taking soil and subsoil and stones with it. Water to deepen the stains on the walls and slate graves, and to illumine the clear gray bark of the beech. Water to release the scent, bitter and spiced, of decaying oak leaves, bitter mixed with a sweet trace from long dormant corms sending up shoots beside the well. Water to drown out all the other sounds of the thaw—of ice breaking up, of the melt running off six eaves, and of far water falling down a granite outcrop. Voices, too, it drowns out, even the raised ones, and the jangling of the harrow—every chink and scrape the tools make.

I want the sound of water right up next to my ears. Once it stops this time, I'm not sure of the sounds I'll hear, not of a disk striking new stone, not of the shovel pitching into earth, not of voices carrying across the spring plantings—"OK." "That's good." "There."

❦

TO REACH HERE I've made my way among countless white birches. The woods are full of their light touch, as they are every May. Their pale new leaves, lit by the sun, glint when they shiver, and—above me—sound mild as subsided water coursing over mossy stones. They're not old. The trunks are all nearly the same—tall, slender, unbranched—and the white bark lights the shadows. In a breeze they sway in unison, as if tethered to the same root, and moved by the same mover.

One oak looms from out of them. So late in spring, and everything about it still suggests winter—a few of last year's coppery leaves cling to its branches, and the ones on the ground beneath it are tough and so slow to decay that I step on years of them. The dry rustle I make as I walk sounds out of time—Canada mayflowers and starflowers are already pushing up through the duff.

As yet, there are no new leaves, and all the limbs are apparent. The lowest ones branch out nearly horizontal to the massive trunk, and every one—there are many—crags as it extends. A few of the limbs are long dead, and on the living, the pale gray bark is furrowed, hung with moss and fungus, and crusted with lichen. The crown itself spreads wide and is higher than any other tree in these woods.

Already hundreds of years old, it may last uncounted more. I know there were others, a line of them that once marked a roadway or skirted the boundaries of a field back when this was a more open land. Maybe spared by sentiment or reverence, this one outlived them all. It's hard to believe that some hadn't thought to fell it at one time or another, since oak had been used for everything from ships to hafts even if it needed long, careful curing to keep it from warping. By now, this one is decayed and crooked enough to have escaped such uses. It keeps its own place, far and rarely seen, nearly forgotten.

Its lower branches are the only horizontal lines on this hillside, and so through the birches I can discern it from far off. I head straight there—it's what I've come here to see. A willful presence that takes all my attention, a tough-minded ancestor checking my enthusiasm. And all that shimmers of spring—the new leaves and their sound and the sweet air—drops away.

In another month the birch leaves will have darkened, and the oak will be in full leaf, nearly blended in with the young woods around it. It will seem more of a kind. My somber, ponderous thoughts will go too, becoming far and faint as the sailors' curses who blamed the oak itself, instead of the hasty seasoning, when their ships fell apart on the seas.

Epilogue

THIS YEAR SALLIE has a cataract in her right eye. She has to finish trellising the tomatoes, she says, before she sees the doctor on Thursday. The twine, all measured and cut, the color of straw in the sun, hangs over one of the trellis wires. She takes a handful and drapes them over her shoulder. As she moves down the row of transplants she pulls a length of twine from her shoulder, stoops, and ties it around the base of the tomato plant. She weaves the young plant around it and ties the twine to the wire above. She trellises two and then takes a step down the row. When I look up from my desk every now and again, I see her incremental progression.

The winter was a fairly steady time with the farm quiet, buried, still, like every other place, but now that it has come to life, I sometimes feel hemmed in by the goings-on. It's hard for me not to feel as if I should be helping out. I stay inside most of the day, and work in my study. Late in the afternoon, when I know Sam is deep in his house, and the help gone home, I walk down the mail road and into the woods, looking for small things—the overwintered partridgeberries or the lady's-slippers beneath the pines.

As I loop back home I sometimes stop in to see my parents. These past few weeks my father has been talking about his asparagus. There are stray beds of it all over the farm—asparagus grows out of my mother's rosebed, on a knoll in the apple orchard, in my Uncle Cal's abandoned garden in back of the farmhouse. There are a few stray stalks amid the grapevines. Old plantings, out of which he cuts a handful of spears every few days. I know he is just glad to be moving around after the shut-in months.

Changes soon, I know. Even if I have nothing to do with the daily workings of the farm, I know I'll have much to do with those changes. When my sister comes down for a visit, we always seem to find ourselves talking about the future of this place. How much you'd have to do to pay the taxes. Who might want to rent the orchard. But such talk has no real weight. It doesn't even begin to touch on the days ahead. Winters are harder on my parents every year. Where Sam is in the future, I don't know.

I can still feel that my returning here has only delayed what was inevitable, and that we've all returned to our old places, which had been set for years: me in my own life; Sam, no matter how stricken, running the farm. Angry, frustrating hours when I feel I've had no choice but to give in. Though there are times, too, when I'm relieved not to be tied to the daily responsibilities of the farm. How could I have handled things here anyway? There's so much I would have had to learn. I would have had to set aside most of the other dreams I have. Dreams which grow more vivid as my plans pare off from the farm. Ideas for books I want to write. More and more time at my desk.

Still, I can't imagine this place as anything other than a farm. Now there is waning light in the trees. The orchard has just been mowed, and everywhere is the scent of cut grass. Out of the long dusk the call of a thrush defines the edge of the cleared land.

FIVE THOUSAND DAYS LIKE THIS ONE

An American Family History

In Memoriam
John Brox
1910–1995

I

. . . whether the hills will look as blue as the sailors say.

—EMILY DICKINSON

1
Afterwards

IT HAD BEEN a droughty summer. The orchard grasses turned sere by late June, the brook beds shrunk to dried mud, and the apples reached no more than half their best size. The worst of them—ones marked up with codling moth scars or scab—weren't even worth hauling to the apple cellar. Strange to see bushel after bushel of Cortland, McIntosh, and Northern Spies in heaps under the trees. Even stranger was the way the sweet smell of those fermenting apples drew the deer out of the woods—more deer than anyone in living memory had ever seen here. They'd forage through the fallen leaves under the bare crowns of the trees, coming more and more frequently as the apples frosted and thawed by turns down the shortening autumn days until they froze through at last and were covered with an early December snow.

Through the first quiet winter storms, the deer stayed

in the part of the orchard that bordered the woods, nosing the snow near the apple bark, raising their heads every now and again, wary, listening. But as the snow deepened they came further into the orchard, and I could trace their tracks from tree to tree. They came at all hours—nine of them once—filing along the edge of the pines at eleven in the morning on a cloudless day. Even my father, who'd seen them all his life, remarked: "Look at that. In broad daylight."

We talked about them every day as I stopped by my parents' house in the late afternoon, my mother in the kitchen, my father at his desk going over the farm year on paper:

"I saw three under the Cortland trees, one couldn't have been more than a yearling."

"Is that so? They must be finding apples still. I hope they don't start grazing on the branches."

"Oh, you think?"

As long as the deer kept to the fallen apples, they stood clear of any concern of ours. Just beautiful things even for the hunters in our family—my father and uncles—all of whom were too old now for the hunt, though seeing those deer brought up the old stories about ones tracked years ago here, or in Maine, or Nova Scotia.

Our talk about the deer kept on, even when my father took sick. In the hospital he couldn't say much because his breathing was rapid and shallow. I sat by his bed—it was far into December by then—and his room, with its beeping monitors and hissing oxygen, was louder than anything outdoors. There were no real words, it felt late, but I was hoping he'd still want to know about the small things, so I told him I'd seen the deer that morning, that they were coming farther up into the orchard all the time. All he could do was blink his eyes, and I couldn't figure out if he knew what I was saying or if he had a question, but a little later, when I asked him if he wanted anything, he smiled as much as he could—I saw his cheek wrinkle—and whispered, "venison." One word that let through his dry wit, since he knew I wasn't much for hunting. One word that comforted me more than all the times he'd answered the usual questions we asked to make sure he still dwelled in time: Did he know where he was? Did he know the day? Or the questions he'd ask us when he could: What's the matter with me? Where's the blood coming from? Have you gone home?

A few mornings later, I was the first in the family to arrive at the hospital and I had to wait outside the room while the nurse finished her care. I knew something was wrong because she wasn't talking to him—always before I could hear the nurse asking my father how he felt, was he comfortable, telling him she was going to draw some blood. She beckoned me in this time and told me he had come down with pneumonia in the night. His hands were cold. It was louder in the room. They'd turned up the oxygen and had given him a larger mask through which to breathe. I could tell right away he was having trouble keeping up. When I said, "Hello," all he could do was raise his eyebrows. That was his last gesture to me in this life, and it's what I keep remembering, wishing for more, thinking of all his years of reserve and, in that last week, all his efforts to say the merest thing.

My mother, my brothers, my sister—all faces fell as they entered the room that morning. And no one dared step out for coffee or to make a phone call. We'd come a diligent way on a narrow trail since the gray early light of Christmas Eve day when I stood in the door frame of my parents' house, and faced the road, listening in the still of the year for the ambulance to come. Now we gathered bewildered around his bed as his breathing grew quieter and quieter. He took his last breath, then his mouth closed. My mother whispered, "No," as the heart monitor slowed to a scribed outline like the low eroded hills, and each reading ended in a question mark.

When the doctor came he listened for my father's heart and his lungs, and then put a thumb to the lid of my father's blind eye and opened it, not knowing it had stopped gathering light years before, though my father always said he could still see the shadow of his own hand. Now, the coin for the journey.

Afterwards—in the days following—I sat at his desk and tried to carry on the workings of his home and farm: changing everything over to my mother's name alone, working out the payroll taxes, the quarterly taxes, all the January paperwork. I was half grateful for the dry figuring of accounts, of the farm year drawn to an abstraction of costs and balances. But such soldierly work couldn't keep grief at bay for long. As I backtracked over the check stubs, I saw how his handwriting had grown shakier down through the year. Like his voice. I'd start to think, becoming gravelly as his lungs weakened. Then I'd notice how quiet the house felt—and he was a quiet man. How did my mother stand it? How would she get through the days, the meals, the evenings? No answering words. Only months later beginning to understand: *never again.*

As I worked I'd uncover keepsakes of his in the drawers of the desk or tucked away among his ledgers and files. His original birth certificate, the death notices of his close friends who had gone before him, his own father's timepiece with its etched copper backing worn fine and the crystal clouded over—mute things that had lost the one who could best speak for them. From now on they'll only be partially understood, same as the stories I can no longer verify that were mostly his alone. "No one believes me," I remember him saying, "but I stood by the Bay of Fundy on the eve of the war and saw apples coming in on the tide. The bay was full of apples. The ships had dumped their cargoes to take on supplies for the war." That's all I know. And no matter how much, I want to know more.

Prayer cards and letters of sympathy came through the mail as news of his death traveled out of the valley. My aunts had been tearing obituaries out of the local papers and sending them to the relatives in Syracuse and Delaware. Friends of far friends wrote and called. I phoned my parents' closest friends in Florida—people I'd never called before—who knew something was wrong the moment they heard my voice. After their weak greeting, a questioning silence into which I poured, "I wanted to tell you my father passed away."

I know their true grief is beyond the formal, scripted sorrow that lands on his desk with every mail. I know my father would have understood their efforts to find words that come near. Near enough, OK, what we settle for while we tilt an ear to the winter air. I feel as if I've been listening—for what?—ever since the wake. It had started spitting snow, and those arriving to pay their respects, though they'd only walked from the parking lot to the door of the funeral home, seemed as if they'd made a real journey the way they stamped their boots and shook the snow from their scarves. They blew on their hands as they cleared their throats on comfortable sayings about the weather: "Sure is cold . . ." and "The roads are icing up." Then, the plush hallways and the floral sprays brought their voices down: "I can't believe it. I thought he'd live forever." Voices that had surrounded us all our lives sounded graver than I'd ever known, murmuring, "Things just won't be the same," "I'm sorry," "Sorry for your troubles," "I had no idea. I saw him just a few days before Christmas and he seemed fine," "It'll be tough."

Eyes, then eye, that saw. Ears, then muffled ears, that heard. When I try to imagine afterwards, I keep coming back to how much my father belonged to this one place on earth. I can't imagine more than all he had in his keeping: three houses, forty cleared acres, a hundred of woodland, and a dozen in fruit trees. *Thou canst not follow me now, but thou shall follow me afterwards.* If *after* is a word that doesn't come near, if what's to come can't be imagined from this life, then why does his farm seem to mean all the more to me now, as I stand in the orchard when the moon is down and watch the comet passing?

His is a New England farm, and for all the stony soil, there's an intimate feel to the lay of the land with its small fields set off by chinked walls and the mixed woods beyond. My father's understanding of this place had accreted over eighty-five years, and at times I know he drove the wedge into wood with the resentment of the responsible son, at times with an effort born only of love. Over eighty-five years the original sound had swallowed its own echoes, and the most he could do was to tell me, "This is where I keep the receipts, this is where I keep the outstanding bills," as he opened the drawers in his office. Not much different from when he tried to teach me to prune the peaches: "This branch . . . here . . . see how the light will get through now?" as I stood puzzling it out alongside him.

So, with the snow falling outside, and the deer lunging through the deepening drifts, I am left to figure the farm from the notes on his desk, from the business cards he had scribbled across. I find a name—*Very Fine, Cal Jennings, Orchard Supply*—and work back from there. The world I'm responsible for is more complex and less patient than the one he was born into. Hospital bills from his last illness are waiting, the pension fund needs proof of his death. Over the phone I have to recite his Social Security number to prove I know him. I have to mail out his death certificate again and again—the form itself, with its raised stamp, is what they want, not the facts that his parents were born in Lebanon and that it was his heart and kidneys that gave out. Even when such work goes smoothly, I sometimes throw down the pen and ask the desk, the walls, and the ledgers why I couldn't have learned all this before.

And then a day comes when I have to erase his name from another account. First it was the checking account, then the Agway charge, and the Harris Seed charge. Sometimes it feels as if I'm erasing him everywhere until his name will remain only in his last place, on the hill, behind the white birch. I hate it, both the erasure and my realization that if we are going to go on I can't make the same decisions he would have made. I'm planning to spend more on repairs to the machinery than he ever would have agreed to. I'm thinking of selling a piece of far land as soon as the market's better.

"Who's going to be farming in this valley ten years from now? Who?" my father once asked. I could say nothing in the face of the long years he had put in. I realize I know little of the work it will be, even though it's not a great deal of land. These acres he has left are almost nothing in comparison to the farms to the west, to farms in general. But it is ours, and one of the last here, and it feels huge to me.

In his safe, among the canceled bank books and the stock certificates, I found the original deed to the farm. In 1902 it had been a thirty-five-acre holding with worn implements and gradey cattle. All the scattered outbuildings are described in detail, and every boundary is fixed: *Thence northerly by said Herrick land as the fence now stands, to land now or formerly of Herbert Coburn, thence easterly by said Almon Richardson land as the fence now stands to the corner of a wall by said Almon Richardson land . . . thence southerly . . . thence westerly by the Black North Road to the point of beginning.*

The contents of the barn's are listed, too: the hoes and shovels, the scythes and hammers, the Concord coach, the jump seat wagon, two bay mares and their harnesses, five dairy cows and one milk pung, about thirty-five hens and all the chickens, one tip cart, and a blind horse. Plus feed for the blind horse. I swear, the worth of every nail is accounted for. For this my grandparents were so far mortgaged they had to cut down the pine grove to meet the payments. And now out of all that has been listed there, what has not been discarded or crumbled to a sifted heap hangs gathering an oily dust in the back of the carriage house or in the bins of the toolshed. The scythes have rusted to the nails they hang from, the leather collars for the horses have dried and cracked.

When all is said and done and we tally the contents of my father's estate, such things will no longer be counted among his worth. What brought us here and forward, the things he started with back in his boyhood where his allegiances began, will be smiled at indulgently and hung back up and considered as nothing alongside the larger things that have replaced them—the Case tractor, the gleaming red harrow, the corn planter. Stand at the door of the barn and breathe in the must of those early things.

Listen for a voice—*hey bos, bos, bos*—calling the cows home. Feel an ache in the worked-out shoulders and cold creeping into the firelit rooms. How else could we have come this way since the April day when, according to the deed, my grandfather, who could hardly write his name in English, made his mark?

2
The Quality of Mercy

AFTER SYNTAX IS gone, and the liturgy, the maxims, the songs, even after no one can read anything of the old alphabet, and the names of things that remain are recognizable only to the few—after all, ragged bits of story still come down from the old country and are told in a new tongue: dry, sturdy, thin as the last weeds to be covered by a January snow. Sometimes those stories feel like tests when they're told. Don't you remember . . . Haven't you heard . . . how she was smuggled into this country under her mother's skirts, how they had to get him out after he'd killed that man in a fistfight, how they wanted to send her back because of the weeping in her eye . . .

I imagine the stories my Lebanese grandparents carried with them had been at first as bulky as a peddler's pack stocked with lace and thread and stockings. My grandmother, not yet eighteen, traveled with her goods along the roads surrounding Olean, New York, and, as she exchanged lace and thread and her own handiwork for pennies, she glimpsed through each opened farmhouse door another life in the offing—the rush of warmth from stove heat, the smell of hard soap, of johnnycake and drying apples. And with each opened door, her pack lightened, and the stories shifted their place in her memory.

After Olean, Lawrence, Massachusetts. At the end of the nineteenth century it was the Immigrant City, the worsted and woolen capital of the world. For miles red brick textile mills lined the banks of the canals and the Merrimack River, and they contained so many looms and spindles that workers came to the city from Canada and all the countries of Europe and the Middle East to tend them. Within the central district you could hear forty-five—some say fifty or sixty—different languages and dialects spoken by dyers, cutters, spinners and weavers, by men who fixed the looms and rigged the warps, and women who felt along the yardage for slubs. Their children hauled coils of soft sliver and breathed in air that was white with cloth dust. Always cloth dust, falling constant as high mountain flurries.

The Italians settled near the commercial district, the Portuguese and Jewish neighborhoods were a little farther north and west. The Franco-Belgians found a place by the Merrimack above the dam, the Poles along the thin, sinewy Spicket River that wound through the north side of the city and eventually joined the Merrimack. The Syrians, as they were then known, settled several thousand strong on the slight rise blocks above the commercial district and the river—several thousand who'd walked through night dust storms and across mountain passes, and sailed the winter gray Atlantic to pass bewildered through ports of entry so as to keep shop, and sell wares, and take up unskilled jobs in the mills where they were among the lowest paid of all the workers. However far the journey, they continued to sing their Mass in Arabic, and their corner stores sold thyme, sesame paste, rose water, and the dried cherry kernels we call *mahleb* that flavor our Easter sweets.

For my grandparents, the life they'd glimpsed in the offing wasn't there among the familiar language or aroma of spices in the neighboring shops. "God knows how they ever got the money together," my father once said, for twenty acres of tillage on the south side, and an early-nineteenth-century farmhouse and fifteen acres on the north side of a dirt road—sometimes known as the Black North Road—five miles west of Lawrence's tenement district and halfway to the city of Lowell.

A hundred years before they arrived in this country, a man named Moses Bailey had raised the first rough buildings on the property: a shed and a cottage he used as his shoe shop and living quarters, which in time became the central link in the long, curved spine of what we think of as a New England farm. Bailey's son extended the cluster of buildings eastward by raising a two-bay carriage house and a white clapboard cow barn topped with a windmill. Westward he added an ell for a summer kitchen, and southward from that ell he built a two-story-and-garret home. It is a dream of the solid and ordinary, the gable roof with its plain sloping sides, the serene rows of windows—two lights over two—without ornament, the two chimneys breaching the roof beam. The eight fireplaces assured that even if the corners and attic remained cold enough to keep the summer harvest, there'd be a spot of warmth in every room.

If you were to slip off the leather latch of a Farmer's Diary from those years and turn the pages past the calendar, past the measure of length, the measure of surface, the measure of solidity, past the apothecary's weight and the diamond weight, the dry measures, the tests for death and the cure for cinders in the eye, you could read the concentrated words of those winter days: *Wednesday, January 4: Finished hauling in ice. Thursday, January 5: Got the sawdust back into the ice house. Tuesday, January 31: 6 below. Pump froze up. February 9: Zero. Sawing wood in the woodhouse. Cold windy day. . . . Samuel Knight died this noon. Jim went down to his sister's. . . . Mr. Knight's funeral. . . . Tough New England storm. Snow flying. Hard wind blow. Sawing wood and churning.*

By the time my grandparents bought the farm early in this century, the chain of buildings Moses Bailey had started a hundred years before was as long and rambling as it would ever be. In a 1901 photograph I can count six different roof heights, the most prominent being that of the broad-peaked house. By then its separate doors led to separate worlds. The daily one faced the work yard and the outbuildings, the place where they made their repairs in spring, where they rigged up the bay mares and loaded the milk wagon. The front door—dressed with an overhang—was the one place that allowed for shade, the one place to shelter a formal visitor. It faced the road to the cities.

With the land and house came much of the life of those who'd gone before, their tools and cattle, their hens and chickens, and whatever they could coax out of the wash and drift from the age of glaciers. My grandfather plowed the richest soils for the garden and the crop of hay: the indifferent, he used for pasture; the worst, let go to woods. The orchard spread across a droughty slope. Over time they came to judge for themselves the particular soils of the country—the sweet or sour, early or late. As they cleared stones and harvested potatoes and recorded their monthly payments to the bank, they would have been too intent on their work to notice that the last local map of agriculture was being drawn.

As I was going through my father's papers in the days after his death I came across what must have been that last map—the *1924 Soil Survey of Middlesex County*. A booklet details the particular, elaborate name and best use of every soil to be found here, and the accompanying map lifts every inch of earth from its dun color and assigns it a rich, saturated one. The colors swirl and eddy—they remind me of Italian marbled paper—with fine distinctions and varying promises: the yellows of Hollis and the oranges of Sudbury loams, the pink of the Gloucester gravelly phase. The pale blue of Hinckley gravelly sandy loam: *derived largely from coarse glacial drift . . . drainage, which is almost entirely internal, is usually excessive to the extent that crops suffer in dry seasons. . . . Crop yields are generally low. Hay cuts from one-half to three-fourths ton to the acre, depending on the season. . . . This land can be utilized for pasture, but it is not well suited to grass.* Merrimac gravelly sandy loam: *Nearly all the vegetables grown for market in this county are found on this soil, but asparagus is the only one grown on a large acreage. A few farmers make a specialty of chickens. Hay yields range from 1 to 1½ tons to the acre on the best-farmed land. . . . The soil is easy to cultivate. It is plowed 8 or 9 inches deep and can be brought into good tilth without difficulty.* Coloma sandy loam, gravelly phase: *differs from the typical soil in having a larger proportion of stones and gravel on the surface and throughout the soil. . . . It can be utilized for pasture land, but it furnishes only indifferent grazing. The rougher areas should be left in forest or should be reforested to white pine.*

The soft hills to our south are marked as drifts of Hinckley loamy sand—*a small acreage is in pasture, a still smaller area is in mowing*—and Gloucester stony loams. The soil of the riverbank beyond those hills isn't deposited in pockets like most other county soils but has washed and settled in narrow bands along both sides of the watercourse—thin stretches, influenced by the drenchings and drainings of mountain water: *brown mellow fine sandy loam . . . not inundated with every overflow but covered by the spring freshets. . . . The land is easily plowed and cultivated, coming readily into good tilth. . . . The overflow in spring is depended on in measure to keep up fertility.* This map is the only local one I've seen where the Merrimack's presence is so slight—really it's almost lost amid the colors of over forty soil types, just a faint blue bend cutting through the reds and greens and oranges in its eastward turn to the sea, as if to suggest we were once a people defined by something other than the river and the redbrick cities that line the drops along its lower course.

Any map is bounded by scope and hand and eye, by the particulars of the world it describes. I don't think the soils of Middlesex County will ever again be so painstakingly described and drawn. Contemporary soil maps consist of data imposed on aerial photographs. Colorless, perfunctory, stiff with information alone, the new maps chart moist bulk density and the shrink–swell potential of soils with exactitude, but the grain of the old language with its good tilth and its indifferent pasture is gone, just as hames and traces and milk pungs are gone from our conversations.

Our farm is no longer a self-regarding world. The barn and the silo have been lost to wind and rain, and with them, the long protective curve of buildings around the east yard that shielded daily life. The house stands more prominent than the remaining farm buildings, and more solitary. A side porch makes it less plain; the maples shading the lawn, less blazed upon by the summer sun. Crocus, then iris, then roses, until the rust chrysanthemums bloom in mid-October.

Moses Bailey's original cottage survives to this day in its place between the house and the carriage house, and we continue to use it as a storage room and toolshed. Its clean white clapboard exterior blends in with the line of remaining buildings. Walk inside, though, and you feel the length of its days. Its windows are blocked from full sun by two cedars so that, even at noontime, it's a pool of cool shadows. The boards of the interior walls are rough-sawn, not once sanded or painted in all their years. The smell of must. On the rafters a hundred nails have been sunk for a hundred hooks—lanterns, oilcans, chains. Look how everything is becoming the same color. The bins full of bolts and screws, the anvil and ax, the hammers, spades, and hoes, even the white pine floor and the hardwood rafters, the paper on crates and the notices pinned to the walls—all the same brown as acorns and fallen oaks.

If the interior of the farmhouse—its fireplaces blocked off, old papers piled in the bread ovens, half the clocks stopped—also feels like twilight, to think of it only as it is now is to miss much of the story, which continues to waken and glimmer in the eyes of those who lived there, in the eyes of one of my uncles remembering seventy years back, and the May evenings he read to his mother as she darned socks and knitted sweaters. She'd want to hear the stories he was learning at college—Shakespeare was her favorite—so over and over again he read her *The Merchant of Venice.* By the late acts the sun had faded back from the dining room table and was retreating from the orchard, his dutiful voice was growing tired. Her knitting needles clicked: *The quality of mercy is not strain'd, it droppeth as the gentle rain from heaven upon the place beneath; it is twice bless'd. . . .*

The story never stopped on the last page. Days, months later he'd hear her tell a clear, clean version to his sisters as they made bread or pies or stuffed grape leaves. A version indiscernible in tenor from the stories she'd sometimes tell that she'd heard back in the dry hills of her own childhood. She'd begin: *There was an Italian . . . ,* and when she wanted to pause for emphasis or to keep up the suspense, she spent an extra moment pushing the silky dough away with the palm of her hand, then drawing it towards her again.

3
Armistice Day

As my father's eyes, hearing, and knees failed ever more certainly, his inner world seemed to grow stronger. In his last year he was steeped in remembrances of his childhood, sometimes of the smallest things—squirting milk from a cow's udder into the cat's mouth as he did the morning chores, or picking blueberries with his grandfather, or planting potatoes in the back field. Other times he remembered epidemics and storms, and the way war made its inroads. Or small events that shook their world—the farm was a remote place then and it didn't take much for something, a wild beehive or a flooded brook, to be talked about for weeks and remembered long into the century.

In part, hearing those stories helped me to feel closer to him in the years since I'd come back to live on the farm. Certainly when I listened to him remember—relaxed in his worn leather chair, his voice unrushed—he was no longer the remote, capable, eternally working figure he'd been in my childhood. Without question back then we were over forty years apart, and now in his late years it was as if these moments of recollection allowed for a breach in the strict chronology of time.

Still, that sense of nearness was now durable, now fleeting. As his past grew more vivid, he also became more placid about letting go of the here and now, about the fate of the farm whatever it may be, as if he knew it was already out of his hands. In those moments I felt helplessly far from him, and almost betrayed, what with me dreaming of a constant future. And in his final months, when there was no stopping the reel of stories, as he began calling up the past all the time, the more he remembered, the more he appeared to be on a rise looking back on a complete, clear life. I knew it signaled his end, and began to feel afraid. Some days, when he started in, all I wanted was air.

Now that he's gone those accreted stories hardly seem enough—a few marks carved by a burin is all—and I find myself trying to make up the world around them. I study them for clues not only to his life, but to the life of those before him. My grandparents spoke other tongues—Arabic on my father's side, Italian on my mother's—and our family doesn't have much of a written past. There are no boxes of letters, no journals wrapped in burlap, not even a Farmer's Diary. My oldest aunt kept notes in her careful hand of all the family names and addresses, marriages, deaths, and children. Beyond that the family record is kept in documents—mortgages, deeds, and citizenship papers. *Be it remembered . . . the said applicant having made such declaration and renunciation and taken such oaths* . . . or, *Know all men by these presents* . . . Tucked in drawers or locked in a steel safe fitted into one of the old fireplaces, such documents are what we have to assure ourselves *they* were back there living their lives.

I know the details of a few often-repeated stories—not many—by heart. Though they may seem random, they hold together what I have like the galvanized nails my father used to repair his apple boxes. The original nails have rusted and bled into the wood, but the ones he himself struck gleam out of the time-darkened pine. Here in its modest blue-white luster is Armistice Day 1918. He laughed to remember it: "That afternoon we must have sung *Marching Through Georgia* dozens of times."

Their schoolroom smelled of ink and chalk and musty readers. In autumn they trailed in sodden leaves, in spring their boots reeked of skunk cabbage. Their cheeks glistened in the falling of a small rain. In winter the room could be smoky and the stove sometimes ran too hot; come late spring a glare spread across the wooden desks. On the walls hung oak-framed gray prints of George Washington and Abraham Lincoln and a map of the world with its old borders drawn in black. White rings stained the sills where jars of frogs' eggs had ranged in spring. The flawed window glass sent tremors through the pines and a steely November sky. A tongue curled over an upper lip, a furrow crossed a translucent blue-veined brow as their hands made ghost strokes above their papers before they drew the first mark.

They were learning the Palmer method of penmanship, and it accompanied them all their lives. I still see old men signing checks at the bank, and they circle the air with diagonal O's to get the right slant and set before setting down their signature as if even their names were thought out and practiced before being written. For my father and his sisters and brothers that motion was more than effort and caution, since each pen stroke brought them a little further from their parents' alphabet, even as it granted them new letters fit for their world.

The eleventh day of the eleventh month of 1918 was a quiet, usual morning. Most still traveled the roads by foot and horse and wagon, so all raised their heads from their work when the sound of an approaching engine became audible, and they grew even more attentive when it idled down. Hastening footsteps, and then the regional school superintendent threw the door open while announcing, "Armistice has been signed! Armistice has been signed! Spread the word!"

"I didn't even know what the word *armistice* meant," my father said. For the Second World War there'd be three stars in the window of the farmhouse, and each brother would know differing wars within that war: the one who'd read Shakespeare was a colonel in the Pacific and has countless stories to tell; another, a soldier in France, has never, as far as I know, said a thing. But in 1918 my father was barely eight, and none in his family had gone over, so his memory of the time is that of a child's, no harsher than a feather falling across his good eye. "What did we know of the war? My mother knit scarves and sweaters at night.

We picked out peach pits from the chicken feed. We had no white sugar."

To spread the word their teacher lined them up, the youngest in front, the oldest holding the flag at the rear, and marched them north down Salem Road, then west on Pelham Road. "These were great distances we're talking about . . ." Small steps on an uneven rutted way:

Bring the good old bugle, boys, we'll sing another song.
Sing it with a spirit that will start the world along,
Sing it as we used to sing it fifty thousand strong,
While we were marching through Georgia.

All the holdings they paraded past were carved out of the same world—a tie-up in the barn, a farmhouse facing south so the yard would be out of the bitter wind and early to warm up in spring, a good place for the first work of the new season. In seeing the way those old houses sit in place I'm tempted to see the same lives in each—there were more than four thousand farms in this county and half the land was under cultivation. But there were many ways of division. As the milk inspectors looked over the farms that supplied Lawrence with milk, they saw a difference, and divided the holdings into the kept and unkempt: *noticeable lack of equipment and method at this farm*, or, *up to date farm . . . common farm buildings having ample means of light for the cows*, or, *insufficient light; no milk room; illustration of an old and cluttered farm*, or, *run by men of foreign birth who have no one to keep house for them . . .*

White pine on the terraces, swamp maple in the lowlands.

Hurrah! Hurrah! We bring the jubilee!
Hurrah! Hurrah! the flag that makes you free!
So we sang the chorus from Atlanta to the sea.
While we were marching through Georgia.

He remembers how deserted the road seemed. Mid-afternoon. The chestnut mare at Wyatt's lifted her head from the feed and turned towards them. Mrs. Timson was so far back they couldn't read her expression but saw a wave of her hand. No one home at the Kennedy farm, or Baileys', Mansur. Only the hired man at the Stewarts' walked towards the road and called out, "So, the Kaiser quit?"

"The Kaiser quit," their teacher answered and marched on past more spare yards containing carts and plows, stove wood cut and partially stacked next to a stump with an ax cocked on it. The windmill fluttered in a diminishing wind. A porch hadn't yet been built and the maple planted to mark the first birth was still a sapling, so nothing softened the frame of the white clapboard house trimmed in white. Their hope was that it would soften with time—that a branch of the maple would shelter the parlor from the July sun, and the lilac would scent the second-story bedrooms. But then it was full of effort, and who would believe the world could come to them on this road, dust-choked in summer, the low places impassable in mud time?

The town—fewer than five thousand then—had sent nearly two hundred young men to serve in the infantry, artillery, batteries, in the navy, the aviation corps. 319th Regt. Field Artillery, Battery F. They were engaged at Chemin des Dames, Seicheprey and Xivray, the Meuse-Argonne Forest, St. Mihiel, Château-Thierry. They'd left a world that, if noted at all, was noted in the same clasped leather book that listed the phases of the moon and the position of the morning stars. And noted in a meted-out hand: *July 3: Mowed the new orchard; July 4: Went down to the Grange for dinner; July 5: Mowed with the horses, got in two loads; July 9: A good rain this afternoon; July 10: Good hay day, got in three loads. . . . Got in four loads.*

Some soldiers wrote long chronicles of the war in France, as if the strangeness of what they'd seen had given them a new eloquence: *As we rode along in the train we noticed old trenches, barbed wire . . . we could hear guns booming in the distance. . . . On that afternoon we went to a nearby ruined village, where to test our gas masks, we put them on and went into a cave where there was gas. . . . We noticed orchards everywhere where the trees had been sawed a few feet from the ground and toppled over. The churches and gravestones had been smashed to dust.*

The schoolchildren marched past all that the men would return to, the grade herds, the milk houses, past barbed wire over stone walls, and then turned and marched back past the schoolhouse to the main road. By then, lamps were lit in the kitchens, herds in narrow file were making their way back to the barns. Weak bells tinkled in the late afternoon. On the uneven road, the teacher didn't notice the children's voices were growing fainter. *A grass widow*, is what my father heard people call her, though at the time he couldn't have understood what was meant by the term. In the winter months she buried heated bricks at the foot of her bed for warmth. She broke the skin of ice that had formed on the water in her china washbasin and slipped her cupped hands in. At the junction of the main road she stopped singing, and for the first time she gathered how half-hearted the voices were behind her. A ragged line of pink in the west, pink on the eastern hills. The moon growing solid as the sky sloped towards evening. She turned towards them to find the flag had been passed forward and the oldest children, who had taken up the rear, had deserted. They had been sneaking off one after another, escaping through the fields and woods, their clothes snagging on damask rose vines along the walls, black alders whipping at their eyes. They gulped the damp woods air, their boots were caked with mud and duff; their faces, flush with their antics. What was left of her parade was four children under eleven carrying the song. Seeing how far the daylight had gone, she dismissed them.

It's likely word spread faster from Paris, France, to Lawrence than it had from farm to farm. In the city the fire horns had sounded off at three in the morning to announce the end of the war, and by noon there was an extra edition of the paper outlining the terms of armistice. The five o'clock edition reported Germany was in the hands of revolutionists. The war machine was already dismantling: all outstanding draft calls were canceled, overtime and Sunday war work orders for the mills were stopped. Even so, the call for liberty bonds continued, as did the three-column ads for Arrowsmith artificial limbs. Retail grocers were asked to call at City Hall to receive sugar registry cards. Private McKeown had been killed in France. *Born in Ireland twenty-eight years ago [he] came here when he was eight years old. At the time of his enlistment he was a wool sorter in the Washington mills and is well known in the city. Private McKeown has one brother who has been discharged from the Canadian army after being badly gassed. Another brother, Peter, is a member of Battery D 45th Field Artillery in France.*

On November 12 there was a long, large Victory parade in the city, which my grandfather took my father to see. Mill workers stood measured paces apart and donned matching hats and aprons and marched along with schoolchildren and veterans and Red Cross workers past the shops and banks of the wide commercial streets. What he remembers of that day is not the parade but the crowds back in the side streets, where the triple-decker porches sagged, where the laundry hanging across the alleyways stiffened in the cold late fall winds, back where the tenements were so close together that the rent collector could save himself three flights of stairs by reaching across the alley from one kitchen window to the next to collect what was due him, and where the official songs of the parade faded into songs of countless languages, and voices talking and calling.

The war—and its end—had separated many of the immigrants even further from their birthplaces, further from the hopes they'd raised with each glass of wine. A common toast among them: *May your return home be soon.* Which was answered with: *In your company.* After four years some of their old villages were heaps of rubble, some a muddy field of the dead. When my father's parents immigrated, their village near Zahlé had been part of Greater Syria under the rule of the Ottoman Empire—my father's birth certificate says his parents were born in Syria. Following the First World War, half the population of their village and its surrounding countryside starved or died of disease. The country of my grandparents' birth came under French rule, then years later became the independent country of Lebanon; it's listed as such on my father's death certificate, as if a ghost journey had been made through colonial determinations, through independence and bitter civil war. In the eyes of the world Lebanon has made a passage from an obscure Middle Eastern country to one whose daily casualties make international headlines; and there, between my father's birth and his death, roams what might have been had my grandparents never put their lives in the hands of steamship agents and translators.

As it was, the loved ones the Lawrence mill workers left behind had seen what they themselves had only heard of. In the alleyways of Lawrence, in a confusion of languages, the crowds burned the Kaiser in effigy—my father remembers seeing that—they hit him in effigy. Eight million had died. The dove, if she comes, will have to rest her breast on the rough road and peck at stones.

The school has been razed, the road paved. The houses that have been built on old fields are oriented to the roads and to the life of the towns and cities. There isn't any need to have a warm east-facing yard for the spring repairs. No significant time is told by almanacs and moons. I have in my house the oak frame from the portrait of George Washington that once hung on the wall of the one-room school. It now frames a watercolor of the farm, which had first hung in my father's office. He told me a woman had come out one day and asked permission to set up her easel in one of the fields, and at the end of the day she gave my father the painting because he had brought her a glass of beer earlier that afternoon.

He used to say she got a good portion of it right: the boards missing on the barn, which has since blown down, and the tilt of the silo, gone too. The tree she painted to the right of the barn by the stone wall was never there, but he didn't have a problem with that. The woman who'd spent that afternoon looking at the side of the barn and house, who'd thinned her colors and dipped her brush into springwater, must have wanted something to balance the picture.

4
Milk

"All you could smell was cows," and he tamped his ash cane on the living room floor as he named down the farms: "Bragdon had thirty cows, Dooley must have had twenty or so. Stevens, he had a small herd . . . Cox, Clough . . ." Neighbors whose places were nearly like his own family's—cattle fed on mash and scraps, and grass in summer. The morning's yield cooled in metal jugs set in stone rooms, in springs, and in wellhouses before it was sold to the mill workers in Lawrence.

In the early years of the city, a farmer would bring his own supply to the central district. He'd load his cans on the back of his wagon and cover them with damp blankets to keep out the dust. When he reached the tenements, he'd pull up halfway down a street and the women would come down the outside stairwells carrying tin pitchers, enamel basins, bowls, mason jars, anything that would hold their day's supply of milk. Five cents a quart, three cents a pint—the first customers got all the cream—until his ladle scraped the bottom of a can and he poured the last blue milk into a mason jar.

By 1910 the three-hundred-acre central district contained over thirty thousand people, and one man ladling out milk to a waiting cluster of women proved too small and singular an act to sustain the growing population. Farms farther and farther away began shipping milk in until the Lawrence milkshed extended farther than the Merrimack River watershed—to northern Maine, to the Canadian border of Vermont. Supplies came down in bulk by rail, and plants near the city bottled and distributed the milk.

The nearby farmers, too, began consolidating their production. Bottlers on the outskirts of the city would collect the milk from area farms, process it, then make door-to-door deliveries. In the predawn hours boys ran up the tenement steps, set down full quarts at every threshold, and retrieved the empty ones. You could hear the glass chime when the bottles jostled against each other. The cream froze in winter and forced off the paper plug. Milk set down in the dark at every door—that custom of delivery lived well into the 1960s—I remember it too—and it's the way I always imagine milk: frosty, clean in its clear bottle. Think of all the kind words and phrases that come from it: *milky, milk glass, milk run, milk tooth . . .*

Milk epidemic. I remember my father saying, "Dr. Batal tells me the undertakers always had a dozen child's coffins on hand—it was the milk." It was also the water, and the raw sewage, and the lack of medical attention. But scarlet fever, typhoid, diphtheria, septic sore throat, dysentery could be spread through milk, and because the supply had started to come from a greater distance, and the plants combined the supply from one small farm with all the other milk of the day, of several milkings, it meant the yield from one herd could find its way into countless households, and you could no longer trace a ladle of it back to its source.

As it was, families in the central city were so squeezed in there was little room for health: their long hours at work, the bad air, the cramped conditions, breakfasts of bread soaked in coffee. For those surveying the state of public health in Lawrence in 1910, the milk supply was as much of a concern as the water supply and sewage and garbage disposal. They recommended tighter inspections of dairies and farms, pasteurization of all milk, and a grading system based on cleanliness. In a world once judged by the cream line, the rule would soon be: the cleaner the milk, the higher the price. Certified Milk, Inspected Milk Heated, Market Milk Heated . . .

Inspectors visited every local farm to check one man's holding against all the others, and to an ideal. *Do you wash your hands before milking?* they asked. *Do you wash the udders? Do you cool the milk immediately after milking each cow? How much ice do you have on hand in summer? When was the herd last tested with tuberculin?* They took the temperature of the well water and checked the milk for impurities. They examined the sturdiness of the barn and the cleanliness of the bedding; suggested separate milk rooms; told the farmers their jugs should have narrower openings to lessen the chances for contamination. Their stables should be as light as the kitchen.

A cold eye on dust and rain in an open pail, rough boards, a cluttered yard. Of course they found cause for concern: a live frog in a milk can, bottles washed in a galvanized tub out in the open, manure piled against the barn windows. To the long-established farmers, it must have seemed as if their very lives were being questioned. Men standing in where their fathers had, working as their fathers had, coming to understand they couldn't teach their children all they'd need to know. Children who smelled of hay and dung and souring milk as they studied their times tables and Whittier's poems in the one-room school on the back road.

The Civil War, the industrial age, the opening of the West, and the played-out soil had already thinned the rural families throughout New England. The rougher fields had gone to woods, and barns had collapsed in on themselves. What survived? Lilacs, the stone walls, the cellar holes, the metal and glass dumped in the dry well. Even with all those lost farms, what must have seemed a paucity then looks like an abundance now. When I spread out the 1911 map of the Lawrence milk supply, I can count over one hundred and fifty holdings within a six-mile radius of the city: *Connelly, 200 quarts; Griffen, 80 quarts; Williams, 450; Boornazian, 130.*

To a world already in the midst of change Lawrence, with its short, dense history, brought even more changes to the surrounding farmlands. Coburn, Cox, Clough—names that had settled the valley—half at least would soon disappear from the farming maps; in their place, the names of Greeks, Armenians, Italians, Poles. No one, not the city planners, not the old farmers, not the immigrants themselves would have imagined that as the settlers moved on to other opportunities, or simply faded out, immigrants would move out of the city to take up farming in unfamiliar soil.

Five hundred dollars bought my grandparents their thirty-five acres and house and barn and herd, and that's more or less what they left to their nine children sixty years later. *Know all men by these presents,* my grandfather's Last Will and Testament begins, *that I, Kalil Brox otherwise known as Charles Brox, Kilil Brox, Kelle Brox, and Kelil Brox* . . . To document his life, he had to include all the official spellings of his Arabic name. The farm hardly appeared different from others on the road: trodden dirt in the barnyard, the brief bloom of the peonies. Wood smoke rising all winter long; and in summer, the sound of a man coaxing a workhorse, or the scythe cutting the dry June grass. The smell of hay and manure. Rather than journey to the Arabic church in Lawrence, my father and his brothers and sisters went to Sunday school at the Methodist church next door. They brought their lunches to the neighborhood box parties and leaned over their primers in the kerosene light at night.

When city relatives came to the farm for visits—the children to spend time in some good air, the adults to pick grape leaves or apples—they walked out from Lawrence on foot early in the day, singing Syrian songs for all five miles, and once here they'd eat under the remaining white pines: roasted lamb, red pistachios, stuffed eggplants, and figs. All day long the closest neighbor walked back and forth along his property line, making sure not one of those visitors stepped over onto his land. Strangers here; and the children, strangers here and in the city. After he graduated from the one-room school, my father took the trolley to Lawrence to go to high school, where a child who had never milked a cow in his life, who hunted the railroad tracks for scrap coal and brought lunch to his parents in the mill, who likely would go on to work in the mill himself after school was finished, a child smelling of cabbage, or of scrap coal, or of kerosene, found the difference and split it. He said with contempt and out of the side of his mouth, "You smell like a barn."

I can't find any record of a milk epidemic having occurred in the valley. There needn't have been anything so drastic for milk to become a losing proposition here. Competition from larger farms in the Midwest, and other possibilities for lives and land took care of that. The livestock books on the shelf in my father's office haven't been opened for years. In order to keep the farm going he turned to growing vegetables for the Lawrence market, for Lowell, sometimes Boston. He loved growing things—in his last summers his whole heart seemed to be in his corn and tomatoes and apples—and I have never much pictured him around livestock, though he says he was a long time in giving it up and kept a couple of cows for years after he'd sold the bulk of the herd.

"Just for the family," he would have said. When the cows freshened, his mother and sisters wouldn't know what to do with all the milk. They were always making *laban*, tart and cold, close to what we call yoghurt now, a vestige brought forward from the desert life, where fermentation kept milk from spoiling. You start a new batch with a remnant from the old, and let it thicken in a covered bowl in the warmest spot in the kitchen. One batch would be mild and the next would make you wince with its tartness. Sometimes it was beaten or thinned so you could drink it. For a hot day, *laban* with cucumbers and mint; in winter, *laban* with raspberry jam spooned in contemplation by the kitchen window.

And palm-sized crumbly cakes of fresh Syrian cheese, they made that too. They'd tie the curd in thin muslin and let it drain into an enamel bowl. On the kitchen counter, in the still of the afternoons, all through the night as the household slept, whey dripped off the cloth—first, resonant on the bowl's hard surface, then dripping more softly into itself, and slowing as the curd compacted. Cheese eaten with olives or wrapped in Syrian bread, crumbled between a thumb and forefinger, or sliced neat with a steel blade. White as the high, far moon. We grew up on it: tangy, salty, dry on the tongue.

5
Last Look

PHOTOGRAPHS—THE OLD ONES—AS brown as barn light. I have several sitting on the table in my bay window, and when I catch a glimpse of them, I settle into a mild contentment, since my parents in them, in the wedding photo, in the antic day at the beach, are nearly unknown to me. So young, for one thing, dressed in the fashions of their time, the vintage cars shining behind them. They feel as far from me as the washed pen lines of old plats are from the current maps of this farm. When I look at them it doesn't seem I've lost the people in them to death or age. Not nearly the sorrow I feel when I look at later pictures, say, of my father in his seventies. The one I keep at the window is in full color. He's standing in the eggplant, snips in hand, smiling easily at the photographer: a picture of the father that I knew, in the shape of the daily life he lived, and I can't quite believe he is no longer here.

It's hard for me to remember anymore when he wasn't slow to get out of his truck—arthritis in his knees. His slowness gave him an air of deliberation as he considered the shapes of the apple trees and weighed which branches to prune. Arthritis in his shoulders, too, and he'd lift the saw in obvious pain, though his cuts were always sure. A few draws of the saw and the apple branch would drop to the grass. He'd work from the bottom to the top of each tree, pruning a few trees, and then driving on down the orchard with his truck door ajar. It would be one of the warmer mornings of late February or early March, and it's likely there was snow in the woods.

In his last years Felix, his oldest, most reliable worker, came back early to help him with the trees. Felix speaks mostly Spanish—he's from Puerto Rico—mixed with a little broken English, and I think if the two of them could have communicated better, my father might have taught him how to prune. As it was, the two of them worked at my father's pace, though it was obvious Felix could have worked much faster on his own; he always walked a little ahead, then waited, arms at his sides, until my father showed him which cuts to make. For the high cuts he pointed with his ash cane, which he'd bought years back at a cattle exposition—all the dairy farmers carried one to hook around the neck of a calf, drawing the face close for a better look at its eyes and teeth. The cane outlasted all the herds here, and my father used it to point out what needed to be done or to help himself down the stairs or to poke at a smoldering log. Some days he simply leaned against his truck and tamped the earth with it.

Touch-up prunings are what the orchard had gotten in those years, so there were shocks of water sprouts atop some of the older trees, crowns too thick with branches, and branches crossing one another. Its mild neglect settled in and became part of our lives, maybe like a little drinking or my mother tossing off the doctor's warnings by saying, "I'm fine, I just have a bit of sugar in my blood is all." Even so, the trees stood sturdy in our winter land. Their basic shape had been set long ago by their early prunings, so the tiers turning off the central leaders have a hard-earned grace about them, and the wide crotches are full of strength. In winter they'd stand up to ice storms and the nor'easters, and we stopped thinking of anything radical changing. Outside, a fine dry January snow falling on those turned limbs; within, my father studying his *Wall Street Journal* in the easy chair, my mother biding time in the kitchen, the layers of glass shutting out the small notes of the chickadees.

And so they stayed until stirred one January by a phone call from someone offering to prune the orchards. I imagine the man who called drove the back roads of Middlesex and Worcester Counties looking for old farmers and old orchards. Here the land values are so high there isn't much of a way to have your own farm anymore unless you were born to it, and those who were not have to find their place among established family holdings. There are large places that need managers and small places where an old farmer has no one to take over or is having trouble with his children.

What is it about our farm that signaled a way in? A local rumor maybe, or the water sprouts on the trees, or he might have noticed how the orchard had gone unmown. High burnished grasses he'd see, since the months had been dry. The afternoon thunderstorms all veered to our north or south or petered out in the blue hills of Worcester County. Every afternoon signs of a storm would build and then fail. The brown grasses—nearly the same color as the hide of a deer—were shot through with purple vetch and white yarrow. I remember staring at those colors for a moment and not wanting it otherwise, it seemed so beautiful, even though I knew it wasn't practical for picking the crop or getting the trucks through, and how easily it could be sparked by fire.

"You can start on the new trees," my father told the man, "and we'll see." If he hesitated in his decision, it didn't show in his voice. Strange for me to hear. He had always loved best that early quiet work of pruning, the chance to start up again in spring, and I just didn't think he'd pass it on so easily. For my mother and me, once we got used to the idea, those were light days for us. Light beyond reason really. The thought of the pruner animated our talk. My mother's voice nearly sang as she recited where he had worked before: "In the Nashoba Valley, for Hanson, for Dawes." He had gone to Stockbridge, he had a license to spray. And my mother's mind and mine started running the same. We imagined how he could maybe head the picking crew as well. It seemed as if it were a move for the future instead of just getting through seasons. My father was more cautious: "He talks good. We'll have to see how he works out."

The man began the pruning on a cold March day. Frost in the shadows, the crocuses just breaking ground. Two trucks made their way down into the orchard; the two of them talked for a while at the first tree, and then the man got right to work. He hung his loppers over a low branch and folded a saw into his back pocket. His shears fit in a holster by his side. Only the pole, which he kept in his truck, was out of easy reach. He started right at a large branch on the first row of Cortlands, walked around the tree and made his cuts, then walked straight to the next one and began again. I thought for sure my father would shadow him for much of the first morning, but after a few trees he drove off.

The history of the trees that I'd heard piecemeal through the years—how the orchard had been planted in an old pasture which had once been the white pine woods my grandfather logged to pay off the mortgage, or how a new orchard blew down in the '38 hurricane and they'd staked the trees back up with downed wires—none of it did he need to know to do his job. What he had to know he could glean from what was right in front of him in the turn of a branch, in the health of the wood, in the cuts that had gone before. It may not have been exactly how my father would have pruned the trees, but there wasn't any arguing with his work. "He seems do be doing a good job," I said after the first day. "Well, those young trees are easy," my father replied. I could see him setting his jaw. A silence. And then he relented: "He knows what he's doing."

The man arrived reliably in the early mornings, pruning to his own ideal, probably thinking ahead to next year, saying to himself he'd have an easier time of it then. He tended to prune more severely than my father ever had. Where he cut the larger limbs the honey-colored wood stood out against a gray land. Familiar small branchings were gone. My eyes had to adjust to a little more sky.

He'd leave his truck door open, the radio going, and the noise made me a little cranky, but I never felt it was my place to say anything. Felix went up and down the rows in the afternoon gathering the cuts into piles—more pruned wood than I ever remembered in tangled heaps on the winter-matted grass. They burned the piles one still day under a damp, leaden sky. A low flame was the color in the land. Brushfires have always been part of the early spring here—all smudge—and I cracked the window open so I could sleep to the smell the way I like to sleep to rain after the long dry spells.

6
At Sea

Months after everything I was driving headstrong into Boston—as abstracted as everyone else on the road—when out of the corner of my eye I saw an ambulance in silent alarm moving steadily through the rush-hour traffic, and all my hasty thoughts stopped. I had to wince back tears as I remembered trying to stay calm when the medic told me not to rush as I followed the ambulance into Lowell.

That late December day there'd been patchy snow on the ground and the beginnings of a still gray winter dawn as I ran across the orchard to my parents' house. It seemed to take forever before the sirens cut into the morning. Then all at once five husky men—four EMTs and a policeman—tossed aside the chairs in the dining room so the gurney could get through, then stood around the bedposts in my parents' room where my father lay in pain. They'd brought in the cold dawn air. They'd tracked in melting snow.

The policeman turned to me and spoke in awkward apology: "Someone has to come for all the 911's—in case there's anything. But there's nothing like that here. Sorry. Jeez, it's been a bad night." And as he started to tell me about how some old man who lived on Hildreth Street had fallen out of bed, I tried to hear around him so I could discern what the medics were asking my father: "When did it start?" "Can you describe the pain?" "Can you get yourself up?" I was too afraid to enter the bedroom or to ask questions myself, so I stood in the entranceway overhearing what I could. One medic whispered to another: "He's extremely, extremely critical."

What do the dying see? A resolving flare of human love? Does a milky eye clear for the last time? From the little I know now, I think they must see what they always have, are true to their natures at the end—those of earth, tied to the earth. My father, who had always been practical, when given his last choice weakly told the EMTs to take him to the hospital in Lowell rather than Lawrence because, as he later explained, the winter roads were better maintained along that route, and we'd have an easier time getting back and forth if he was in for a long stay.

And it was a straight way there past the house he'd been born in, the orchards he'd planted, and up the hill that had once mired horses in spring in the time when the route was known as the Black North Road, which you can find on maps hundreds of years back, before our family's knowledge of this country, back past the oxen teams and wheel tracks to a footpath crossing the wilderness north of the Merrimack. Now it was a clear gray paved road for a small entourage just under the speed limit: a squad car, an empty rescue truck followed by the ambulance carrying my father—my mother sat up front with the driver—and me following last.

No other traffic that winter Sunday morning, so no siren, and only the merest slowing at intersections. The deliberateness of the ride was a strange foolish comfort as I told myself *it mustn't be too much of an emergency*, even as I knew for sure his kidneys were so weak he could not possibly survive a problem with his heart, and this was the closest he'd have to a last look at where he'd lived his life as we passed the neighbors, the town offices, the Grange Hall, passed all the goodwill and any grudges and misjudgments, the red emergency lights sweeping across everything like a lighthouse beacon falling back across those sleeping in an innocent mist among low coastal hills who, since they are not the ones caught in a fogbound open sea, sleep unawares. Whoever we'd become in those hours, we were different from those on dry land.

Once in Lowell we skirted the river and its bridges and passed the brick mills and Pawtucket Falls where tawny winter water poured over the boards onto icy rocks at the bottom. Then we turned away from the river road, and the ambulance docked at the emergency entrance to The General.

The intensive care unit there is a circle of rooms around the nurses' station. The waiting area is an outer circle embracing the patients' rooms. You can walk around and around through the day and see the same groups of families camped in that outer circle—the ICU has no restrictions on visiting hours: exhausted men and women who'd come from their shifts, smelling of oil or of ink; old women sitting straight-backed, their eyes closed with strain, or maybe prayer; children sleeping at their parents' feet; spent coffee cups and Coke cans and snack wrappers scattered about. The ones who've just arrived hardly leave at all. Those there long enough might start thinking about spotting one another, coaxing one another to go home for some rest: *Come on, you have to keep your strength up* . . . You never think it could be so plain and small, but there it is: the best hope in that place is a long haul.

For nearly a week we were one family of the eleven waiting there in that circular hall, talking with several surrounding families, comparing progress or the lack of it, reading the slightest movement of a hand or the strength of a sigh for a sign. It seemed we were all on the same journey—odd comfort—and when my father turned quickly, and one afternoon six days after being admitted, his fire drowned in its own ash, we were set adrift on our own small raft again. How strange it was to have lived day and night for a week with those people, and now to hug them one last time and leave them forever to their own fates. I knew that by evening my father's bed would be filled by another.

7
What Is Kept

OF THE ARABIC my grandparents spoke, even my father knew next to nothing. I heard him slip in a word or two now and then, when he talked to his brothers about something he didn't want us children to understand, or when he greeted family—*Messiah am!*—as they stepped through our threshold at Easter. And he'd repeat a common saying of my grandmother's, which translates: "A thousand, but not another thousand," meaning, I guess, the earthly years we have left. My aunt at ninety-four says hoarsely, "We weren't like the children in the city—we didn't hear Arabic every day. All our neighbors spoke English, so we spoke it, too."

If my father knew next to nothing of Arabic, I know even less. Really, just the names of food—kousa, shish kebab, baklava, hummous, tabbouleh—pronounced flatly without the gutturals anyone comfortable in the language would lend them. As a child I was taught to make traditional Lebanese dishes; to hollow out kousa—a pale green, sweet summer squash we stuff with lamb and rice and steam over a bed of scrap bones; to cover phyllo dough with a damp towel—the unbaked pastry was thin as gauze and ruined by the open air—as I built up buttery layers of baclava. To drizzle the hummous with oil, and sprinkle it with paprika, and ring it with slices of red onion before serving. I learned from my mother, who knew nothing of Lebanese cooking until she married and moved out to the farm where she was taught by her mother-in-law to make the things my father loved.

By the time I was school-aged, my mother was an expert at it. She and I would sit at the kitchen table and hollow out one kousa after another. The squash was soft and slender, and the shell had to be thin and even in order to cook well. I was a child, impatient and awkward with the corer. My mother went over my work after she'd finished hers, paring more squash off the thick places and patching my holes with bits of the inner flesh, before she filled all the kousa with lamb and rice.

Chopping, paring, coring, rolling, simmering. Slow, attentive work—without connotation or inflection—more easily passed along than language, with less to misunderstand. And once learned, something that comes back after years, long after words have been lost. Still, measured by detail and consistency; and never think such work wasn't judged at every family gathering in the resistance of a fork, in the crumbs that were left behind.

A few years back, before my mother tired of cooking altogether, she began to talk about trying out new recipes.

For decades she had more or less cooked the same dozen or fifteen dishes for the family: some Lebanese, some Italian, some drawn out of the New England past of franks and beans and clam chowder, the American present of hamburgers on the grill. She looked to me for advice: "What could I do with the roast?" she'd ask, not in a determined way, but quiet and speculative, as she unpacked her groceries. Or, "You know about herbs. What could I add to the chicken?" I felt the same sting as when I first notice a burnish on the September leaves. "You know plenty of ways to cook chicken," I'd said awkwardly, feeling I'd traveled too far. I walk so much faster than her now, and arrive everywhere first.

I have to slow my pace consciously as we walk up the steps of the Melkite church for my father's forty-day Mass. We're here more for the gesture to the old Lebanese community than any church vigilance on my father's part. He'd have trouble sitting still for this, especially since, for the convenience of the relatives who have to travel a ways, we've settled on the late Mass, which is sung in Arabic. As I enter the church, I can smell the incense from the earlier services. My eyes are drawn upwards to the iconostasis that divides the altar from the people: a row of paintings of the angels and saints—straight-nosed, almond-eyed, heads tilted under gilt nimbuses—Byzantine, accomplished without perspective. Flat and shimmering, made of wood and mineral, egg, alabaster, and pigments ground fine as the dust of relic bones, they're meant to illumine another world, and separate us from it.

The priest shakes the censer and the incense begins to sting my eyes. The prayer candles flicker. Three men to the right of the altar begin to sing in Arabic. Their voices start low and move up to an extended nasal waver, blending and carrying the music forward, voices singing as they have over the long line of the dead, ours and others, no matter the soil. Once the priest begins the Mass, the three men lead the congregation in Arabic responses. I can't say a word, but I welcome that language—incomprehensible to me on the page, in the ear—more beautiful and comfortable than the translation of the stern dogma I'd long ago turned my back on. The priest, I know, is singing of the unrelieved suffering of this world, and how everything exists for the next. I remember how my father loved the things of this earth. I close my eyes and hear only the pure beauty of those resonant voices. Long after they stop, their sound continues ringing in the small church, ringing, then decaying like anything held and left to itself—a whole note, love, or belief.

II

The sun rarely shines in history, what with the dust and confusion.

—THOREAU

8
In the Current

MY FATHER ALWAYS remembered a hot, muggy inland summer sometime in the twenties. The riverbank beyond the hills to the south of the farm was dry and dusty. The trees and grass, coated with dust. A boy—white-skinned, arms sinewy from work—grasped a rope swing and let himself fly out over the water. At the highest, farthest point of flight he flung himself away—trees blurred in the rushed descent, the river swallowed him, then brought him back to the surface. When he climbed back up on the bank, he was covered with brown-black oil. "He looked like another creature," my father said. During those minutes on shore countless ephemeral things in the current glimpsed by and were gone. The boy reached for the rope again and flung himself in the air, then into the oil and dye and scour-streaked Merrimack.

9
White Mountain Snow Dissolved

IT WAS ALREADY *the water of Squam and Newfound Lake and Winnepisiogee, and White Mountain snow dissolved, on which we were floating, and Smith's and Baker's and Mad rivers, and Nashua and Souhegan and Piscataquoag, and Suncook and Soucook and Contoocook, mingled in incalculable proportions, still fluid, yellowish, restless all, with an ancient ineradicable inclination to the sea,* wrote Henry David Thoreau of the Merrimack River, which he and his brother, John, traveled upon in the late summer of 1839.

While reading Thoreau's account of his journey in *A Week on the Concord and Merrimack Rivers*, I've wondered again and again how so much could have come and gone in the hundred and fifty or so years since they sailed home free with a September wind carrying them. The farms that defined the banks of the river, where they stopped to buy bread or melons, where they slept at night, are almost all certainly gone. The barges hauling lime and wood and brick—gone too. And whole worlds that had barely emerged when they sailed have flourished and passed. The textile industry that Thoreau witnessed in the incipient city of Lowell in 1839 eventually spread to cities on nearly every drop along the lower river, cities that brought immigrants here from the Middle East and Europe and Canada whose names and languages still swirl in the valley air. The redbrick factories of that industry were so imposing and insistent that we still live with the idea of them though the looms have long since fallen silent, and the cities have become other cities, and the Merrimack itself claims a far more modest place on the map of the country than it had when Francis Cabot Lowell and the Boston Associates scoured its lower length for industrial possibilities. More modest and far, too, from the river that William Wood, chronicler of the New World, described for future voyagers in the early seventeenth century: *All along the river side is fresh marshes, in some places three miles broad. In this river is sturgeon, salmon, and bass, and diverse other kinds of fish. To conclude, the country scarce affordeth that which this place cannot yield. . . .*

As the Merrimack gathers its waters and its snowmelt, it descends through the worn slopes of New Hampshire's White Mountains past granite outcrops and hemlock, cone-heavy white pine, birch, maple, beech, through the cities of Concord, Manchester, Nashua, and into Massachusetts, where it bends eastward at the city of Lowell, and then, for the last thirty lateral miles of its journey to the

Gulf of Maine, traverses the low rolling hills of a coastal plain. The bend in the river lies to our west, and this town is set among the hills on the northern bank, just past the inner crook where the Merrimack turns towards the sea.

On the early maps of New England—*being the first that ever was here; cut and done by the best Pattern that could be had, which being in some places defective it made the other less exact: yet doth it sufficiently shew the Scituation of the Country, and conveniently well the distance of Places*—the Merrimack appears as a central spine rising out of ship-laden waters, rising then branching off beyond the northern lakes and into the wine hills. The bend in the river is hardly discernible, and this town doesn't yet exist. You can clearly see how the first colonial settlements in the lower valley had spread out south and west of the river, the broad waters of which had been a barrier to the settlement of the north bank.

The Pawtucket Indians, early inhabitants of this place, called it *Augumtoocooke*—the wilderness north of the Merrimack—a name it was known by until its seventeenth-century mapping, when surveyor Jonathan Danforth's plats divided twenty-two thousand acres into reserved lands and official grants for a handful of families and their descendants. The southern boundary was marked entirely by a calm stretch of the river. The first faint paths of settlement led to and from the ferry crossings.

Drawcutt, Draycott, Draycote, Dracut. 42°41' latitude, 71°19' longitude. Straight lines drawn across fishing places and hunting grounds. Boundaries marked by white oaks and brooks, by granite blocks and heaps of stones. Their plowing and planting and clearing, the modest homes they hoped would be permanent, the plat itself, marked the passing of an earlier world, though by the time Danforth drew his map the Pawtucket Indians were already a remnant tribe. Explorers and fur traders had, years earlier, made their way up the river bringing European disease with them. Between 1614 and 1617, more than three quarters of the Pawtuckets died of a sickness, still guessed at, that killed within three days of its first symptoms: *The living were in no wise able to bury the dead. . . . hundreds without burial or shelter, were devoured as carrion by beasts and birds of prey, and their bones were bleached by the sun.* They who'd trod paths along ridges and across the valleys, who'd found the shallow places to ford streams and the safest route to the sea, their paths were narrow—*unmapped, unmarked except in the atlas of memory*—and worn so deep by years of use that some are said still to be visible in parts of southern New England.

Brown ink washed out on parchment, black lines on bond. All the maps and all the dissolutions of time that render them inaccurate—scouring rains, disease, invasions, wars and floods. Since the Danforth plats, the wilderness north of the Merrimack has been mapped and mapped again with topographic surveys, road maps, charts of the watershed and the milk shed, soil maps that follow the geological logic of the ice ages and have no center, assessors' maps squaring off the place to account for the ownership of every inch, zoning maps dividing the plots for purpose and use. Each illustrates a small part of the story and none begins to tell the whole.

Thoreau wrote of his own unnavigable Concord River: *it was thought by some that with a little expense in removing rocks and deepening the channel, "there might be a profitable inland navigation." I then lived somewhere to tell of.* Well, if profit and industry mark a place, then redbrick cities—count westward and then northerly against the current: Haverhill, Lawrence, Lowell, Nashua, Manchester, Concord—that stand at each drop on the Merrimack make this somewhere to tell of. The same water flows through each city, and by the mid-nineteenth century the same was diverted and harnessed to turn the wheels of the Amoskeag in Manchester, and harnessed again for the mills of Lowell, and, after a brief, nearly noiseless journey along our southern border, again for the mills of Lawrence where my mother's father spent and finished his years as a weaver. The city of Lawrence covers an area smaller than any of the surrounding towns, no larger in circumference than one could walk in a day, yet it once wove the world's worsteds, and still in the air is the litany of mills that came and went over a hundred years of speedups, slowdowns, fires, accidents, and strikes: the Pemberton, the Arlington, the Ayer, the Everett, the Wood, the Pacific . . .

Even if we've come to accommodate them now, the short dense histories of the Merrimack cities hadn't grown out of the local needs of the farming communities and didn't fold easily into the land or the life of the surrounding towns and farms. Though the cities were cut out of town properties, they were conceived as centers for the export of fabrics all over the world, as investments for the fortunes of Lowell, Cabot, Appleton, Abbot, Lawrence—names that mark our maps again and again, having mixed with settlement names: Andover, Methuen, Boxford. And Algonquin: Merrimack, Passaconaway, Wamesit, Wannalancit. One day the mill sites had been mapped as open and green—just as here. Nearly the next, a grid of streets worked out from redbrick banks. In 1820 the world at the bend in the river was made up of scattered houses and barns, pastures and stone walls, a sawmill. In the late summer of 1823, the place was called Lowell, and the first cotton mill began operation. By 1837 eight textile firms employed six thousand women and eighteen hundred men who produced nearly a million yards of cloth a week. By 1840 it was the second-largest city in Massachusetts. Twenty thousand souls.

Each of the early cities on the river seemed to sprout up at least as quickly. When Thoreau sailed near Amoskeag Falls where the city of Manchester, New Hampshire, was being built, he and his brother made *haste to get past the village here collected, and out of hearing of the hammer which was laying the foundation of another Lowell on the banks.* When he set down the account of his journey years later, he recalled: *At the time of our voyage Manchester was a village of about two thousand inhabitants, where we landed for a moment to get some cool water. . . . But now, after nine years, as I have been told and indeed have witnessed, it contains sixteen thousand inhabitants. From a hill on the road between Goffstown and Hooksett, four miles distant, I have since seen a thunder shower pass over, and the sun break out and shine on a city there, where I had landed nine years before in fields to get a draught of water.*

Because we are set between the drops in the river, because there hadn't been power to exploit along these wide-curved miles where the river deposits more than it carries away, ours has been a quieter story, one less marked in time, especially in the eastern part of town where we live, and which, even now, has retained some small vestige of its agricultural identity. What do I know of this place's history before my grandparents came? Little more than I can read from the graves or look up in the local books. Little more than the facts of who settled where, the names of those who died in the wars, where the churches and schools once stood, where old pastures had been walled in by stones. A world made of daily life more than history, really, the world I know best by artifacts I pull from the earth, by glass bottles mottled with prismatic colors, by the blades and tines of old metal tools whose handles have crumbled into the soil.

I imagine the life here didn't differ significantly in its underpinnings from the life of hundreds of other New England towns. During the seventeenth and eighteenth centuries, colonial settlement followed the smaller rivers and streams into the backcountry of New England. A fall in a watercourse of only ten or fifteen feet could power an overshot or breast waterwheel for a small mill: sawmills; grist mills for grinding meal, malt, and flour; fulling mills that dressed homespun cloth after it was woven—washed, shrunk, felted, then the nap raised, and sheared. Fuller's teasel grew along the banks—the dried heads were used to raise the nap on cloth—and still grows on some of the old sites. Such mills almost never operated year-round—the bulk of the fulling mill's business, for instance, came in autumn, finishing wool that had been sheared and washed in the spring, and carded, spun, and woven in the summer.

Northeastern Massachusetts is so congested now that, except in a few conscientious towns, the original settlement centers are lost in sprawl. A white steepled church here, a barn connected to a farmhouse there, stand as specimens preserved among new developments and restaurants, stores, and strip malls. They're hardly enough to give a sense of an earlier time, but claim their place the way a wolf tree does in a regrown woods: the wolf tree's long branches reach out and up, its crown having spread out in the full sunlight of an open field in the years before the younger trees sprouted up around it. It stands as a marker of time passing and gone, and a share of its beauty is drawn from the fact of its survival, from the fact of its standing still beside the sway of supple birches and the sweep of pines. They startle you, those wolf trees, when you come upon them in the woods. Alone always, and as arresting as the clear yellow eye of their namesake.

But follow the roads that follow the rivers back into the north hills of New England today and you still see the small towns as they almost were: wooden and stone shells of old grist and fulling mills and the rapids roaring beside them. If you walk out and up above the town until the sound of the water falling onto granite is beyond your hearing, and then look back, you can see the way the circumscribed world of stone and brick and painted wood, off plumb as it is, spreads out from the bright, rushing falls. Smoke from the chimneys gets lost in the first light snows of November. The measured flow of water is half-caught and half-free.

The early lives in such towns and nearby farms were adorned with a plain weave, a linen warp, a woolen weft. Stripes, checks, bedtick, damask, web shirts. Women and girls made clothes, towels, sacks, bags, bedding for the family and also sold and bartered their woven goods at the local stores. They broadcast their flax seeds in spring and sheared the wool from their indifferent breeds of sheep whose fleece, if it didn't come near the quality of that from Spanish Merino, French Rambouillets, and English Southdowns, sufficed. Girls grew into young women, then into mothers themselves, spinning by the fire and weaving in the shed rooms and attics. They dyed their cloth with vegetable colors: indigo, madder, the bark of red oak, the petals from iris that blossomed in the June meadows and gave a light purple tinge to white wool.

With the coming of the Revolutionary War they wove uniforms for their sons, fathers, brothers, and husbands. As one woman wrote: *Our hands are soldiers' property now; jellies are to be made, lint to be scraped, bandages to be prepared for waiting wounds. Embroidery is laid aside and spinning takes its place. Oh, there is such urgent need for economy.* Dracut claimed about eleven hundred inhabitants at the start of the Revolution, including women and children. Four hundred and thirty-nine soldiers went off to fight—proportionally more soldiers to citizenry than any other town in the colonies—a fact marked down again and again in the local histories. Those who remained sent a pair of shoes, a pair of stockings, and two shirts to each enlisted soldier from the town. After the war, in the new days of Independence, homespun cloth became a fashionable mark of self-reliance, so much so that Harvard and Yale voted to wear homespun suits at commencement. Prizes were given for spinning and weaving. Vows were made to eat no lamb and wear no imports.

Today as you enter the Museum of American Textile History in Lowell, the first thing you see on the second-story corridors on both sides of the entry hall is the largest collection of spinning wheels in the world. Hundreds of wheels have been carried from singular places beside hearths and in ells and are now set one after another in clear glass cases lit from within. Sturdy, square homemade wheels nudge fine shop-made specimens. Small wheels for spinning flax; walking wheels for wool—a girl might cover three or four miles in the course of a day's spinning. Some are adorned with chip-carved table ends or a beaded edge or chamfered legs. More than a few are elegant, without ornament, crafted in one of the Shaker villages while the scent of lavender drifted through a raised window. Wheels collected from foreign countries: one, carried out of Asia Minor, had spun silk and is decorated with hammered turquoise; another—a Charka wheel, built in a case so as to be portable—can fold up to the size of a book. It follows Gandhi's design and is meant to be used sitting on the ground. It had been purchased by a collector on the streets of Bombay.

But most of the wheels haven't traveled all that far, having found their way here from the Green Mountains, the Ozarks, or an island off the coast of Maine. A few are painted a deep butter yellow or bright red or ebony; others retain the warm luster of pine and maple, or the dreaming grains of oak. On some the years have flattened the wheel between the spokes or the wood has been gnawed by rats trying to get at the grease around the axle. Move your eyes just a bit and they can rest on a perfectly preserved parlor wheel—it might have once been a wedding present—adorned with ivory bells and finials.

For each one preserved, countless others sit in neglect somewhere with worms in the wood or a disciplined, turned leg in splinters. What is the difference between the lost and the saved? Between the ones left to attics and kindling and the one bundled and hauled over in steerage by German immigrants? Or carried back from a Black Forest village by a soldier returning from the Second World War, or salvaged from a chicken coop, the manure painstakingly scraped off? Is it the same as the difference between a word lost and a word caught as a sentence trails off? How can there be an accounting of the voices drowned out? What of voices that were strong when they lived the story, but in the long, later telling uttered reedy and hesitant syllables?

They say a humming wheel rises to a sound like the echo of wind in a storm. Almost every New England girl must have known it, busy as all the fates spinning out the only line of destiny she knew: *I have got the most of my wool spun and two webs wove and at the mill and have been out and raked hay almost every afternoon whilst they were haying.... O Sabrina how my western fever rages. Were it not for my father and mother I would be in the far West ere this summer closes but I shall not leave them for friends nor foes! Mary and Elias say Liz dont get married for you must come out here. I shall take up with thare advise unles I can find some kind hearted youth that want a wife and mother, one that is good looking and can hold up his head up. Then when all that comes to pass I am off in a fit of matrimony like a broken jug handle.... Tomorrow I have got to wash churn, bake and make a chese and go over to Daniels blackbering. So good night...*

The next day, too, she had to wind her spun yarn on a reel to make skeins for washing and drying. The making of cloth was one thing before the next—just like syntax—and through all the hours of dreams, resentments, lost thoughts, and song, everything would come out even and of an equal tightness, a handsome selvage, a smooth plain weave. She'd cross-stitch her initials or sign her maiden name in a small hand in the corner of a towel. She might pen in four or six or eight lines of verse. Her warp of linen, her weft of wool contained the same durable root as the web of words.

Though the practice of spinning has been nearly lost, though many of the wheels persist as artifacts tangled under a kindly light, surely it meant something, surely there's a gathered power in all the solitary hours of work those wheels contain. I've never seen a crowd congregate under the bright corridors of the spinning wheel display, though an occasional scholar wanders through, and groups of schoolchildren line up and then move on past the wheels into the exhibit at large, where one room after another displays the journey from homespun through the modern textile industry. Each room is dark to begin with, and as you arrive at its entrance each lights up and stays lit for as long as you look at it. When you step away it darkens again, and the next one lights up, and so on, as you pass through counting rooms and storehouses and mock-ups of the early factories. You make your way into the factory weaving and spinning rooms of the twentieth century, when the spinners, by the time of the strike at Lawrence in 1912, were working fifty-six hours a week for six or seven or ten dollars in their pay envelopes. Always the roving had to be damp and warm to keep from breaking. Lint swirled through the humid air. Such conditions were a breeding ground for the "white plague," pneumonia, and tuberculosis. In Lawrence, a third of the spinners wouldn't survive ten years of such work, and half of those who died hadn't reached the age of twenty-five.

The cities in the valley, like much of the industrial world, could trace a path back to Manchester, England. England, in order to protect its primacy in manufacturing, forbade skilled British textile workers from emigrating and forbade the export of plans or models of the mill machines. When Francis Cabot Lowell, whose Boston Company founded the first mill city on the Merrimack, toured Manchester he memorized the plans for the high-production textile machinery he saw there. What he took away in his memory has become a large part of the lower valley's story, as did what he sought to leave behind: the degrading working conditions, the lives overwhelmed by exhaustion and poverty, the air permeated with cloth dust and anthracite.

He toured Manchester in 1810, at a time when the place of girls and young women on New England farms was perceptibly changing. Within a generation after the Revolutionary War, Spanish Merino sheep had been introduced to New England, improving the quality of native fleece and making it more marketable to the family-run factories that had begun to appear on the smaller rivers of New England. The larger and more profitable market encouraged farmers to turn to pasturing sheep all up and down their hillsides—farmers burned Mt. Monadnock bald trying to drive out the wolves that preyed on their flocks. This larger market also meant a less essential place on the farm for young women. Factory cloth was finer and far cheaper to produce than homespun. As the price of coarse cloth fell, the home spinning wheels and the looms in the ells and sheds fell silent, then gathered dust and were put away.

Francis Lowell reasoned that by employing New England farm girls in his textile city he would be assured of a cheap, dependable workforce and the girls would make more money in his factories than they could earn selling their weaving to a local store or their spinning to an agent. By regulating their living quarters and lives, he believed he could prevent the dire conditions of Manchester from taking hold along the Merrimack. And such regulations would help assure farm families of their daughters' safety. Three-quarters of the first workers in Lowell were young women from New England farms. Their wages were half those of the men.

By the time Pawtucket Falls, where Lowell now stands, was dammed to create a millpond that would assure the factories of reliable power, the river was already a far cry from the one the first European explorers saw as they rowed against the current up a course that crossed broad fresh marshes, its waters teeming with fish. By the early nineteenth century several worlds had been built up along the river, and had waned. Thoreau noted the passing of cargo trade on the waters: *Since our voyage the railroad on the bank has been extended, and there is now but little boating on the Merrimack,* he wrote. *All kinds of produce and stores were formerly conveyed by water but now nothing is carried up the stream, and almost wood and bricks alone are carried down, and these are also carried on the railroad. The locks are fast wearing out, and will soon be impassable, since the tolls will not pay the expense of repairing them.*

Before the world of cargo, and before the coming of European explorers, the lower Merrimack valley was the world of the Pawtucket, the Wamesit, the Nashaway, the Souhegan. "River of sturgeon," the Algonquin word *Merrimack* means—or swift water, or strong place. The inland tribes, who depended for their stores on the reliable run of salmon, bass, and sturgeon pushing up the river at spawning season, gathered every spring at even the smaller falls along its course. But Pawtucket Falls, with its drop of thirty-one feet to the granite rocks below, was the most renowned gathering place of the Algonquin world.

The old town history of Dracut says that Passaconaway was the leader of the area tribes at the time the wilderness north of the Merrimack was divided into land grants and holdings for colonial settlers. If, prior to settlement on his lands, Passaconaway had counseled resistance, within a year of the first grants he could see the future, and it was at Pawtucket Falls he chose to speak for the last time. His last conciliatory speech comes down in common books—this speech alone. The words are variously recorded by various Englishmen: *The English came, they seized our lands; I set me down at Pawtucket. They fought with fire and thunder, my young men were swept down before me when no one was near them. I tried sorcery against them but still they increased and prevailed over me and mine, and I gave place to them, I that can make the dry leaf turn green again. I who have had communion with the Great Spirit dreaming and awake. . . . The oak will soon break before the whirlwind—it shivers and shakes even now; soon its trunk will be prostrate. . . . Then think, my children, of what I say; I commune with the Great Spirit. He whispers to me now—Tell your people Peace, Peace is the only hope of your race. . . . these meadows they shall turn with the plow—these forests shall fall by the axe. . . . We are few and powerless before them. We must bend before the storm; the wind blows hard; the old oak trembles; its branches are gone; its sap is frozen. . . . Peace, Peace with the white men, is the command of the Great Spirit—and the wish—the last wish of Passaconaway.*

The area tribes never again gathered at Pawtucket Falls. Today water brims over the flashboards of a dam and sometimes splashes, sometimes cascades onto the rough rocks below. Its color, as it falls, runs to old ivory; in hard winters that falling water forms into a ragged freeze. Except at snowmelt, the river at the foot of the falls is so still and low you'd think it couldn't possibly recover enough strength to flow out of the city. Graffiti on the granite. Some trash. Weeds and grasses have taken hold deep in the riverbed. Sometimes I see a man fishing from the bank. Who doesn't wonder as they pass where the power of the Merrimack has gone, the Souhegan and Suncook and Soucook and White Mountain snows dissolved?

Once I drive beyond the Pawtucket Dam, my perception of the river changes in an instant, as if dissonance had collapsed into one long clear note of reconciliation. Suddenly I'm above the flashboards and running on a road out and away, the river beside me is wide, brimming, dark, holding back gathered strength from the north. Steely, impenetrable under an overcast, and glittering under a clear sky, there are days the river there is graced by compensatory white sails. The difference above and below the dam is far greater than the difference between being half-caught and half-free—it's a difference of scale, accumulation, and history. The dam, when it was built, not only exaggerated a geological break in the riverbed. Here in the valley it separated agriculture from industry.

At first, the sight of so many bands, and wheels, and springs, in constant motion, was very frightful. She felt afraid to touch the loom, and she was almost sure that she could never learn to weave; the harness puzzled, and reed perplexed her; the shuttle flew out and made a new bump upon her head. . . . Here was a new place in the world, and it was lit differently. Whale-oil lamps hung on pegs by each loom. The windows, nailed shut, let in a slanting light to rooms that were sprayed with water to keep the humidity high so the yarns wouldn't break. From a distance, a roomful of looms had the same rhythm as a loud heart, but in the weaving rooms the machine noise was an undifferentiated envelope of sound. The weavers sucked thread through the small hole at the end of their shuttles and inhaled lint, sizing, and dyes. That shuttle design—used into the twentieth century—came to be known as "the kiss-of-death shuttle" because of the efficiency with which it spread TB among the workers.

Most of the girls stayed on at the mills for no more than a few years while they awaited marriage, or looked to increase the money they could bring to a marriage, or tried to escape their households. In *The Lowell Offering*, a literary magazine written by and for the mill girls, Harriet Farley gave a sense of that world in the guise of a letter home: *There are girls here for every reason, and for no reason at all. . . . One who sits at my right hand at table, is in the factory because she hates her mother-in-law. She has a kind father, and an otherwise excellent home, but, as she and her mama agree about as well as cat and mouse, she has come to the factory. The one next her has a wealthy father, but, like many of our country farmers, he is very penurious, and he wishes his daughters to maintain themselves. The next is here because there is no better place for her unless it is a Shaker settlement. The next has a "well-off" mother, but she is a very pious woman, and will not buy her daughter so many pretty gowns and collars and ribbons and other etceteras of "Vanity Fair" as she likes; so she concluded to "help herself." . . . The next is here because her parents are poor, and she wishes to acquire the means to educate herself. The next is here because her beau came, and she didn't like to trust him alone among so many pretty girls. . . . Many, who are dissatisfied here, have also acquired a dissatisfaction for their homes, so that they cannot be contented any where, and wish they had never seen Lowell. But tell Hester that I advise her to come. . . .*

The dry goods stores in Lowell sold gauzes, lawn, and muslin for frocks and baby clothes; Marseilles quilting, Florentine cloth, and swansdown for vests. Corduroy, velvets, damask. Fabric shot with silk for fine dresses and shawls and skirts. Fabrics named after places in the Far East: nankeen, harrateen, shalloon. Fabrics befitting the glinting city Charles Dickens described in 1840: *One would swear that every "Bakery," "Grocery," and "Bookbindery" and every other kind of store, took its shutters down for the first time, and started in business yesterday. The golden pestles and mortars fixed as signs upon the sun-blind frames outside the Druggists appear to have been just turned out of the United States Mint; and when I saw a baby of some week or ten days old in a woman's arms at a street corner, I found myself unconsciously wondering where it came from: never supposing for an instant that it could have been born in such a young town as that.* In the mills they wove plain goods out of coarse cotton for farm families in the West and black slaves in the South. A heavy sheeting—durable, reliable, inexpensive—weighing less than three yards to the pound. What they wove was called negro cloth.

The girls and young women were used to exhausting hours of work at home, and the factory workload, which seems incomprehensible to us now, was sometimes shouldered matter-of-factly: *You ask if the work is not disagreeable. Not when one is accustomed to it. It tried my patience sadly at first and does now when it does not run well; but in general, I like it, very much. It is easy to do and does not require very violent exertion, as much of our farm work does.* Still, at home there'd been breaks in work and a seasonal variety. Now their days were confined to hours and bells, and they no longer worked at their own speed, but were subject to slowdowns and speedups beyond their control, subject to the demands of the machine itself, closer to the machine than the human: *The Overseers . . . are to see that all those employed in their rooms are in their places in due season. They may grant leave of absence to those employed under them, when there are spare hands in the room to supply their places; otherwise they are not to grant leave of absence, except in cases of absolute necessity.* They had become operatives. In winter they were summoned and released into darkness. The first bells rang at 4:30 a.m., the second at 5:30, the third at 6:20. The dinner bell rang out at noon and rang in at 12:35. The evening bell rang out at 6:30 except on Saturday evening.

Still, in a certain moment, their plainwork had substance for them: *I have sometimes stood at one end of a row of green looms, when the girls were gone from between them, and seen the lathes moving back and forth, the harnesses up and down, the white cloth winding over the rollers, through the long perspective; and I have thought it beautiful.*

Many of the young women had come from large families, but they had never experienced communal living quite like the life of the boardinghouses: *Chairs, chairs—one, two, three, four, and so on to forty. It is really refreshing, sometimes, to go where there is only now and then a chair.* The particulars of their world were gone, particulars they sometimes allowed themselves to remember: *It is now that I begin to dislike these hot brick pavements, and glaring buildings. I want to be at home—to go down to the brook over which the wild grapes have made a natural arbor, and to sit by the cool spring around which the soft brakes cluster so lovingly. I think of the time when, with my little bare feet, I used to follow in aunt Nabby's footsteps through the fields of corn—stepping high and long till we came to the bleaching ground; and I remember—but I must stop, for I know you wish me to write of what I am now doing, as you already know of what I have done. Well; I go to work every day. . . .*

When they climbed into their boardinghouse beds at night and pulled the blanket over their shoulders they may have heard a few whispers near to the timbre of their own voices softly rising out of the same room. Absent were deeper, older voices muffled in a downstairs room, voices separated from their own by years and experience and time, yet familiar and confiding, there when they woke in the morning, there at night, accompanied by the ringing of a cup set down on the plank of the table, or the sound of a log shifting in the fire and falling into cinders that, in breaking up, sounded like the tinkling of glass.

The gaze of the larger world was turned away from those hearths, from the acid soils and the fields turning up stones every spring, from those places keenly recalled by young women in the factories. The farmers' children, if they weren't heading for the factories, had begun to head west for the deeper, richer soils: *I also wish you could see a prairie. You would feel as you never felt before. You would feel as I once did, when for the first time I stood upon the edge of the prairie upon which I now reside. It was about noon of a beautiful October day, when we emerged from the wood, and for miles around stretched forth one broad expanse of clear, open land. I stood alone wrapt up in that peculiar sensation that man only feels when beholding a broad rolling prairie for the first time. . . . Fancy upon a level smooth piece of ground free from sticks, stumps and stones, a team of four, five, or even six yoke of oxen, hitched to a pair of cart wheels, and to them hitched a plough with a beam fourteen feet long, and the share, &c. of which weigh from sixty to one hundred and twenty-five pounds, of wrought iron and steel, and which cuts a furrow from sixteen to twenty-four inches wide, and you will figure the appearance of a "breaking team."*

The children of New England between 1820 and 1840 were born with knives in their brains, wrote Ralph Waldo Emerson. I love the furrowed mystery of his words and have wondered again and again how much and what he meant by them, laden as they are with suggestions of introspection, possibility, and violence, of a heretofore unimagined world opening as young women walked into the crowded world of the factory and the communal life of the boardinghouses. A world where they sometimes found a kind of camaraderie as they took over for each other and covered for each other during absences in work. As they reached for an articulation and understanding of their own lives beyond days of staring into the machine, they papered the panes of the factory windows with reading material, wrote poems and stories of their present and past and dreams, and attended lectures: *Lowell Hall was always crowded and four-fifths of the audience were factory girls. When the lecturer entered, almost every girl had a book in her hand and was intent upon it. When he rose, the book was laid aside and paper and pencil taken instead. . . . I have never seen anywhere so assiduous note-taking.*

The durable cloth they wove into countless white bolts was carried away on wagons or shipped by rail to the new reaches of the country, was unrolled across a table where patterns were pinned and chalked onto the plain weave, and the cloth cut and sewn into work shirts. The last hint of salt from the young women's hands mixed with red southern soils and deep brown prairie loam as whole families stooped to plant freshly broken fields. However durable, soon enough the cloth was worn and bleached and frayed by time and effort until it was patched and threadbare, and at last cut up for quilts or rags or a child's toy, after which it all but disappeared.

Yet fragments of those young women's free hours survive in the letters, poems, and stories they wrote, which come down to us in brown script or in lead type, in accounts of childhoods and workdays, and in dreams of a new social order, like that of Betsey Chamberlain's, which she published in an issue of *The Lowell Offering*:

I had closed my book, and sat ruminating upon the many changes and events which are continually taking place in this transitory world of ours. My reverie was disturbed by the opening of the door, and a little boy entered the room, who, handing me a paper, retired without speaking. I unfolded the paper . . .

1. RESOLVED, *That every father of a family who neglects to give his daughters the same advantages for an education which he gives his sons, shall be expelled from this society and be considered a heathen.*

2. RESOLVED, *That no member of this society shall exact more than eight hours of labor, out of every twenty four, of any person in his or her employment.*

3. RESOLVED, *That as the laborer is worthy of his hire, the price for labor shall be sufficient to enable the working-people to pay a proper attention to scientific and literary pursuits.*

4. RESOLVED, *That the wages of females shall be equal to the wages of males, that they may be enabled to maintain proper independence of character and virtuous deportment. . . .*

Betsey Chamberlain lived in the Lowell of 1841, where the working conditions were in decline. Inside the factories, cut backs in hours during slack times and the slowing of machines meant young women earned less during longer hours. In times of increased production they had three and four machines to tend, and those who fell behind found their wages drastically reduced. They were harassed by overseers who'd begun to receive, as incentives, production premiums. Factory hands were working seventy-five-hour weeks. Wages for piecework had dropped. There were four annual holidays. Such conditions would eventually drive the young women from the factories, forcing owners to rely more and more heavily upon Irish immigrants fleeing the potato famine, the French Canadians hoping to put the hardships of rural Quebec behind them, and the countless others seeking to escape poverty or the past, debt or conscription.

All still to come. In the moment that held Betsey Chamberlain's dream, as the tailrace waste waters rejoined the river, and the factory bells rang across the water, the last Indians living in wigwams in this part of the valley were put on a train north.

10

Five Thousand Days Like This One

SPEED ANNIHILATING TIME. Time annihilating time. By 1912, Lawrence spindles produced more cloth per employee than any other textile city in the nation. *At the time I was just about fourteen years old. We were changing the empty bobbins, take the full ones off and put the empty ones in and then start to fill her up again. . . . And you know what we had to do? Keep on going and going till night, keep on doffing all the time, fast and fast. "Come on, the boss is to come." "Come on, are you still there?" "Come! We got to keep the thing going."*

In winter, to keep warm, they'd stuff pieces of cloth in the chinks where the mortar between the bricks in the factory walls had disintegrated. They'd bring in pieces of cardboard to place in the broken windowpanes. In winter the streets weren't plowed, only the sidewalks, since everyone walked to work. Food was scarce and expensive. What gardens they had were gone under frost. Coal: some carefully laid it aside all summer long, storing it in their bathtubs. Others bought it in small, expensive quantities throughout the winter. Children hunted for scraps along the railroad tracks.

Some of the immigrants had been born in remote places where, when the snow fell at night, they couldn't see it, though they'd felt its wet flicks on their cheeks and lashes—light, thoughtless—and all night long they'd sense an accumulation in the way the already-small sounds of the world tamped down. *La neva, neige, schnee, snow, snow, snow.* In Lawrence during the strike of 1912 you could see the snow falling through the sweep of the searchlights casting off from the mills along the river. Borne out of the sky, white, startling, furious—by the time it settled on the granite sills and the caps of streetlamps and at their feet, it had begun to mix with soot and ash.

When they left their old countries, the accounts say, their greatest dread was not of any season, but of being denied entrance, of being turned back because of trachoma, viruses, and infections. How do you begin to fear cloth dust falling like snow that never dissolved, swirling in the moist, hot air, whitening their lungs, and burying them from within? *And it was hot. Even in the winter it was real hot. . . . And those old shoes we wore, walking in there, and the floors were oily, and you'd be breathing that lint. Your eyelashes would be all full of cotton. But we knew we had to make a week's pay in order to survive.*

It had been illegal for American industries to solicit workers from Europe, and it was never proven the industrialists advertised, though many insisted that's how William Wood of the American Woolen Company—the Wood mill in Lawrence could process a million pounds of wool in a week—and the owners of the Amoskeag in Manchester attracted workers; others claimed the steamship lines advertised in the hopes of filling their holds: *They had advertised in the newspapers and put out flyers about the wonderful opportunities for weavers, spinners, and dyers in this country. The advertisements they put up were like circus posters. . . . They showed a man just coming out of the millyard with a wallet in his hand full of money, and he was going up to the bank. A lot of people never even saw a bank in those days.*

Even with meager belongings—papers and change stowed in a breast pocket, a folded envelope of dried spices—when a family reached Lawrence, they'd find there wasn't enough space in the tenements for a world that had once spread across a valley. They strung up their laundry in the kitchen or between alleyways, and their chickens scratched at the basement floor near the coal bins. Goats, too, and pigs kept in the tenements. Young children lit the kerosene lamps, started the stove, and waited for their parents to come home from the mills. And if, while waiting, a child looked out the alley window, instead of an immense and uninterrupted night, she'd see herself reflected in the glass.

Some had intended to make their money and return, others turned on their old life forever, remembering nothing to return to but smoky houses and hard snows. A good many Poles and Armenians and my Lebanese grandparents wanted nothing more than to leave the industrial cities and return to farming even if on unfamiliar soil, and many managed to buy up the dying marginal farms in the valley. Others weren't satisfied anywhere after their crossing. They set sail back to Europe and as soon as the ship left port they began to miss what they knew of the textile cities. When they returned to their old places they might be known as *Americani*, as if the journey could never quite be gone back on, as if in the sheer act of crossing and recrossing they'd lost their place forever and could never be home-staying people again.

Manchester, Lowell, Lawrence, Nashua, and Haverhill each had differing proportions of immigrant groups. A woman from the Amoskeag in Manchester, New Hampshire, remembers: *The spinning room was mostly French, the card rooms were mostly Polish, and the dye house was mostly Scotch. In the worsted dye house, it was mostly Irish. The French people would bring the French into the spinning room, and the Scotchmen would bring their friends into the dye houses. It was the same with the Polish people that worked in the card room. That's the way it worked. . . . The French people were probably 50 percent; the American people—like the Irish-Americans, Scotch, English—would run probably 20–25 percent; the Greeks would run 10 percent.*

In Lawrence, the mill managers throughout the city saw to it that no one nationality composed more than 15 percent of the workforce in any one mill. The older immigrants, mostly English-speaking, held the better-paying jobs. The new immigrants from southern Europe held the lowest, making six to seven dollars a week, so low that the entire household needed to work if a family was to survive. With enough able-bodied adults to bring in money, there might be enough. Young families had their difficulties, with a mother forced to work and maybe only being able to put in half a week while she left the children with neighbors. Pay and half-pay couldn't feed a family and heat a winter home.

All entered and left the mills together at the sound of the bells. Within, there were countless divisions. The young and the old had their assigned places. Older women, cheap labor, were employed to mount the empty bobbins. Young boys swept floors or cleaned bobbins or delivered materials from room to room. The caustic world of the dyehouses and bleacheries was a man's world—men wrapped in scraps of cloth to keep the acids from penetrating to their skin—where you needed thick hands and strength to wring out the material. Precision and experience in the wool-sorting rooms, which were filled with English immigrants whose hands could feel at a touch differences in length, thickness, curl, and softness of the fleece. A sorting room had its own danger: the raw wool could sometimes transmit anthrax. Young girls were needed for finish work such as burling that demanded precision, good eyesight, and fine, sensitive fingertips. Though it was relatively quiet in the burling and mending rooms and where the woven bolts of cloth were inspected, the work took a toll on the eyes: *Everybody got only so much black cloth because it was hard to mend. That day my eyes were getting watery, and the cloth inspector gave me my third black roll. She was English and I guess she didn't like the Germans, I don't know . . .*

The accounts of the warp rooms and weave rooms tumble one on top of another: *There were so many black frames and so many maroon; and on the other side of the room where the black dust wouldn't fly, they had the white. Those who worked on the black got a cent more in their pay, when we got out of there we'd look like real Negroes we were so black. . . . My work was handlooms—four shuttle, four color. Some weavers weave all their lives and never get the hang of it. . . . I preferred to weave because it paid better. . . . We tied each end with a knot to a black-topped pin, then we take these pins and tie it on the beam. Each one, even, even, even. So that everything was close. We had to look all the time. Sometimes the thread breaks, and we had to piece it up. We'd stop the beam and make a weaver's knot, just to make it even. We start the machine again until the beam was full. . . . When the beam was full, there would be eight hundred to a thousand yards on it. . . . You had to watch for when the ends break. You had to know where and why and if the spool was empty. . . . If your hands sweat, we had a little bag of white chalk, because it would stain the silk. . . . You had to be on your toes all the time. Watch your clock; watch your silk. You got hell if you wasted time. . . . They never wanted to see a loom stop, because once you stop a loom you're gonna make a bad mark, you know. . . . And it was gloomy. I think they had twenty-five watt.*

The noise in the weave rooms was loud enough to break the sleep of earth. The whole place vibrated and the workers would be shaking, their ears ringing as they left at the end of the day. The money was small. The same thing over and over as they labored under the threat of stretchouts and speedups, of grudges that could slow the work down and affect pay—if the loom fixer had something against you, he'd skip over you and go on. The mill bells, the mill gates closing at 6:00 a.m. sharp. Piece rate. There was a premium system under which, once the required amount was earned, workers received additional wages only if the work was performed in an unbroken four-week period. A day's absence canceled their premium.

Even well-intentioned laws passed by the state could make a winter colder. In January 1912 a Massachusetts law decreased the work week from fifty-six to fifty-four hours for women and for children under eighteen. The law made no provisions for compensatory pay for the loss of hours, unlike the previous act of the legislature, which in January 1910 had decreased weekly hours from fifty-eight to fifty-six but had also increased day and piece rates to make up for the hours lost.

On New Year's Day 1912, the American Woolen Company announced that workers should expect reduced pay in their envelopes. On Thursday, January 11, the weavers at the Everett mill—mostly Polish women—opened their pay envelopes and saw the money was short. They stopped their looms and stood by their machines for a while. When questioned, they simply said: "Not enough pay." When the new legislation was explained to them, they said: "Not enough pay." When they were asked to leave quietly, they threw down their aprons and marched out of the mill calling "All Out. Short Pay!" As other workers joined the protest, some slashed belts and broke harnesses as they walked away from their stations. The violence increased as the strike progressed from mill to mill. By nine the next morning the riot bells sounded from the tower of City Hall. Two hours less meant the workers had lost between sixteen and twenty-five cents out of their weekly paychecks. A loaf of bread for each hour lost.

Brot, bread, *pain*, *pane*. *Khoubz* in Arabic. Blessed, thanked for, kissed—like a child's hurt—if dropped. Sliced against one's woolen vest, sliced against the galvanized tabletop, or torn to sop up the last of a thin sauce. Once stale it was softened again in a salad of tomatoes and onions and oil, or broken into milky coffee for breakfast. Eaten with every meal, or sometimes the meal was bread alone. In the central district of Lawrence there may have been as many variations of making bread as there were languages: a slack dough with olive oil added to it: caraway, butter, raisins; the loaves shaped into rounds or like slippers; braided, flat; the tops scored or sprinkled with seeds; glazed with milk to soften the crust, with egg white to give it a shine. But such variations are always smaller than the whole: flour, yeast, water, salt. When the corner bakeries were going full force, the air smelled simply of bread—yeasty, faintly sour—in the German neighborhood, the Irish, the Italian, the French Canadian, the Syrian. Not much escaped and roamed freely other than the aroma of bread drifting into the households, here and gone and here again. *Brot*, bread, *pain*, *pane*, *khoubz*. The word alone in their native tongue. And theirs alone. Who among them believed *pane* and *khoubz* really meant the same thing?

❦

By mid-January, the governor of Massachusetts had called in militias from Lowell and Haverhill to help keep order in the city. The Boston militia prepared to go to Lawrence as well. What had begun as a disorganized effort among people with uncommon tongues, no more articulate than the two short sentences, *All Out. Short Pay!* had become a strike involving twenty-five thousand operatives, forty different nationalities, the skilled and the unskilled, demanding not only an increase in wages but changes in working conditions as well.

Swept along with the deliberate will of the strikers, and taken for supporters, were some who stayed on the sidelines, too quiet to get involved, and some who couldn't go to work because there was no work, or because they were afraid. *You didn't dare be in the mills at all. . . . They'd even come around to your house, and if there was a light early in the morning at six or seven o'clock, they came along and they rapped at your door, and they'd tell you put out the light, you can't go to work.*

With so many languages, at times it was hard to tell a deliberate act from a misunderstanding, as when a Syrian man who was buying milk for his child in the morning was told to go back in the house by a member of the militia. Who can tell whether he disobeyed deliberately or simply didn't understand? As he continued on his way the militiaman struck him across the face and broke his cheekbone. The incident has always stood for resistance, as if once the strike began there could be no small action.

Likewise, there could be no small death. When John Rami, a sixteen-year-old boy, was stabbed in the back with a bayonet during a strike demonstration, his anonymous life became a historical moment; his funeral, a cortege of thirty cars and wagons. A few months after his death he is remembered as "a young Syrian" in records of the Congressional hearings on the strike. During those hearings, conducted early in 1912, one worker after another gave testimony of the conditions in the Lawrence mills. Those interviewed ranged from the most experienced to the youngest of bobbin boys and pieceworkers—boys and girls fourteen, fifteen, sixteen years old who'd never before been asked what they thought, who'd brought their pay envelopes home and kept twenty cents for themselves after giving the rest to their parents for household expenses. The questioners sometimes had difficulty hearing the questioned: *Say that again . . . You have said that so fast I could not understand . . .* In the Immigrant City Archives, Lawrence's Historical Society, the bound volume of those hearings is losing its spine. The pages are brittle. The voices fall through the years:

You asked me whether I supported my family out of ten dollars a week. Of course we do not use butter at the present time. We use a kind of molasses; we are trying to fool our stomachs with it. . . .

It is a bad thing to fool your stomach . . .

Yes sir; there are always some people who are healthier. Those who have come recently from the old country are healthier. . . . They have red cheeks, and so on. They are apt to find it pretty hard for the first few weeks, because they are not used to such machinery. They call them "devils" and not machinery. After working for a while they are getting used to it, but they say that in England and France they did not do as much in a week as they do in three days in this country. . . . When the Italians get their pay and see that they never have enough bread, and still their wages are cut down, their wages getting smaller, and two loaves of bread in that pay envelope . . . the Italians, going down, you know, they made a little noise; they sang songs and they said, "We are going to fight them for more bread." . . . Probably if they would throw a piece of ice or snow, it was because they were desperate. . . . They were good people; and as soon as they ask for a piece more of bread they told them they are foreigners and that they are all kinds of things.

But would you like to have the law changed so that boys could not go into the mill until they were sixteen?

I would; but what would we eat if I go to school?

How long have you been working at that mill?

About two years. I have been working a few months in the Washington mill and in the Everett mill, and two years in the Wood mill.

Are you sixteen years old?

Sixteen in a few months, or sixteen last July.

Talk louder.

Sixteen last July.

Yes, sir; had to work barefooted there, with only overalls and a small shirt on.

Why did you have to do that?

You would fall; and you would have sore feet if you worked with shoes on . . .

Were you hurt in the mill?

Yes, sir, part of my right thumb was cut off.

Part of your thumb is gone now?

Yes, sir.

You have been working without shoes in the winter?

Without shoes; it is wet there the same as in the cellar.

Why didn't you wear shoes?

Because we can't stand it; it is too hot.

You could have had shoes on if you wanted to, could you?

Well, then we would not be able to work . . .

You say you would not be able to work with shoes on?

With shoes on.

What I want to know is if you had shoes?

If I wanted to put my shoes on, all I had to do was to put my shoes on . . .

What furniture have you in the house?

Oh a couple of beds; that is all.

Have you carpets on the floor?

I guess not. I guess some horses lives better than we do.

Well I would rather you answer my questions nicely, John, and not try to be funny; it doesn't pay. We just want to get at the exact conditions with all of you people that live in that city and work in those mills.

Right along that line, Mr. Wilson. I want to ask one question.

Yes.

I have heard quite a number of you people talk about living on bread and water. Has there ever been a time when you were compelled to live on bread and water?

Yes, sir.

How long were you compelled to live on it?

Well sometimes we did not have enough money to buy bread one day or two days.

That is not very often, is it?

No.

The organizers of the strike knew they couldn't rely on language alone for success. After they secured representatives and translators for every group—for the Italians, the French, the Polish, the Syrians—they gathered the workers on Lawrence Common so that all could see the strength of their numbers. From the side streets and tenements the strikers converged on the spare winter green in twos and threes, alone. Many who remember the meetings remember the songs—everybody singing in a different language, songs that ran around like ragtime, old tunes granted new words: *In the good old picket line, in the good old picket line . . . The strikers will wear diamonds in the good old picket line.* If you ask a group of those who were there what they sang, many will recall "The Internationale." They start off hesitant and slow—*I forget it now to tell you the truth . . . If I heard it I could sing it.* But the words come soon enough to one—in French, and another translates behind him: *It's the final fight . . . Let's get together—it's tomorrow . . . The international world will be the human world . . . Let's get together—it's tomorrow.*

The Lawrence Common is strictly a common, without gardens or ornaments. A modest, circumscribed respite lying at the heart of the central district. Once it was graced by elms. Now its scattering of maples isn't dense enough to keep you from noticing what surrounds it—all the squat brick and stone buildings, and the churches: the stone-and-red-door patrician Episcopal church with its stained-glass windows designed by Tiffany and LaFarge; the Hope Congregational built with granite blocks left over from the construction of the Lawrence dam; St. George's Orthodox with its plain, blonde exterior brick sheltering an interior of gilded and iridescent icons. And beyond the Common are what seem like countless churches, more churches than mills—Sts. Peter and Paul, a modest wooden building first constructed as a mission for the Portuguese; the Holy Rosary, with its angels carved out of Carrera marble; St. Mary's, as old as the canals, first for the Irish who built the canals.

Joseph Ettor, arriving from New York, must have known that the city of woolens and worsteds and dyes and scourings and forty-five languages was also a city of alabaster, marble, and stained-glass angels, of prayers to be shielded from suffering, and prayers to be given unfailing strength to bear it. Ettor, along with Arturo Giovannitti, represented the Industrial Workers of the World during the strike, and he is sometimes remembered as an outside agitator, or as one who aggravated the talks between mill owners and workers. The old strikers remember him as having *the eloquence of an Italian and the cunning of a Syrian. . . . You couldn't get away from him when he was speaking.*

Whatever else, he articulated the vast gap between what the city's textile workers endured in their daily lives and what anyone would dream their lives to be in the moment he held up a pay envelope and spoke: *This human being, an image of God, gets six dollars and forty-three cents for his week's work. This man has a mother, a wife, and four children to provide for. . . .*

The strike continued through much of the winter. Ice on the river broke up, and the snows melted back from the streets. The workers relied more and more on the soup kitchens—the *laban* and lamb and bulghur of the Syrian cooks, the beans and macaroni and tomato sauce of the Italians. After so many weeks on strike, who could afford the groceries advertised in the papers—the Tunis dates and Camembert cheeses? The children of strikers were sent to other cities to be cared for—an old European labor dispute practice—in order to conserve the dwindling food supplies, to bring the nation's attention to the city, to keep the young out of the fray: *I sent my child away because I did not want my child to see what is going on in the city.*

The Congressional hearings continued. A picture of workers staring down the militia appeared on the cover of *Harper's Weekly*. They say, in the end, it was the wider scrutiny that finally resolved the strike. On March 12, the American Woolen Company agreed to the strikers' demands. The workers had gained an increase in wages of one to two cents an hour in their weekly paycheck, which could be counted as four loaves of bread, and a little more.

It hadn't the strict measure of an organized parade, their victory march, which appears now, in photographs, as a fluid human line in caps and winter coats. Everyone is walking easily, some are talking among themselves, others must be singing. It's a moment of reprieve in a world of workers who carried with them nearly a hundred years of clothmaking. So much had changed since young women began weaving cheap cottons for farmers in the West and slaves in the South as their brothers and fathers worked the first and second mow in the hayfields and the forests grew back over abandoned pastures and Betsey Chamberlain dreamed that the price of labor should be sufficient, the laborer being worthy of his hire.

Unimaginable, the yards of cloth woven through all the quiet times, and during wars and financial panics and moments of prosperity, as the first cottons—the shirting, sheetings, drillings, and osnaburgs—gave way to cambrics, linens, piqués, and lawns, and to blue wool for Civil War uniforms, which were succeeded in turn by corduroys, moleskins, velveteens, silks, and chintz. By the end of the nineteenth century, the sample books displayed hundreds of choices in florals, plaids, stripes, and saturated colors for merchants and clothing manufacturers to ponder. Yet there was little time left after the strike of 1912—maybe time for storm serge, time to supply soldiers with khaki and olive drab and blankets for the cold of Europe—before synthetics came in, and the factories went south for cheaper labor rather than refurbish the old machines for a new cloth.

Much in the making of cloth is a dream of precision. Even, even, even. In the remembering, it's less so. More than the building of the Lawrence dam, more than the daily lives, and wars and epidemics, the ten weeks of the strike of 1912 is the most written about moment in Lawrence's history. A hundred historians have tugged at the same set of facts and statistics to gain their perspectives, to tell labor's version, the feminists' version, management's version, a version roused by speech, a version roused by singing. The Strike for Bread and Roses is what many call it now.

Work created the Merrimack cities, and work was the reason so many came. Still, sometimes the avenues were lit for evenings out. Caruso sang, Charlie Chaplin performed live on stage, Lillian Russell. In winters when the sluggish waters above the Lawrence dam froze solid, there were carnivals on the ice, with ski jumps and toboggan runs and large circles cleared for skating.

When my mother remembers her father she says: "He was a weaver in the Wood mill, he'd tend bar at the Sons of Italy on weekends, and bicycle to his garden plot in Pleasant Valley." I've asked and heard more—she knows more—I imagine he talked more—about his garden than I have of his work in the mills. "He would have loved to have been a farmer," my mother says. By the time my grandparents arrived in Lawrence, the mills covered more than three hundred acres of the city, and the tenements were set as thick as they'd ever be. The demand for housing was so great that the landlords had filled in previously open spaces where small gardens, arbors, and trees might once have been with more tenement housing. What gardens the mill workers had were pushed beyond the city limits to the banks of the river in a place called Pleasant Valley. With its alluvial soils rich and clear of stones, there are farms on the land to this day.

Among the fruits and vegetables my grandfather planted in his plot were some of what he knew from the village he was born to halfway between Naples and Rome, where the growing season was warm and long under a Mediterranean sun. Everything—peppers, eggplant, tomatoes, basil—was frail to the frost and needed to be set in warm earth. He had to start the seedlings indoors or under glass. He had to protect them in the cold September nights. He had to bury his fig tree every fall, and unearth it every spring and set it upright once again.

They weighed hardly more than cloth dust, those seeds, but they were of a different substance entirely than the world he inhabited all week: the narrow way where the web of carded fibers turned to soft strands of sliver, which were drawn and twisted, and shuttled through the shed of a warp, and beat into their final place. Something to grow and taste and smell of the life he desired.

My grandfather died when I was four and I only remember an old and ill glimpse of him. But I can imagine a Sunday supper—*Sunday was our day* the mill workers liked to say—in the lingering twilight at the end of summer. The late sun, a sheen on all the galvanized things, the smell of the season's grapes, the last of the wine in a cup, his own tomatoes and peppers and eggplant simmering in olive oil. Minced green onions, some basil and parsley. After so many years he still spoke with the rough accent of Caserta Province, the place he'd left before a quarter of his life had passed. After so many, he raised a toast in his old tongue: *Cinque mille questo giorno.* The same every time, a four-word shorthand for *May there be . . . may you have . . . may we all have five thousand days like this one.*

11
Influenza 1918

In ordinary times, the bankers, lawyers, and mill owners who lived on Tower Hill opened their doors to a quiet broken only by the jostle of a laden milk wagon, the first stirrings of a wind in the elms, or the quavering notes of a sparrow. It was the height of country; the air, sweet and clear. Looking east from their porches they could survey miles of redbrick textile mills that banked the canals and the sluggish Merrimack, as well as the broad central plain mazed with tenements. To their west was a patchwork of small dairy holdings giving over to the blue distance. But for the thirty-one mornings of October 1918 those men adjusted gauze masks over their mouths and noses as they set out for work in the cold-tinged dawn, and they kept their eyes to the ground so as not to see what they couldn't help but hear: the clatter of motorcars and horse-drawn wagons over the paving stones, as day and night without ceasing the ambulances ran up the hill bringing sufferers from the heart of the city, and the hearses carried them away.

It had started as a seemingly common thing—what the line storm season always brings borne on its wind and on our breath, something that would run its course in the comfort of camphor and bed rest. At first there had been no more than six or eight or ten cases a day reported in the city, and such news hardly took up a side column in the papers, which were full of soldiers' obituaries and reports of a weakening Germany. As September wore on, however, the death notices of victims of the flu began to outnumber the casualties of war. Finally it laid low so many the Lawrence Board of Health set aside its usual work of granting permits to keep roosters, charting the milk supply, and inspecting tenements. The flu took up all its talk—how it was to be treated, how contained, how to stay ahead of the dead. The sufferers needed fresh air and isolation, and their care had to be consolidated to make the most of the scarce nurses and orderlies. So the Board took a page from other stricken cities and voted to construct a makeshift tent hospital on the highest, most open land that offered the best air, which was the leeward side of Tower Hill where a farm still spread across the slope.

At home the mill workers breathed in the smells of rubbish and night soil that drifted up from the alleyways. Where they lived was low-lying, so such smells, together with smoke and ash, hung in the air. Their heat was sparse. They were crowded into their rooms. The flu cut right through, spreading ahead of its own rumors, passing on a handshake and on the wind and with the lightest kiss. No spitting. No sharing food. Keep your hands clean. Avoid crowds. Walk everywhere. Sleep with your windows open.

They slept to the sound of rain—rain pouring from their gutterless roofs, turning the alleyways into a thick mud, rain on the wandering hens pecking at stones in the streets, rain on the silenced pigeons puffed and caged in their coops. At times the rain was hard, driven from the north, like mare's hooves on their roofs, drowning the parsley and oregano set in enamel basins on the window ledges. Other times it fell soft and fine out of a pale gray sky, making circles fragile as wrists on the surfaces of the canals before being lost to the brown, frothy water there. And sometimes it was no more than a mist that settled in the low places, obscuring the bottoms of the stairwells and the barrels and the piles of sawdust, only to shift and reveal the same world as always. Then the rain would gather its strength again, seeming to rake their lives all that much harder. Scrap coal couldn't keep away its chill.

A doctor may as well have gone house to house down Common, Haverhill, and Jackson Streets, so numerous were the cases. He'd knock, and often his knock would go unanswered since it wasn't the family who had sought him out. More likely the sickness had been reported by their landlord or neighbor—afraid that the influenza would spread—so the doctor heard a sudden silence within and a face at the window disappeared into shadow. What kept the families from opening the door was the fear that the doctor would tack a card to their home warning of the infection within, and the greater fear that their sick children would be ordered to the tent hospital. Once there they wouldn't be seen again until they were dead or cured.

When the doctor finally gained entrance—at times with the help of the police—he could find whole families had been laid low, and the sick were tending those who were sicker. They had sacks of camphor around their necks, or mustard spread on their chests, a cup of chamomile by the cot. Whiskey. Garlic and onions weighed in the air. Some sufferers lay in windowless rooms where the damp had kept in the smoke from a low coal fire, and what light there was wavered from a kerosene lamp. Almost always the disease had gone beyond a cough and aches and a runny nose. There was blood mixed in with their phlegm, and they couldn't move from their beds. In the worst cases their skin was tinted blue.

One doctor could see hundreds of cases a day, and in his haste to complete his records, he sometimes left out the ages of the victims and often the names. They come down now in the *Influenza Journal* distinguished only by their address or their nationality: *four Cases, 384 Common Street (downstairs).* Or, *Mother and Child. Baby Rossano. Father and Son. A Syrian fellow. Polish man.* When the rain finally let up and days of mist lifted to bring on clear dry air, the number of influenza cases still didn't slow. Every woman who gave birth, it seems, died. The elderly, schoolchildren, and infants, yes—but strangest of all was how it took the young and healthy, who had never been sick in their lives. Just yesterday they had worked a full day.

The entrance to the tent hospital on Tower Hill was clotted with ambulances arriving with patients and standing ambulances awaiting their dispatch orders. Many were still horse-drawn and the mares stood uneasy in the confusion. The motorized cars idled, and choked the air with gasoline, the tang of which overlay the warm, familiar smells of hay and animal sweat. Everyone wore gauze masks, and there was no talk but orders. *Don't back up. Bring that one over here.* Nurses checked the patients' pulse and color and listened to their lungs. *We need more masks. Find me a doctor. Help me with this one.* The gate was patrolled by a military guard to assure that only the sufferers and those who tended them went beyond it. Waiting black hacks stood three deep.

Every day at 5:00 a.m. a soldier blew reveille. The quick, bright notes parted the confusion at the entrance, and gleamed above the hospital grounds—a far call from a country those patients no longer came from. The general din at the gate may as well have been the sound of a market day in a port city, and they, drowsing on a ship that had pulled away. They didn't stir. It was no concern of theirs, each in his or her own tent, the tent flap open to the back of a neighboring tent. Tents were arranged in rows, in wards, and in precincts, making a grid of the old hayfield. Its crickets were silent. Its summer birds had flown. Electrical wires hung on makeshift poles, and you could hear them swaying in the storms. The soaked canvas flanks of the tents ballooned in a wind and settled back on their frames. Boardwalks had been laid down between the tents, and footfalls, softened by the drenched wood, came near and receded. The nuns' habits swished. What country was this? A cough. A groan. The stricken tossed in their fevers. Their muscles ached. One moment they had the sweats; the next, chills. In forty-five different languages and dialects they called for water and warmth.

Many were cared for in words they couldn't understand. The student nurses and the Good Shepherd Sisters spoke English and French, but to the Germans and Italians and Syrians their voices may just as well have been more soft rain. A face half-covered with gauze leaned near their own. They were given water to drink. Cool cloths were placed on their brows. They were wrapped in blankets and wheeled outside for more air. Someone listened to their hearts, and then to their bogged-down lungs. A spoonful of thick serum was lifted to their lips. Their toes and fingertips turned blue from lack of oxygen. In many pneumonia set in.

It was the same suffering in each tent, in each ward, in each precinct of the hospital. And the same in the surrounding country, in all cities, in all the known nations of the world. It struck those already stricken with war, in the military camps, the troopships, the trenches, in the besieged villages along the Meuse, in the devastated plain called the Somme, in the Argonne Forest. It struck those who knew nothing of the war—all the Eskimos in a remote outpost, villagers in China. Some died without having given it a name. Others called it "the grippe," the flu—influenza—meaning "under the influence of the stars," under Orion and the Southern Cross, under the Bear, the Pole Star, and the Pleiades.

When care failed in the Tower Hill hospital the Good Shepherd Sisters closed the eyes of the dead, blessed the body in the language that they knew, blessed themselves, and closed the tent flap. The sisters on the next shift said a last prayer in front of each closed tent and turned to the living.

In the central city, those who were spared became captive to a strange, altered music. All the sounds of their streets—voices and songs, teams hauling loads over paving stones, elm whips cracking the air and animals, bottles nudging one another in the back of a truck, the deliberate tread of the iceman on their stairs—all these were no longer heard. Or weren't heard as usual. Survivors strained at the absence as if they were listening for flowing water after a cold snap, water now trapped and nearly silenced by clear ice. Schools and movie houses had been ordered closed and bolted shut; public gatherings were curtailed. Workers, their numbers halved, walked down Essex Street to the mills in a slackened ribbon. Their tamped-down gossip touched on only those who had been stricken, who had died in the night. They traded preventions and cures, some wearing masks, others with garlic hung around their necks. More pronounced than the usual smells of the fouled canals or lanolin or grease were the head-clearing scents of camphor and carbolic soap.

The flow of supply wagons slowed as well. There was no commerce in bolts of velvet, silk puffs, worsted suits, or pianos. Bakers who used to shape one hundred granary loaves a day—split and seeded and washed with a glaze of milk—took to preparing fifty or sixty unadorned loaves. In the corner groceries, scab on the early apple crop spread, grapes softened, then soured, and pears turned overripe in their crates.

The absence filled with uncommon sounds. Children with nowhere to go played in the streets and in the parks as if it were another kind of summer. They sang their jump rope songs and called out sides in the letups between rain. The pharmacies swarmed with customers looking for VapoRub, germicide, and ice. And all the carpenters—whether they had formerly spent their days roughing out tenements or carving details into table legs—had turned to making pine boxes. Their sawing, and the sound of bright nails driving into soft wood, could be heard long into the night. Even so, coffins remained scarce and expensive.

The streets running up to Tower Hill rushed with ambulances, police cars, and fire engines. The alleyways and side streets were clogged with passing funerals. Meager corteges were everywhere—there, out of the corner of an eye, or coming straight on. In hopes of slowing the spread of the epidemic, the Board of Health had limited the size of the funerals to one carriage. They prohibited church services for the dead, and forbade anyone other than the immediate family to accompany the coffin. So, a black hack or a utility wagon with a loose knot of mourners following on foot behind was all. Some of the grieving were sick themselves, some barely recovered, and they had trouble keeping up if the hack driver was proceeding faster than he should—there were so many, had been so many, and someone else was waiting for his services. The processions appeared to be blown by a directionless wind down home streets past the millworks and across the bridge to the burial grounds on the outskirts of the city.

The mourners entered a place starred with freshly closed graves and open graves with piles of earth next to them—clay, sea-worn gravel, sodden sandy loam. The gravediggers kept on shoveling—they had long stopped looking up from their work. Even so, they couldn't stay ahead, and most of the coffins were escorted to the yard and left near the entrance along with others to await a later burial. Few of the processions were accompanied by ministers or priests. The parents or children or sisters of the deceased bowed their heads and said their own prayers. Perhaps they threw a handful of earth on the set-aside box. Maybe they lay a clutch of asters on the top. So plain and unsacred, it may just as well have been a death in the wilderness. Small. A winter spider crawling across an old white wall.

"We knew it was serious, but we didn't know how serious," my father said. The farm is less than five miles to the west of Lawrence, but by the time news reached here, it was muted and slowed—no more than a rumor on the sea wind biting in from Cape Ann. Their eastward view was open then, and they could see the leeward slope of Tower Hill, though it was far enough away to appear plainly blue.

On the first of October 1918 they woke to see the flanks of those white canvas tents set in columns and rows across the hill. And that night the horizon was so crowded with lights that it must have seemed as if the heart of the city had grown closer.

As in the city, whole families on some farms were stricken, others spared. His family was spared—all he knew of the flu then was white chips of camphor in an old sock around his neck, and his mother whispering to his father in the evenings: "You'll bring it here." His aunt and uncle, who had a nearby farm, and his cousins all came down with it in their turn, until the whole household was confined to their beds. No doctor came. My grandfather, after he had tended his own herd, saw to theirs—to their water and feed, and to their milking. He drew water for the house and brought them bread. He'd light the fires and bring in a day's supply of wood. Even so, with the windows open, the rooms felt cold as quarried granite.

The youngest boy died. The parents, still weak, were slow to perform the offices of the strong. They washed the body and had to rest. It seemed to take most of a day to make a respectable, small pine coffin. They cleaned the front room, set the coffin in the bay window, and took their turns sitting beside it. Not even small things were the same. Not the rust-colored chrysanthemums blooming against the kitchen door. Not the lingering fragrance of thyme and mint in the yard.

And the large things to be done—the work that had waited all through their sickness—waited still and weighed heavier. It was late enough in the year so that the weeding didn't matter anymore. But carrots, potatoes, cabbages had to be harvested and stored. Wood to be gotten in. The late apple tree was laden with fruit—the Ben Davis apples would cling to the branches all winter if you let them. Enough work to fill their days for as long as they could foresee.

There are two small, walled-in graveyards in the middle of our farm, and they seem odd and adrift now among our fields and woods, though in the early part of this century there had been a Methodist church adjoining them. It was pulled down for salvage sometime in the forties, and its granite steps are now my parents' doorstone. My father would sit on one of the pews when he pulled off his work boots. Now he's buried among those graves, just up the hill, behind a white birch. But in those years only the names of the settlers—Richardson, Coburn, Clough—had been chiseled into the stones. It wasn't a place for recent immigrants to be buried, so his uncle's family walked behind the coffin to Lawrence and set their child beside all the recent victims in the city. The mounds of earth beside the open graves were compiled of heavier and stonier soils than any they had cultivated in the arid land they had been born to. Impossible to return to that country now, though they said their words in Arabic before turning west out of the gate.

For another week after the funeral they could still see the tents, white in the new days, just as yesterday. Then at the end of October the epidemic broke, the fires were banked.

The tent hospital was taken down in a driving rain, and the stricken were moved to winter quarters at the General Hospital. At night Tower Hill once again appeared darker than the night sky. Predictable quiet returned to the neighborhood of mill owners, bankers, lawyers. The schools opened again, then the theaters. The policemen and firemen took off their gauze masks. On the twelfth of November, the Red Cross workers marched in the Victory Day parade. When the city looked up they counted more dead of the flu than of the war.

The winter of 1918 was so cold the water over the Lawrence dam froze and had to be dynamited. The following spring, the field where the tent hospital had stood was seeded in hay. It was mown that summer, and winds swept the timothy and redtop. Here, after the child had become bone, a liturgy was said for him. A child whose life is no longer given a name or a length, so short it is remembered by the one fact of his death.

On summer evenings, my father would sit on his porch, looking into our own horizon. The long simple line of the hill was gone. Pine and maple had grown up, and buildings squared off against the sky. Once in a while out of nowhere he'd mention the lights of the tent hospital as if he could still see them, strange and clear.

12

The Pure Element of Time

. . . one shared it—just as excited bathers share shining seawater—with creatures that were not oneself but that were joined to one by time's common flow. . . .

—VLADIMIR NABOKOV

FOR SOME THE mill years were a curse. The hurt silenced them, or their spirit. Others will say *that was our life*, though the longer and larger purpose of all their work was so their children wouldn't have to live the same. Sometimes it's those children who remember, and the anger rests in them: *I'd be walking to Lowell High in the morning. It was right across the river from where I lived. And the steam from the mill would be floating over the bridge. And in the winter it was so cold that the steam would form tiny little snowflakes that would fall over me. And every time I'd walk across that bridge and saw those snowflakes falling on me, I'd think of my mother in those wretched mills.*

The late voices, the last voices most commonly set down, the ones closest to us, won a certain strength after the ten-hour movement and the strike of 1912, of 1919 in Lawrence, of 1922 at Amoskeag . . . After all the speedups and slowdowns and recalibrations of the years, they had arrived at a humane place, and then the economy carried it away. A few of the last can be mournful—whatever else, those mills finally put enough bread on their table. And when they remember—speaking not to each other but to a future—they speak into winter-clear air with its shining ring: *You* go *up and down the aisle, and you say, "I used to have all these looms to fill."* *Now you don't have anything. Now it's so empty you can almost hear the stillness come across the room. You go through a section where a lot of them are running, and then you come to where it's awful quiet. Only a few are running. And it's a lot colder, too. People don't say very much when they're leaving. They're sad, and a lot of them cry. It's a bad thing when there are no jobs to be had.*

History no longer runs along the river—commerce and industry have turned to the highways—and some of the most densely crowded parts of Lawrence's tenement district—the places where kitchen utensils hung on the outside clapboards and were shared among families—have been lost to urban renewal. My grandfather had insisted on living on the third floor of the tenement because of the better air, and now that third floor where he drank his coffee and dipped his bread in olive oil, where my grandmother made *pizzele* and *zuppa di scarola* while my mother studied the rules of English grammar, is all air.

Along the waterways of Lawrence, the mill buildings, like the old farms of the valley, stand in stages of use and disuse. Some—their windows freshly glazed, the bricks newly pointed—house computer industries, outlets, warehouses, and even a synthetic textile trade. Others are spalling brick and broken lights. New immigrants continue to come—from Asia and the Latin countries now—and are settling in differing proportions in each city along the Merrimack. Most Cambodians have settled in Lowell, while currently over 40 percent of Lawrence's population is Spanish-speaking, mainly from the Dominican Republic and Puerto Rico.

My father once said if our place is still a farm years from now, a Cambodian will own it. But the land values are simply too high for this farm to be bought by anyone as a farm, so when it changes hands it will become something else. Still, however different from the world of nineteenth-century immigrants, however much the landscape has changed, sometimes the desires can feel nearly the same. I remember a man coming by the farm looking for work at the start of the year—I strained to understand what I could of his Spanish and his broken English—and I was startled when he said, *I hope to move to Lawrence in April.*

Older residents of Lawrence keep the city they knew in the atlas of memory. The languages and dialects—the forty-five or fifty or sixty—that they'd spoken in the weaving rooms and dyehouses aren't heard anymore. The banks, department stores, and restaurants they knew have gone to suburbs or fallen to chains. They say, *Downtown is gone.* Essex Street, where they shopped for clothes and secured their mortgages and drank their coffee, is half boarded up. And on the side streets of the city is the main street of another life, bright with its own wares, lit for its own evenings out, and the old weavers and loom fixers can't read the storefront signs on the ten-footers—just above eye level—written in Khmer and Chinese and Spanish. *They don't remember the mills . . .* That particular *they* of the other, you hear it all the time. Even in praise: *You know some of them are really good workers.*

But the old people don't have to look to newcomers for incomprehension. The history here has moved so swiftly one generation seems to share little enough with the next. Their children may still understand the first languages but don't speak them beyond the household, and sometimes talk as if their parents weren't there: *They live such a restricted life. I told them they should go to Florida for the winter.* We've moved beyond our parents, as was wanted. *When I talk about my work in the mill,* says a woman who worked in the Amoskeag in Manchester, *to my daughters especially, they think it's a story; and when I say something about living on a farm, to them it's a story. They don't believe it's true that we were that backward.*

The smell of bread still drifts through the streets. Flour, water, yeast, salt. In the Syrian bakeries the fires are as white hot as they've always had to be to form pockets in the flat loaves. The old wood-fired ovens, fed with second-growth New England forests, must have burned far more wood in a year than many outlying farmhouses together. In the intense heat of the fires the loaves puffed up hollow in the center. Once out of the ovens, they collapsed as they cooled, and the baker wrapped the bread in towels or muslin to keep it soft enough to fold around an olive or fresh cheese or a slice of cooked lamb.

By the early 1960s, when my mother made her trips to the Lawrence bakery, the ovens were heated with gas or electricity and the bread was kneaded with a dough hook in an oversized steel mixer. And now, as the bread moves through the ovens on conveyor belts, the children of immigrants name their bread with adjectives: Irish soda, Syrian, Italian, Jewish rye.

My mother used to buy a dozen plain loaves of Syrian bread, and one loaf each of fancier flatbreads still eaten mostly by Syrians and Lebanese alone, and called by their old names: *simsim* and *zaatar*. *Simsim* means "sesame," a bread sweet enough to love right away. Its top is spread with seeds and drenched in a sugar syrup that's been flavored and scented with rose water. Mild and assuaging, it perfumes your mouth with its sweetness, and leaves a sticky mess on your hands and lips. Syrup pools on the plate.

Zaatar is topped with sumac. The dried red berries are crushed fine and mixed with olive oil, thyme, and oregano. Oil stains the waxed white bakery paper it's wrapped in, the spices and dried herb leaves on top are black from the baking. Dark as good earth, the taste of earth. Sumac, thyme, and oregano—used sparingly elsewhere—are used in abundance on *zaatar*. One strong flavor on top of another, acid compounded by bitter, the oil binding them but not subduing them. *Zaatar* smells stronger than the yeast itself, and the air is full of its scent long after it's finished baking. Though there's no one in our family who doesn't love its rough flavor, *zaatar* is a long-acquired taste, and I don't see it traveling much farther than the Arabic language has in this valley: a Mass for the dead, the old folks talking among themselves. What few words we speak anymore, we pronounce softly.

❦

YOUNG CAMBODIAN BOYS swing out and drop into the flat tawny waters of the canals, one after another, again and again. Dominican men fish for whatever lives in the Merrimack below the falls. When I see this, I can't help but catch my breath. In my bones it's the river we turned our back on. When I was growing up in the sixties and early seventies, the Merrimack was still an industrial tool to carry wastes to the sea, and far from swimmable. Always there were jokes about tetanus shots and skin dissolving on contact with the water, and fires racing across the river's surface. *I remember I slipped in off the riverbank once. Christ, when I pulled my leg, out it was purple. Purple for a week!* Those who could went elsewhere to be near water, heading north, away, against the current, to clearer, smaller, incipient streams or to camps along the lakes near the mountains—quiet, shuttered, muggy days away from daily life—up to where the Merrimack is fed, to Squam and Winnepesaukee and Newfoundland.

Since the Clean Water Act of 1972, industrial wastes no longer pour into the river. Its pollution now comes from storm drain overflows, salts and oils from road runoff, fertilizers and pesticides from farmland, acid rain, acid snowmelt, and metals leaching out of long-spoiled grounds. For most of its length, the Merrimack is classified as a grade B river, which means you could swim in it if you wished. On a clear day, with a glittering surface, it can seem like the past has settled with time or has washed away. Crew teams, low on the water, grace the quiet stretches. Speedboats leave long wakes down its center. There are public landings from which you could slip in a canoe or a kayak. I've been startled by something simply beautiful: a flurry of white sails on the river above the dam one high summer day. To be startled at all is a reminder to me of how much I still see the river as I knew it in childhood, a reminder that in spite of the years of change here, and in spite of my own goings and comings, this valley's legacy has worked into my mind as sharply as the keen of the northeast winds.

The newer immigrants carry their own dangers with them. A Cambodian woman who came to the farm to pick beans and blueberries in the height of the growing season wouldn't go into the field without a dagger sheathed at her waist. Her fear was that soldiers would come out of our mild, cutover woods, and she wouldn't listen when we told her it was safe.

These days, once salmon reach the Great Stone Dam in Lawrence they are guided up a narrow channel, caught in a large hopper, then hoisted eighty feet over the dam. There's a window inside the Lawrence Hydro plant—the present-day controllers of the dam—where a man sits day in and day out during the runs and counts. Mid-May to mid-July, the spawning season, twenty thousand fish work their way up the Merrimack. He counts one hundred salmon, which are captured and taken to a fishery to propagate.

It had been taken for granted, the teeming river William Wood saw in the seventeenth century when he set down his observations about its fresh marshes and its bounty of fish. What beauty and ease we've regained over time must be guarded by the continual testing of waters, the dry weather surveys, the wet weather surveys, the search for a balance or forward movement that eludes with every overflow and every melt of the snowpack. At times I think of a man at the end of his tether doing only as much as he needs to stay alive. The electricity is off, his supply of wood is gone, he is keeping warm by lighting one match after another. We say *the wildlife is coming back* as we stand cold and counting each single thing that passes—the hundred salmon, and the twelve bald eagles feeding on them.

One rainy winter afternoon at the Immigrant City Archives I listened to a taped interview in which Ernie Russell recounted his life on the river during the sixties and seventies. He'd built a camp on eleven-acre Pine Island, which lies in the river between Lawrence and Lowell, not far from the six-lane interstate. I imagine he could hear the steady, surflike sound of traffic as, on the highway beyond, all of the north sped into Boston, and Boston fled north. He floated and rowed over the lumber he needed. He stowed seed for the songbirds and scraps for the crows. At night he worked and read by Coleman light, and in winter he kept warm with a woodstove. He saw a bald eagle grappling with a heron, and the cormorants settling on the broken ice of March. He heard owls calling across the February dusk.

I felt I was listening to a man from far away: *Below the Lawrence falls I don't know well, above the Lowell falls I don't know that well, but in my section there isn't anything I don't know: every rock, the rapids*... He could name the creeks running into the river. He knew there was no pickerel left. There is a world—sometimes large, sometimes fleeting—his voice seemed to suggest, that exists in addition to the obvious one of our own making. I thought of the Bolognese painter who composed still lifes of the same bottles and vases, before, during, and after the Second World War. Painting them again and again, seeing steadily with a changing eye, fixing a dense, earthy thickness, affirming a line until at the end of his life the forms shed their solid lines and are no longer objects but the ghosts of objects. I imagine he loosed the boundaries of form not to negate the hard facts of the time, but to come down through the years to tell something, too, of the life that continues.

IN THE SLEEPING cities, once in a while something is raised from sleep. You can see the Ayer clock, at the heart of the Lawrence mill complexes, from almost all points in the city. The largest mill clock in the world, it once called the city to work. It broke down sometime in the 1950s after the textile trade had entirely declined, and for decades its four faces were stained dark at night; by day they were frozen at one hour. Its bell had disappeared.

The clock was restored in the early 1990s and stands prominent and true over a city where everyone keeps his or her own time, where no one again will tie a small, smooth weaver's knot that will disappear into the nap of the cloth. To raise part of the money for restoration, a committee sold the four clock faces for five thousand dollars each, sold the hour and minute hand, and each hour of each day. The children and grandchildren of the old textile workers bought hours in memory of wool sorters and weavers. The hour the workday started was bought, as was the hour the workday ended. A toll for the lunch hour and a toll for the final hours.

These are low cities, and the Ayer clock stands beautiful and complete above ruined mills as if history were born whole with its new tolling. As if history were entirely the overexpanded dream of a nineteenth-century textile city, it claims a place beyond traprock and soils, grazing sheep and fishing grounds. Its clock faces, too near to be moons, are tied to earth with red clay and lime. *You will only get so far,* they say, *you can only reach halfway.* These are the hours, a tin resonance, low on the register, more precise than time measured by distance, not subject to weather like blossom time with its petal fall swept on a wind. Unlike time told by a candle burning down, or a glass of sand turned, they are the same hours again and again with no end. Here is time as countless men and women have remembered it, its Roman numbers shining into the night. Here is history standing foursquare above us, stopping us, insisting on facts. And we help it by building such permanent things, meant to last, to tower over us, bearing their responsibilities, as if those founding years are what we are tied to forever.

Who can bear the thought of the wind bringing the fallen leaves in? We who have chosen our colors—dove gray, dove white, a cool green—who've chosen what to keep, what to change, we who take such pains to make of our brief moment a place in time, who insist that what sufficed for the old lives—the brittle metals, the darkened varnishes—will not suffice for ours. Who isn't afraid of living in a city built by the dead? Who isn't afraid the place that created us will turn its back on us?

❦

RARELY, A WRECKING ball knocks a mill down wall by wall, and afterwards crumbled brick litters the yard, glistening when it rains and paling with dust in the sun. More often the lost mills burn. And when they go, they go fast. The beams are wood; and the wood floors are soaked with industrial oil. Years of oil have dropped off machines and belts. It was a young boy's job to run between the spinning floor and the weaving floor monkeying around in the gears, oiling one machine after another, since oil could keep the crippled works going just a little longer, just enough to keep the pace up.

When a spark ignites the years, and alarms sound in the winter night, the city fire trucks are joined by those from the surrounding towns, and for the larger fires volunteer departments are called in from the coast, from the broad hills of New Hampshire, from small farming villages with one truck, to join with firefighters from the cities and suburbs near Boston in raising their ladders and pumping water onto the flames. Just behind the yellow police line, the sleepy, silent workers are wrapped in their blankets looking at their jobs disappearing before them. And neighbors—theirs are the old wood triple-deckers—are anxious to have their roofs hosed down.

It takes no time for those fires to become the talk of the city and the surrounding towns. When Malden Mills, which employs fourteen hundred workers with over fifty nationalities among them, burned in the winter of 1994, onlookers gathered on the spine of hills that cuts across the valley at the edge of Lawrence. From there they could see the whole city on the floodplain spread out before them. They talked and watched, their breath forming clouds of mist in the night air. *It was just a matter of time . . . The place was a matchbox . . .* When something large went down the whole hillside gasped *Ahhhh! Jeee-sus!*

I live five miles away from Malden Mills, and I heard of the fire the same way I heard of the San Francisco earthquake—on the national television news. Only then did I look west to find I could see the blaze even from the farm. I pulled a chair up to a second-story window to watch, and I looked across our fields and woods to an orange glow far on the eastern horizon. And then sparks, and the glow flared skyward—I guessed a wall had fallen. I smelled smoke, though what I was seeing and imagining seemed a world away, what with the snow-covered fields in front of me, and the dark silhouettes of the white pines. The road was quiet; the night sky, full of star clusters and planets and nebulae, a night so dark you could see Betelgeuse for the red giant it is.

At Malden Mills they make Polartec fleece, which—though an entirely synthetic fiber, sometimes manufactured out of recycled plastic bottles—is meant to look similar to the shorn coat of a lamb. It's produced in an array of bright colors—magentas and teals and French blues—and intricate patterns based on Aztec and Nepalese and Navajo designs, just as the hikers and mountain climbers who wear it—and those who want to look like hikers and mountain climbers—would wish. The feel of the cloth and the method of production may be entirely modern, but the ones who lost their jobs in the fire had the old fears of a world without work: *It's been tough, I saw grown men almost crying this morning . . .*

In its aftermath the fire became large. The story of the boiler that had exploded with enough force to knock out the sprinkler system, and the winter winds that reached forty-five miles per hour that night, and the police evacuating the homes surrounding the mill, moved well beyond the valley, was told and retold all over the country, and told again when the owner of Malden Mills, Aaron Feuerstein, vowed to rebuild and keep work in the city. Senators and Congressmen came to tour the smoking ruins. The president briefly nodded Lawrence's way in an hour as bright and bewildering as the fire itself, for who had turned towards them until that moment?

III

It is like what we imagine knowledge to be:
dark, salt, clear, moving, utterly free,
drawn from the cold hard mouth
of the world . . .

—ELIZABETH BISHOP

13
Baldwins

APPLE VARIETIES, LIKE anything else, have their years, and Baldwins—along with Greenings, Russets, Winesaps, Sheep Nose, Ben Davis, Astrakhans—drift now on the edge of sleep. The trees survive as solitaries along what was once a fence or at the far end of old orchards that stretch across gravelly New England hillsides. A heart-lifting surprise to see a hill like that in winter: craggy, evenly spaced trees with staunch, gray-barked trunks and bare branches crazing the air, the contained red life in them glinting in a cold sun. Commercial orchards won't look that way again. There's no economical way to harvest trees that are so big, and most of what hasn't been lost to housing has given way to dwarf or semi-dwarf trees that bear more quickly, are easier to pick, and can be densely planted to give a higher yield per acre.

And so many kinds of apples, fallen to market pressures, have been replaced by more uniform, evenly colored varieties. In 1920, the prominent Massachusetts apples were McIntosh, Baldwin, Wealthy, Red Delicious, and Gravenstein. Of those, only Macs and Red Delicious are commonly known anymore, persisting in larger markets among newcomers such as Empire, Mutsu—developed in Japan—and disease-resistant Liberty. Of course, it is better that some of the old varieties have nearly disappeared. My grandfather had to practically give his Ben Davis away—cottony, tasteless—their virtue, if it is a virtue, being they clung to the branches all winter.

But spicy, juicy Baldwins are another story. By late September the apples have deepened in color to a brownish red with a rusty splay at the stem end. Ripening as the fall itself slopes towards its close, Baldwins—a pie apple, a keeper—taste better after the frost. They're picked in October when the orchard grasses have already turned and morning frost lingers in the shadows. As the winter progresses, the skin of a Baldwin wrinkles in storage, but its flavor and crispness hold, and its wine-dark smell fills closed-in cellars and refrigerators.

Here, the remaining Baldwin tree is framed in my bay window. The late light backs it in all seasons, and I watch its changes as I work, and read, and eat my breakfast and lunch. The man who planted this tree also built my small white farmhouse—he repaired with scrap, insulated with newspaper, saved string, lived a more frugal life than I could ever imagine. Who knows why, but it's this tree that reminds me of his effort and economy and the rough stone over his grave.

Baldwins bear every other year, and the fall feels different to me in the years the tree is laden. This past September, when I looked up one mild afternoon and noticed how the branches were bowed by the weight of their fruit, I felt my first sting of the coming cold season. It's hard to predict what will fist up your heart—maybe the smell of decay in the garden, or the clearer light, or the leaves whirling in front of the car as you hurry home at night. After that first sting the bright peak weeks follow, then the yellows become brown, the rusts deepen, and the laden branches are what you see every day until the harvest is over and the oaks alone have it. Afterwards, bare branches against a big sky, and the light and forms of the world are too hard to dream into, so you get used to the spare, smaller life, and what once chilled your heart no longer has its old power. What is withheld becomes what is beautiful.

And what of things revived? Of all that is hauled forward so self-consciously? The catalogs that come in my mail are full of old utilitarian things elevated, and as I flip through the pages I'm tugged by the orbit of seamed, thick glass milk bottles, wire egg baskets, and galvanized buckets set apart from their workaday purposes and their strength. Dowdy, plain, taken for granted in their former life, in the warmer rooms of our time such square and solid forms have gained a grace and even seem seductive. And nothing is more seductive than the Baldwins offered for sale in every Williams-Sonoma catalog I've received this fall. Along with Spitzenbergs, Arkansas Blacks, and Winesaps they top off a weathered half-peck basket. Six pounds—about twelve apples—are thirty-two dollars, plus shipping. Available in the catalog only. Each apple, polished to a still life, is more perfect than any Baldwin I've ever seen—larger and redder, no russeting from mites, or sooty mold, or frass at the calyx. Called heirloom or antique apples now, they seem to say, *See how beautiful the old life was?*

And it does take time before I hazard to ask how much of that old life I would want with its aches, its silence and remoteness. The ones who lived it can't afford the price of these goods, nor would they ever pay it. More farmhouses than farms remain. The interiors of the ones my oldest neighbors live in look nearly the way they did eighty years back. A salt box and tea canister by the stove in the kitchen. A plaid wool coat hangs on the coatrack in the hall. The last pansies of the year are set in a milk-glass vase on the table. "The younger ones just don't understand," Mrs. Burton says. "They want me to *adjust* since Carl's died. But the world is so different now. I wouldn't know where to begin." Her husband had had a long illness and had left off work months before his death, though they waited until after the services to sell his herd. She pares away the toughened skin on one of her windfall Baldwins, going a little deeper where there's a bruise from its drop, and nicking out the rough spots in the flesh with the tip of her knife—apple maggot, curculio, codling moth, all the troubles apples are heir to. It's true, the past is a different country.

In this one a late-fall storm gathers its strength over the Atlantic and sweeps in a warm rain. It lashes the bay window and mats the mottled leaves on the ground. No moon, no stars. The bark of the Baldwin tree is silver in the wet night, and its resistant branches toss against each other in the gusts.

14

By Said Last Named Land

As we glided over the broad bosom of the Merrimack, between Chelmsford and Dracut, at noon, here a quarter of a mile wide, the rattling of our oars was echoed over the water to those villages, and their slight sound to us. Their harbors lay as smooth and fairylike as the Lido, or Syracuse, or Rhodes, in our imagination, while, like some strange roving craft, we flitted past what seemed the dwellings of noble home-staying men, seemingly as conspicuous as if on an eminence, or floating upon a tide which came up to those villagers' breasts. At a third of a mile over the water we heard distinctly some children repeating their catechism in a cottage near the shore, while in the broad shallows between, a herd of cows stood lashing their sides, and waging war with the flies.

—THOREAU

THE TOWN OF slight sounds and home-staying people Thoreau imagined as he sailed past on the cusp of the Industrial Revolution, and all the towns we've been since the first lands were granted to a handful of families and their descendants—a spare farming community separated from the world by a quiet stretch of river, a waylaid place between two mill cities, a promise glimpsed in the offing—have slipped into history. Their records and maps, showing lands bought and sold over the nearly four hundred years of settlement, are stained with thumbprints and oils, and the corners are feathered from countless exchanges and handlings. Read closely the earnest markings of those documents and you read a world of boundaries marked by stakes and stones, by the trees and the fences that then stood: *The lot of tillage land on the southerly side of said Black North Road and bounded: Beginning on the south side of the said road at the corner of the wall by the old grave yard; thence southerly by said grave yard and by a wall to the road leading from the Methodist Church to the Whittier Brothers; thence by said road . . . thence westerly by said last named land by a fence to the corner in the wall by said Richardson . . . to the point of beginning. Said lot contains fifteen acres, more or less.*

To believe we are still a small place apart—to believe in the word *town* in its root sense as a world protected and enclosed—is as perilous as believing the old maps suffice, with their hand-drawn lines describing assessments made with lengths of measured chain, with one eye squinting, tracing stone walls, defining borders by neighbors, knowing in some places the map would be defective and make the whole that much less exact.

The river no longer divides us from the world or brings us the world, which comes in by the road and leaves by the road, and distance everywhere is cut through by time. We are a town surrounded by two interstates. We are a reasonable commute to Boston. What had once been prime farmland, the best going for somewhere around eighty dollars an acre in the early decades of the century, is worth multiples of tens of thousands, and the map we live by now is the zoning map of the town, with its broad, direct lines designating solid areas of residential, business, and industrial lands. The shaded or hatched or dotted overlays, which obscure all delineations of ownership, old usages, soil types, elevations, the milk shed, the watershed, and the grid of roads, describe a world where old terms have slipped their meanings. In early documents sometimes the landlocked woods of the town—old pastures let go to pines, the pines now standing a hundred years—are notated as *vacant land.* Such a term has no place on the zoning map, just as *more or less* hardly fits our requirements for an exactitude greater than paces and strides and weighted observations can measure. Nor can *tillage* exist alongside the current concept of highest and best use, where, for final worth, the assessors reimagine fields and orchards and woods as house lot parcels of one acre, no less.

Such efforts as we've made in these years—on the current map all the land surrounding our farm is zoned for industrial use—will take a long time to go to woods again. Hardly a mile away from the farm is an abandoned Esso plant that had been built—if I read the soil map right—on Gloucester sandy loam: *the weathered surface of a comparatively thin glacial-drift. . . . In cleared areas the soil, to a depth of 6 or 8 inches, is brown, mellow sandy loam. . . . one of the important soils, as it represents the best farming land in the region. . . . hay, the most important crop, yields from one-half to one and a half tons to the acre, depending on the season and seeding conditions.*

The rusting tanks of the old plant have been hauled away, but the flat-roofed brick and concrete buildings remain. Moss is creeping up the concrete walls, and mildew stains spread like shadows. A weedy, paved-over lot is scattered with rubbish. It remains a kind of no-man's-land warning the world off with its chain-link fence topped with coiled barbed wire. Signs reading FOR SALE and

NO TRESPASSING have been up for years, and the place is passed over by developers for old woods every time. I imagine investors are afraid, in spite of assertions that the land is not contaminated, of the future liabilities of building on such land, of what might be uncovered, afraid the work of cleaning the soil may be like cleaning up the river—long and laborious and, even so, not enough. It's easier to break new ground as long as there is new ground. Even the poorest soils keep a time different from human years, the way they build up out of the work of lichens wearing away the glacial rocks, the work of earthworms and sun, of voles and bacteria.

FOR AS LONG as I can remember there's been a utility corridor along the eastern edge of the farm. I've never known where the high-tension wires link up, only that to our north they disappear into rising hills, and to our south they span the river and continue across the undulating plain towards Boston. This coming year Portland Natural Gas plans to put a six-hundred-and-fifty-mile underground line through the corridor that will bring vast undersea gas reserves from off Sable Island, Nova Scotia, to New England. They'll cut an eighty-foot swath across part of the Maritimes, Maine, and New Hampshire, cut through the southeast corner of our property, and then link up with a national pipeline grid that begins in Texas and works north. Now that six of eight nuclear power plants that once produced electricity for the Northeast have shut down, everyone connected with the project easily says, *You know natural gas is the fuel of the future.* The Canadian reserves will last a hundred years.

Further north in Maine, where the plan veers off the corridor and the proposed gas line crosses farmland and woods, some of the landowners have protested: *How can something be good for Richmond if it's not good for a group of people who have lived there for years? . . . If they ruin the land they'll ruin something we love. . . . I am fighting because I don't want to die in bitterness.*

Because the gas line here goes over a small portion of land already utilized by power lines, it will not encroach on us the way it will encroach on those in Maine. Yet, I keep thinking of those protesters in Maine, though I don't know what to think myself, as the land agent from Portland Gas and I walk the utility corridor at the edge of the farm, finding our way through the brush by following a path made by deer and deepened by dirt bikes. He is a quiet, even man, with all his attention on the job in front of him. Nothing about him specifically to make me uneasy. But with my father's death has also come the unfamiliar responsibility of making decisions and negotiating with the outside encroachments on the land, so even though I don't know what to think, my every word is guarded. Guarded maybe because one of my father's cautions to me was *play your cards close to your vest.* And a little because really I'd just like to refuse every encroachment now that development upon development has been coming in so quickly around us—trucking firms, concrete manufacturing, more trucking. It just seems like too much too fast, and this, too, is part of my wariness.

"We all think, perhaps too much, of the piece of land where we were born and of the blood our ancestors gave us," said Jorge Luis Borges after the Falklands War. "In ancient times the Stoics coined a word which, I think, we are still unworthy of; I am referring to the word *cosmopolitanism.* I believe we should be citizens of the world." I'm always arrested by Borges's words when I read them, and, however much attachment I feel to this place, I also imagine I agree with him. What place, a citizen's duty? What place, the heart's affections? Where does the greater good lie? In a regional decision? In a local want? What choice—to stand with the Mainers or simply let this one pass—I wonder, would a citizen of the world make here and now?

The land agent and I talk about the lumber that they may need to cut, the restoration of the woods, and the price they'll pay for their easement. Orange flags mark the route of the future gas line, and we walk from flag to flag through the marsh across the brook to higher ground. The vegetation that has grown up along the corridor is not like any I see elsewhere here. The electric company keeps the woods back for the sake of the lines, but the land isn't tended or cultivated. Clumps of sun-loving sweet fern—its head-clearing scent rising when you walk on it—run scattershot through the highest dry ground. There's a stand of hazelnuts with their tiny nutmeats encased in elaborate husks. Sweetfern and hazelnuts both will be uprooted when they cut the trench for the gas line. They plan to replant the swath with grass. "They like to plant grass," the land agent tells me. "They know there's a leak when the grass dies."

We come up the crest of the hill just at the old mail road that runs through the woods. We stop and stand on the two discernible tracks where wagons once traveled across this hill towards the river. In the woods on either side of the corridor trees have encroached on the old road, and branches lean over the last of it, which has narrowed to a footpath. Only underneath these power lines and on the stretch where our tractors still use the road can you make out the two tracks defined over the years by wagons and oxcarts. The road had been on maps hundreds of years back, fifty years back, but on recent maps is no longer marked. "This is where our property ends," I tell the agent.

"I have you down as owning the other side of the road too."

"This is where our land ends," I say, "I'm sure of it." I remember my father in the lamplit circle of his desk showing me this boundary, telling me about the lumber they hauled out of the near woods after the hurricane of '38. In his last year, when I knew he was dying, when I knew many of the responsibilities he carried would fall to me, we hardly ever dared bring up the practical things. Our own deeds are old and inexact and we probably should have had the land resurveyed before he died. But the most we managed was that sometimes he'd unroll one of the maps he kept stored in the ceramic crock in the corner of his office, and he'd show me the boundaries of our property and fall into a story about how he came to own it: *I bought it because we needed wood for the winter. We'd go through fifteen cords. A French Canadian used to come out from Lawrence—carried his own ax—to do some of the logging. I paid him a dollar a day.*

When I think about all that went unsaid while he sat there remembering, I know that part of me didn't want to disturb his stories; it was easier—for us both—to abide their wanderings than to think about the future. If we'd had better maps or had spent more time going over what we had would I be any more prepared for what I've had to face? Always the future is its own bewilderment, and has its own rewards and sorrows. Not long at all since my father died, and already I sometimes feel he wouldn't fit back into this world, that we've arranged our lives beyond him.

"I'm certain this is the property line," I repeat to the agent. "We've never owned that land." He begins flipping through the papers on his clipboard, back past the aerial maps of the property where I can count the trees in the orchard, can see the pattern of waterways through the woods and the clear edge of every field, back past the earlier hand-drawn maps with the stone walls painstakingly set down in a precise hand. He looks exasperated. He'll have to search through the old deeds to find the absent owner of the adjacent land. He begins to flip through more papers. As I wait I look north up the long corridor where the power lines chase the retreat of the glacier and cut across every sea-flowing thing.

After a few minutes he says, "I'll have to get back to you." Then we begin our walk back down from flag to flag to the road below.

"You'll put all the stone walls back?"

"They have to restore everything."

A fine November rain is falling, and the dampness makes everything clear as rain falls into the last color of the season, bringing up the red at the branch tips on the birches, and the patchy green of the grass in the warmer places. I see black alder lining the low land along the brook, its silver-black bark shining, its red berries brighter than anything else on the land. Winterberries they're called, because they stay on the branches until spring.

Late every year my father would take his truck down the back field and come back with a dozen cut branches for my mother, a little color at Christmas to arrange in vases and pitchers. He called them foxberries—I don't know for sure why, I've never heard anyone else call them that, but it seems right. Foxberries. By the end of winter they're bird-pecked, and shriveled, nowhere near as brilliantly red as they are in this November rain, but the light is softer and longer then and what color remains is still apparent in the smoky March days.

Surveyor's chains and steel tapes sag over distances. They expand in the heat and sun, so that you have to wait for a cool cloudy day for the best readings, and have to adjust readings done in full day to compensate for the inaccuracies. You need to keep an even stride over a rough road. The fog comes in and you can't see benchmarks or the other members of the surveying team. You go back over the same ground to check and remeasure the work. It is an old trade, and I imagine its earthbound sounds—the tools clinking and creaking—will grow ever smaller and more distant, quieter, quieter with each successive map made, until they are nearly silent—hardly louder than key clicks.

What will determine our map of the future? The twenty-four satellites orbiting the earth that constitute the Global Positioning System? Each satellite emits precisely timed radio signals. An observer at any point on the earth's surface receiving signals from at least four of the satellites can figure the precise horizontal and vertical coordinates of any point on earth. The waves from the satellites can be read in snow and rain and fog. You can correct errors that have stood for centuries. You can map featureless lands. The only inaccuracy is a slight corruption the government insists upon for civilian use, which will confound attempts to exactly locate sensitive military areas. The map you can compute from satellite readings can be freed from the eye and the subjectivities of the eye, the hand and the limits of the hand. A spatial database has no limits on the density of information. You can add the element of time and can capture the three-dimensional. Some farmers already use the system in conjunction with grid soil samples to determine the precise nutrient needs of every square inch of their fields. They use it to figure potential yield. Precision farming, farming by the inch. *Once you see a yield map you never quite look at that field in the same way again.*

Surely one of those satellites is above us now, tumbling through soundless space, while the land agent and I make our way down the path beneath the power lines. As we approach the road the sound of traffic swallows the hum of the wires. The light rain has stopped. Towards the west clouds have lifted, and the branches and grasses and fox-berries are shining in a brief moment of sun. I can't yet see a single star, just that incalculable deep blue of near twilight, vast and peaceful.

15
Storm

THE FIRST REAL snow of the season came on a warm wind from the coast. Large wet flakes swirled through the brief December afternoon, settling on bare branches, the winter-brown grass, and cold barn roofs. At first the snow dissolved as it hit the black roadways, though by nightfall it started to accumulate there, too, a snow so heavy the wind couldn't shake it loose from even the most supple birches, and its clinging weight began to fill in the narrow crotches of the oaks, it bowed the slender birch trunks and tugged down the upturned, light-loving boughs of the white pines. Only the stout branches of the apple trees refused to yield—they're pruned to bear the weight of a heavy crop—and the snow merely ghosted their limbs.

Inside, the storm built its cocoon of silence until late that evening when the weaker branches began to give way under the weight of the snow. Then I could hear the dry cracking of wood everywhere, and enormous muffled thuds as pine, oak, and maple branches landed on the whitened ground and roofs, on the power lines and iced-over roads, on the polished granite gravestones on the hill.

The power went down about ten that night and in the greater quiet candlelight brought, I began imagining trees falling in the woods beyond the farm. In their long, slow descent some would snag in the branches of sturdier trees; others, scatter their load of snow on decaying boles—pecker-fretted, crumbling to earth—and on lichen-stained uprooted trunks, rousing burrowed, sleeping life as they toppled onto the last trace of the logging road and fell across stone walls, brooks, old cattle bridges, and frost-withered Indian pipe. All night a falling while my house grew cold, and I slept.

The next morning was clear, bright, and every bit as silent as the night before: we still had no power. I'd never known the electricity to be down for more than seven or eight hours at a time, so I couldn't help but think the power would jolt back on at any moment in spite of such damage: I saw maple branches in splinters. The linden on my mother's lawn had lost three of its limbs—and all its grace—and would have to be cut down. Though not a twig had fallen in the apple orchard, the less sturdy peach trees suffered: some of their east-facing branches had snapped and were hanging by threads of bark. The trunk of the last Elberta had split clear down the middle. The white pines seem to have been hit the worst, especially right at the edge of the timber stand along the old mail road. Where the pines face the woods, half in shadow and crowded by other trees, they're short limbed. But where they face the open road they bear long outstretched branches that curve upwards and feather out to get the most of the sun. Nothing's more handsome than the spreading shape of those branches, but such a shape also makes them vulnerable to heavy snow, since they take on too much weight. Now countless limbs were tossed on the ground, one after another, all down the mail road. And above, where they'd been torn from the trunks, I saw exposed sapwood—warm, honey-colored, out of place against the cold gray tones of the winter bark. I'd see that sapwood all winter long, and it would remind me of this storm, as would the downed pine branches, which remained supple and green, and stirred in the wind as if they still had life in them.

As I made my way across the fields I was as surprised by what survived as I was by what had fallen. There is a solitary standing deadwood rising out of the near woods—a maple, I think—its limbless bole is a sober mark across our summer landscape, its punky bark shreds at your touch, it smells of decay. It had survived intact—a great refusal raised towards the winter sun.

Word was the greatest damage had been here in the valley. To our south they'd had all rain, to our north the snow was drier and lighter, and had caused far less destruction. Here, the radio said, over one hundred thousand people were still without power, and the linemen hadn't seen anything like it in thirty-five years. Even with help from crews coming in from Canada and Pennsylvania, it would take days to get to all the repairs.

Luckily it wasn't cold enough for pipes to freeze, but the thought of being without power for any extended time made other cares stick in my throat. My mom and aunt—what if their houses grew too cold? The old Glenwood cookstove at the farmhouse where my aunt lived had long been given away. It was clear she'd have to spend the day at my mother's house, where we still had a working woodstove, though it hadn't been lit since my father died. Even that last winter he didn't often build himself a fire. That was a sure sign of his failing, since he loved the spot of heat a stove gave, especially after coming in from the cold. As I splintered old shingles and gathered cordwood from a remnant pile on the porch, I tried to remember when it was he'd built his last fire. Sometimes I think of his death and I feel as naive as a child, the way I can't get past the thought that he no longer hears or sees, or feels the bracing January air on his face. *It is strange that a man should be sewn up in a sailcloth and should soon be flying into the sea. Is it possible that such a thing can happen to anyone?*

At ninety-four my aunt has outlived her sister and two of her seven brothers, and I know she fears outliving the others. She had never married, had never—for more than a month or two—lived anywhere other than the farmhouse she had been born in. Now when one of her brothers suggests she spend part of the winter in Florida, she answers, "The days pass here just fine. I sit by the window and before you know it the sun is going down." Though she refuses them simply, she means it entirely. It's nearly all she insists on anymore, having years ago let others take on making her appointments at the Lahey Clinic, filling her prescriptions for her heart medicine, paying her bills, balancing her checkbook. *Well,* folks say, *she took care of them for so long, now it's her turn to be taken care of.*

She shakes her head to think of those years when the house was so full everyone couldn't sit at the table at once to eat, then from the drainboard she lifts a cup and plate and sets them at the far end of the long dining table which she keeps covered in white oilcloth. Morning coffee and toast, jam on the second slice. In the center of the table are silver spoons in a silver creamer. Behind her, the cupboard is full of fine china; the closet, of everyday dishes that have been stacked in place for years. As a child, and then as a woman, she'd spend the afternoons at her own mother's side, slipping proofed loaves of Syrian bread into the Glenwood, or mending workclothes, or shelling peas. She mashed potatoes and turned them into an ironstone bowl, whisked the lumps out of the gravy, sliced the roast, boiled carrots and beans, set out plates and silverware and called her brothers for dinner. I can imagine they'd run straight in after leaving off picking up drops in the orchard, they'd trail in from the barn smelling of hay and dung, from the fields smelling of new earth, having jabbed the spade into the soil to mark their place. They'd drop their homework mid-sentence, or clamber down from the upstairs rooms to crowd the table with its steaming platters that she filled and filled again.

Who would remember anything of childhood without those smells of bread and gravy, without the clicking of knitting needles, or the even strokes of a rolling pin on a wooden counter? And it hardly seems enough now to pray for her easy passage or to say to each other the way we do, "She's doing well for her age," "She's lucky she can still get around," "I can't think of what she'd do if she couldn't live in that house." The house had been standing a hundred years when they were young, and I doubt it felt old then with all the life of the place swirling through. Now it's down to one light, and the clocks all tell a different time, and every change feels monumental. "Why don't we paint the TV room come spring?" I once suggested. "Wait until I'm gone," she answered.

Although the power outage had quickly brought the temperature in the rambling drafty farmhouse down, my aunt—long a creature of her routine—was reluctant to spend the day at my mother's. I had to marshal her out, and as soon as she settled into the chair in front of my mother's stove she promptly asked, "They've no idea when they expect the power back on?"

"The lines to the power company are jammed—I can't get through, I don't know how long it will be," I answered.

She reframed the question half a dozen different ways, and asked it every ten minutes or so all through the afternoon. "When do they expect the power back on?" "You've no word?"

"Don't know, no word," I'd answer again and again, trying not to sound exasperated, as I walked in and out of the house, getting in more wood, checking on things, joining them for a while, then leaving to take care of my own house and to do a little work while there was daylight. All that day it never occurred to me to drive anywhere—somehow I thought the wider world was shut down, too. Later I'd read that pockets of the nearby cities had full power and the hardware stores and supermarkets were jammed with people buying candles, batteries, and groceries for next time while we drew closer to the cast-iron stovebox and waited for the refrigerator to kick back on and the dark lamps to suddenly stain the walls with radiance. Every moment we believed it could be any minute now. A belief answered with the sound of a candle guttering or a log falling into ash. As we sat, I imagined it was the same for all families up and down the road. All plans set aside, lighting what stoves and candles they had, wondering if they had enough, thinking about the food spoiling in the freezer, whether it be frozen dinners, ice cream, sides of venison, or the corn it took them a whole steamy August afternoon to blanch and scrape off the cob.

In the failing afternoon light we hunted up more candles—the nubs of old tapers and half-spent Christmas pillars. As I warmed up some canned soup on the stove I was reminded how my father would cook chestnuts and popcorn on its top. "Dad would have enjoyed this," I said softly to the air as I stirred the pot, and I saw the pain of remembrance flicker across my mother's face. I don't know if silence or remembrance is best, but I was longing to press a hurt simply to remind myself it was there.

Yes, he would have enjoyed this, and he would have been the one to take care of everything for my mother and aunt during such a power outage. It's been easier for me to make peace with farm and business responsibilities than with this. I've heard again and again over the past year: *Your mother's so lucky to have you*, and no matter who says it—the lawyer, the insurance broker, my mother's oldest sister—they use exactly those words, and their tone is always the same. No other words can make me feel more like everybody's sentimental idea of a good daughter, old girl—patronizing, instructive, always finding a way to surreptitiously check my mother's pillbox, to make sure the doses are measured out and taken.

Sometimes I wish she wouldn't tolerate my care. I wish she could break out of the place death's aftermath has consigned to her. I remember right after my father died I thought she'd want to step out into the world for herself, so I set up a checkbook for her and tried again and again to get her to take a trip with a church group. But she has wanted none of it, not the finances, not the travel. Often when she speaks about her life she slips into the past: *I've had a good life.* My desires for her are selfish, too. I'd love to feel like simply a daughter again. To walk in the door and sit down for a cup of tea with no other thought than to ask first about her day rather than to be thinking of the bills I have to look over and the calls that need to be made, even though at times, when her loneliness feels beyond reach, I hide in those duties.

I'd foreseen nothing of the places we'd all assume after my father's death. Maybe I'm surprised most of all by the way my mother has begun to grow closer to her earlier life, as if all other destinations have given way to old original paths. She won't venture far alone on the highway or up north to see her grandchildren, but she'll wind her way through the old tenement districts of Lawrence—though they are hardly the streets of her childhood—to do her errands at the Italian grocery and at the Italian bakery. Lawrence may be the place she still knows best in spite of nearly fifty years on this farm, where she kept to the house and yard. I don't ever remember her going for a walk into the woods or along the orchard here.

The stove, having run full all day, gave off a radiant warmth that spread through half the downstairs rooms. I filled it with wood and left my mother and aunt sitting beside it wrapped in their afghans. I bundled myself up in scarf and hat and parka, expecting a winter cold had come in with the setting sun, but as I stepped into the open night, I was surprised by the mildness of the air.

Safer than houses, I thought, the quiet road, the windless trees. As I crossed the peach orchard, the green glass votive I'd carried out with me cast a flickering light and made long shadows of the frail peach branches, and longer shadows of the white pines beyond. Shadows so distorted you couldn't tell the damaged trees from the unscathed ones. I turned full circle in the middle of the orchard, and I could discern in the living room window of my mother's house the faintest glow from their candles. Besides the votive I carried with me, theirs was the only human light I could see. The hills of Lawrence were dark. Gone, the lamplit neighboring windows and the long curve of streetlights going west and east. I couldn't even make out the red radio beacon that I usually saw on the southern horizon.

I have lived in remote places and have seen night skies full of the weight and lift of stars—so many stars there didn't seem to be an inch between them—but I had never seen such a sky above our farm. Here I was accustomed to nights paled by city and suburban glow, and I had a moment's trouble finding constellations—defiant Orion rising, the Pleiades, Cassiopeia—I thought prominent in our half-starred sky. Now with their familiarity drowned by the other stars, they seemed smaller and part of a greater whole rather than the distinct gatherings I'd always taken them for. Wonderful to see for once. I felt we lived in a wilder, more remote place, and for all the mildness of the night there seemed a greater distance between outside and in. I couldn't help but think I stood in a world closer to what we truly are, that with our lights we have only thrown up the thinnest scrim, and tenuously disguised how near we are to the night.

My own house was soaked with the cold. I wrapped myself in throws and odd blankets to read for a while amid all that was dumb and useless: the oven, the faucets and sinks, the lamps, the radiators, my computer, stereo, TV. Shapes no longer having a function in that quiet. Funny to think if I were to take them all away my rooms would be half-empty. And without the sounds I'd long since become used to—the chuffs of the furnace and refrigerator, the heat crackling in the radiators—I knew what I'd taken for silence was just an imitation of silence. All I could hear now were the few cars going by, fast and steady and purposeful—going somewhere—and I felt an immense distance between them and me. *There are those of the world,* they seemed to say, *and those not.*

After a few hours I returned to my mother's house to feed the fire. I called out as I entered, but, no matter how loudly I whispered their names, they lay too deeply asleep to hear. I bustled as much as I could to try to rouse them—I could be anyone!—and they didn't stir, not even when I opened the creaking stove doors to throw more wood on the fire. And though the house was far warmer than my own, I left them there, sleeping so faithfully, and returned to my bed where I dived fully dressed under the covers, my wool hat on my head. It was still dark when I heard the furnace and then the refrigerator click on and saw light from a downstairs lamp shed into the room. Almost instantly all the exaggerated feelings of the last day and a half fell away and I was swallowed into the world again.

The storm had left no more than six inches of snow on the ground, yet we spent all winter cleaning up its debris. In every break in the weather folks hauled out their tools and set to work. There was a run on chainsaws at the hardware stores, and anyone with a chipper had small jobs for months. In the cemetery, I saw families taking turns with a hacksaw to clear the downed branches from the graves.

Come spring what's left of the scattered debris—the small branches and needle clusters, which seem so conspicuous and unsettling scattered on the snow—will begin to molder into the warming earth, and soon will be indistinguishable from the usual world. But I know the time will stay with me. For those thirty-two hours our own place had seemed so strange, but also so clearly defined as home. Now when I think of those hours, it's like the recollection of a tender dream.

16
White Clover

AFTER MIXED NEW England woods are clear-cut for pulp, lumber, and cordwood, sometimes from a seeming nowhere the land sprouts pokeweed, bramble, white clover, or fireweed. They say the plants are from seeds that may have been lying dormant for a century. The place would have been a field then—pasture or meadow—and when the pines and maples overtook the open land, their shade hampered the germination of such light-loving plants. I imagine them small as dust, the dormant seeds of clover, cradled in mineral earth as the forested world passed into the soil—the fallen trunks greening with lichen and softening as insects and mushrooms work in, bringing them down to mix with pine needles, abandoned bark nests, and deer bones gnawed to dust. White clover, which had been buried along with the daily light of agriculture, now sun-warmed after so much time, sprouting through fresh wood chips, leaf litter, and mold.

In 1919, with half its land under cultivation, Middlesex County in northeastern Massachusetts ranked fifth in the United States in the value of vegetables grown for market. There were also twenty thousand milk cows, and almost everyone knew the difference between Holsteins, Guernseys, Ayrshires, Jerseys. Sour or sweet soil, late and early land, pasture, meadow, tillage. They worked with two-horse turning plows and teams of oxen, though farming had begun to enter its shiny, new stage: tractors also broke the spring soil and chemical fertilizers were coming into use. Yet by 1920, there were already a thousand fewer farms in the county than there had been in 1910—a decrease of over sixty-five thousand improved acres of land. Cultivation had been pushed farther and farther back from the terraces and drumlins of the Boston Basin, where market gardening had begun, and where there'd been fine soil for growing vegetables.

The decline of agriculture here is partly a story of the best land being lost to development, partly a story of the worst land being abandoned. And within those stories are all the singular stories of those caught in work. The blurred couple in front of their hay wagon hardly had time to see the future. They were becoming obsolete with their particular tasks and endless chores: plowing, planting, haying, hoeing, cultivating, harvesting potatoes, storing carrots, thinning beets. The milking, mucking out the stalls, repairing the buildings and plows, shoeing the horses and oxen, clearing stones, clearing trees, the wood hauled out, cut and split and stacked, mending fence. Cooking pies and bread, three hot meals a day, canning tomatoes, beans, peas, peaches, making cheese, boiling sap, bottling milk. The clothes and beds, heating, testing, pressing the irons into the Sunday best, the floors to be scrubbed, the kitchen garden to be weeded, the children, the elderly, the sick. I remember still the day my aunt said, *I brought in so much wood as a child, I don't care if I ever see that Glenwood stove again.*

As the old-field woods are cleared, white clover sprouts its way into a world where farming is seen not in a daily light but a kindly one, like a language nearly lost and so practiced with all intention, and apart from a native understanding. Their chores have become our lost arts to be studied in schools for agriculture and classes for spinning and pie-making. Now that farms in the valley are vestigial, glimpsed in passing between industrial cities and suburban sprawl, the ones that are left have taken on a reverent sheen, sometimes precious and self-conscious. I know those driving by our farm see a landscape to admire and dream into, something settled and steeped with a history of family and work. *Don't ever sell*, they plead. *It's like an oasis.*

Our settled houses, the composure of our cared-for land and the woods beyond draw people's eyes because they remain. But how can we stand up under such a gaze? At times there seems to be such a wide gulf between who we are and who others see us as. And I don't want to let the last years stand for the whole, to let so many years of strength remain unremembered and unrecalled. At the accumulative end there should be voices insisting on the worth of all the days, even if—few remember anymore—farming was a life most of the country had walked away from.

On this second April after my father's death—the family having made it through one farming year without him—there are days I'm even hopeful for the future of this place, since David—a man who, all through his school years, worked for my father—will be coming back, at least for the coming season, to take care of the orchard and some of the fields and the farm stand. I had written to him in Ecuador where he was working in the Agricultural Program for the Peace Corps. We'd made arrangements for the coming year by long slow-traveling letters—his written longhand by failing light, mine on a computer—or by telephone when he came down out of his village every once in a while. It was hard to believe we were in the same century when I talked to him. It felt like a leap of faith across more than miles—like a leap across time, when I imagined, or tried to imagine, where he was calling from: the cold, high country too steep for tractors, and too poor. Animals still working across the terraced hills, and travel between places largely on foot.

I don't know how long he'll stay on—he could do a thousand other things with his life. But he's here out of an old abiding loyalty, I think, to the farm he worked on all through his teens, all through college summers. *The best worker I ever had* I remember my father—not one to mete out praise—saying.

David wouldn't be returning from Ecuador until late in spring, so I was left to order some of the seed myself. For a few of the smaller crops I found my father's old orders scribbled on scrap paper: Avalanche cauliflower, Howden pumpkins, and Comet broccoli. I know he chose the cauliflower because it was self-blanching and the Howdens because of their thick skin and good weight, and I copied his order in both amount and variety. I felt safe in the measure of such things, imagining the long consideration of quality and yield, appearance, size, sweetness, and keeping qualities.

I know he stood by Lady Bell peppers for a long time, and with good reason, but this year there's been a nearly complete failure in the seed crop, so each customer has been restricted to one quarter ounce of seed. I spent hours this past January at the end of the day as the gray afternoons turned to dark, flipping from one catalog to another—it's no help that the commercial catalogs offer more choices every year: hybrids, improved varieties, disease-resistant strains—looking for how to make up the one more ounce of pepper seed we'll need, weighing the attributes of blocky or long peppers, thin-walled against thick, looking for old names I'd heard—anything familiar—comparing the days to germination between one variety and the next, and marking down which ones are early to color red. Who was I that I worried over a packet I could pour into the palm of my hand, over something so nearly weightless I could hold it out for the wind, or simply blow it away like fine cold snow?

In the end I launched into what future I could imagine and ordered eight kinds of peppers. There was safety in variety, I reasoned, so I chose three different kinds of blocky green bells: North Star, Yankee, King Arthur. I ordered jalapeños and Thai hot peppers, along with yellow and purple bells and cubanelles. Then I went on to order other things we'd never grown: tomatillos, baby eggplants, romanesco broccoli, and tiny pear-shaped yellow tomatoes.

Not so much, these decisions, especially when I try to imagine all the new seed that has taken root while clover lay dormant. Seed carried in a bale of hay or caught on fur, or dropped by birds. Winged seeds—light as chaff—blowing in on a salty wind. Loosestrife, rank in the low places, snagged on sheep's wool and washed into the river from the mills. Tomato seed brought in the pockets of Italians. The Polish walnuts given by a Polish neighbor. All the years of my father's Blue Hubbard squash carried down from what he'd chosen and dried and saved every winter through fifty years. A jar of the last, from two Novembers back, sits on his office shelf, labeled in his own hand, a part now—and reminder of—the simple past: *They were.* They were the ones who slept over cellars with carrots buried in sand and onions hung from a rafter, who kept whatever they judged best from one year to the next.

However much of the story disintegrates, won't some part survive in shards and remnants? Aren't there other dormancies—lying in wait among white clover and fireweed—which light alone won't raise? In some soils where flints and stone knives wash onto dirt roads after the rain, the story is almost beyond us. In others, it sours the land: French farmers turn over spent shells, barbed wire, and bones from the trenches. Our own fields turn up bits of ceramic—the handle of an ironstone cup or the clear blue-patterned rim of a porcelain plate. Glass from medicine bottles, too—cobalt blue or pale green—which have clouded like sea glass now. All minor notes and half steps with a few surges in the song. Enamel splinters from a painted red chair, cinders from doused fires, newsprint thinning, tearing, breaking up into smaller and smaller increments, until single letters mix with ash and clay and sand.

17
Twilight of the Apple Growers

THE EARLY NEW England farms all had their apple trees that grew along a fence line or in a small block: stalwart, long-lived, their bearing limbs pruned to a craggy, turned grace. Apples—old workhorse crop—crushed for cider, dried beside the fire, and simmered into sauce. The late-ripening storage varieties—Baldwins, say—touched with frost, were packed away into March, and as the days lengthened their skins toughened, then wrinkled, their flesh softened, and the dank, stone-sealed air of the apple cellar deepened to a winey depth.

In 1920, over a quarter million bearing apple trees had ranged across the drumlins and eskers of Middlesex County—far-spread full-sized trees producing more than a million bushels of fruit. We are so far from an agricultural economy now, I can't help but think at first that a million bushels of apples must have meant the early century was a propitious time for farming here, yet by then agriculture in New England had been in decline for years. Even by the mid-nineteenth century Henry Thoreau, walking his own corner of Middlesex County, could see the falling off: *None of the farmer's sons are willing to be farmers, and the apple trees are decayed, and the cellar holes are more numerous than the houses, and the rails are covered with lichens, and the old maids wish to sell out and move into the village, and have waited twenty years in vain for this purpose. . . . lands which the Indian long since was dispossessed of and now the farms are run out, and what were forests are now grain-fields, what were grain-fields, pastures.*

By the time my father planted his last orchard in the 1980s, apple growing in Middlesex County had diminished to little more than a thousand acres of orchards spread across seventy-four farms. In 1992, eight hundred and nineteen acres remained. Fifty-three farms. And they—tucked in among spreading suburbs and industrial complexes—are without context in the county or in the eastern part of Massachusetts. In tough weather—a droughty midsummer, say—when growers wonder how the apples will ever size up, they listen for news of rain in the beach and boating forecast, the rush-hour weather, and weekend weather.

Our own farm claims little more than three hundred of the county's remaining apple trees. The main orchard, sloping behind the barns and houses and bordered on its far side by a stand of hundred-foot white pines, is hardly noticeable, except in blossom time, to a passerby. Most of the year our farm is dominated by the row crops—fields of corn and trellised tomatoes and vines of cucumbers and squashes, yet, even so, when I think of the possibility that all of it may pass away, it's the apple trees I can't imagine going—maybe because through all the changes here, there's always been an orchard.

In 1901 along with the house, the gradey herd of milking cows, the barn and its contents—pails, hammers, plows, scythes—my grandparents gained ownership of the Baldwins, Gravensteins, Red Astrakhans, and Ben Davis. A few of those trees hang on—thickset now, and crusted with lichen—in the far corners of the farm, but by mid-century most had been replaced by my father's plantings of McIntosh, Red Delicious, Cortland, and Northern Spy. By then he had sold the herd and had begun to concentrate on growing fruits and vegetables for the nearby city-dwellers, hauling his produce to the corner markets in Lowell and Lawrence until the cities emptied of their prospects, the corner stores failed, and the world around us turned suburban.

As new ranches and split-levels were set down on old fields and in old woods, the traffic on our road became constant and we sold nearly all the produce we grew on our roadside stand. By then the varieties of apples we offered included Macoun, Jonah Reds, Paula Reds, a redder strain of Cortlands—red had become important in marketing—as the last orchard my father planted started to bear. Those trees cover only a few acres, but their wood is grafted to semi-dwarf rootstock, which means easier-to-pick smaller trees that produce higher yields per acre. Not so long-lived, they'll be replaced after a quarter-century or so, and will never look as tough and wind-staunch as the old Baldwin trees.

When most of the row crops have been plowed under and the fields are deep in rye, apples—like winter squash and pumpkins—bring in money after the killing frosts. Though they may extend the season, my father never counted apples among his most productive or lucrative crops. He always said corn was the draw. Even now, come July my mother will return from Sunday Mass saying, "All anybody wanted to know was when the corn would be ready." The stand opens with corn, and with the first sparse pickings everyone crowds the table and watches anxiously as a young boy tosses bushel after bushel on the table and packs the pile down into dozens and half dozens, they watch the dozens disappear and the bushels empty, afraid all will be gone before they get their turn. The stand—last stop on a Friday afternoon before the road to the beach, to the White Mountains, to the lake country in Maine—runs with corn all through the summer and the demand hardly lets up until summer itself lets up.

Never such a story for apples, which appear after corn, then tomatoes and peaches have contented everyone, and even the first astringent smell of the early Gravs and Paula Reds and Jerseymacs are lost when set beside the sweeter smells of peaches and cornsilk. Apples gain their true place in the cooler, drier air after Labor Day—the world back on its axis—when those who stop do so on the way home from work, or during a weekend morning in the middle of errands. *I love the smell of fall* some will say, as they breathe in the cold, sharp scent of apples. The first red leaves are falling one by one. The days are already growing brief.

If the scent of apples closes one year, pruning the orchard opens the next, bringing a kind of relief on the other side of winter restlessness: work to shake off the quiet contemplative months in a time not crowded by countless other chores and a shortness of time. Come late February or early March my father would take out his snips and saw and begin to work down the rows of the orchard. I like to remember him alone under the wide winter sky, studying the shape of each tree before making his cuts. He always liked to be doing something, even after the arthritis in his knees made it difficult to walk without the help of a cane. In his last years he'd drive his truck nearly from tree to tree. I can still see him as he'd pause after turning out of the seat to brace himself for the sure pain that would come when he put his weight on his knees. Down to bone on bone the doctor had said. Even so, he stayed with it—that was always his advice to me: *stay with it*—until he was eighty-five, his last spring, when he hired the man to prune the orchard for him.

After my father's death—the fields buried in snow—it was the dormant trees with their gray, turned branches that loomed large to me and made the farm feel like too much to care for. In the weeks afterwards, as I tried to square his affairs, I didn't give a thought to the summer row crops. I just wanted—beyond reason really—to see the orchard pruned as it had always been. I looked through all the cards and the several list-finders on his desk in an effort to find the name of that pruner he'd hired the spring before, and once I found the listing, heart in my throat, I called.

"I'd be glad to do it," he answered. "You've got a nice setup—an old romantic orchard. They don't make them that way anymore." He was glad to do it, and though his price was higher than my father ever would have ever settled on, I didn't know what else in the world to do but agree. In those months I'd probably have paid anything just for things to be the same. It was worth it to see him show up, snips and loppers in hand—strange with his radio, sunglasses, and wide-brimmed hat—on the first open day in late February. Even with all the corn snow still on the ground—we'd had over ten feet of it that winter—he worked deftly up and down the rows, and the shapes of the trees clarified in his wake.

❦

HERE IS ANOTHER spring. It's a late Wednesday afternoon and David and I are traveling west through the broad length of Middlesex and into more rural Worcester County on our way to the first Twilight Meeting of the year for apple growers. The interstate belies any sense of distance or towns-traveled-through as it cuts across the rolling, wooded hills of our region. The grasses are just starting to turn green, and the light feels a little milder as the sun slants toward the western hills, though the cold comes in quickly still.

We've both been pruning the orchard this year—Dave has taken a chainsaw to the tops of the older trees, while I've been working from the ground, pruning the lower limbs the best I know how, looking into the tangle of the crowns, trying to clear out what's growing down, or in, or crossing another branch. The watersprouts, the winter damage, the deer-bitten branch tips. You could hear me mumbling to myself: *That won't do any good—that should go—this one, maybe this.*

We still have a way to go with the pruning even though the buds have been swelling for a while. As we left today for the Twilight Meeting—named so because it's held at the end of a workday—the buds, having already passed dormancy, silver tip, and green tip, were at the stage called half-inch green, when you can see the leafy folds just starting to break out. In the weeks to come will be tight cluster, pink, full bloom, petal fall, fruit set, then the June drop. Through all the variants of the April days—the warming trends, the cloudy, cold, gray setbacks, and the freak snows—the buds push forward towards their blossoming, softening the harsh winter forms of the trees. All the long months I've looked out on the severe turns of the long-pruned branches, and now at dusk, especially, the orchard feels full of peace, with the haze of incipient silver-white and green floating around the crowns of even the oldest, craggiest trees. For the moment they seem to be made only half of substance, and hardly bearers of fortunes and tradition.

The trees are clearly on our mind as we head into the Nashoba Hills. "I can't imagine the orchard paying for itself," I say casually.

"I know," Dave answers.

"Who will you get to pick all the apples?"

"I have no idea."

At the meeting there'll be apple growers from the northeastern region of Massachusetts, which includes Essex, Middlesex, and Worcester Counties. Each meeting is held at a different host farm every month of the spring and early summer. The owner or manager leads a tour of his orchards, packing houses, and storage areas and afterwards the Extension agents from the University of Massachusetts Tree Fruit Program report on recent field trials and the insect migrations and hatches. They give out certification points for those with pesticide applicator's licenses.

The host farm this month is in the Nashoba Hills, which has always been the prime apple-growing region in the area and is where the largest orchards remain. The Soil Survey of 1924 praises the Charlton loam of these hills: *derived from glacial-till deposits which are commonly from 10 to 40 feet thick over the bedrock. This soil occurs on low, smooth, rounded, oblong or drumloid hills, and in many places caps the tops of ridges and hills having more or less stony hillsides. . . . Drainage is thoroughly established, but the soil has an excellent moisture-holding capacity, and crops rarely suffer even in dry seasons. . . . Apples do exceptionally well. The trees make a healthy growth, and the fruit is of good quality.*

As we turn off the interstate and approach the farm, the road is nearly swallowed by high round hills on either side. Slender dwarf trees cross them in soldierly rows, trees light and airy after their spring pruning. Though it's higher and colder here than at home, the green on these buds is breaking out, too. We pull into the parking lot of a broad, beamy, closed-for-the-winter farm stand. The windows are dark and bare, a fading SEE YOU IN THE SPRING sign covers one of the doorpanes. The parking lot is filled with pickups. Some, high-riding shiny new four-wheel drives with elaborate detailing; others, rusting fenders and slatted wooden sides on the bed to give it some depth. On the doors, the orchard names: *Wheatley Orchards, Barstow Farms, Farms, Farms, Farms.*

We meet up with the others behind the farm stand in the apple storage area. There may not be many left but the apple growers range in prosperity and experience from a bewildered couple who've just bought into an old neglected orchard to third- or fourth-generation farmers who manage state-of-the art operations. Most are men in their fifties and sixties, some in their seventies. April is the time to finish up what pruning is left, and—just from work—they're dressed in work pants, jackets, peaked caps, and boots. They've already been working outside for months, so their cheeks are rough and red from the spring winds. There are a few women, and a scattering of younger men. We, with our five or six acres of orchard, half of which are aging, standard trees, belong with the smaller enterprises. Much of the meeting—the discussion of the breakdown of senescent fruit, comprehensive spray programs, and the shipping market—is banked towards larger growers and will be beyond our immediate concerns. Still, it feels like something we can't afford to miss.

Maybe there are forty or fifty of us gathered in a long hall that runs alongside the refrigeration areas. As I sit, I suddenly get a scent—round and deep—of long-stored apples, and, with the chill in the air, I forget the spring and feel for a moment as if winter is still to come. I remember the way the smell of apples filled our car when I was a child. On longer trips my father always traveled with a bag of apples on the floor behind the driver's seat. Northern Spies. When he pulled in for gas, if he had started in talking to the owner or attendant, he'd reach back and offer him an apple as he paid up.

I come around again to realize it's the scent of last fall's crop I now smell. The host pulls out a couple of stored samples to show how they've held up all these months. Beautiful, large, streaked with red. I remember the pruner telling me *Massachusetts apples are renowned outside Massachusetts—they're exported all over the world—McIntosh sell for a dollar apiece in England.* This time of year you'd find few in the nearby supermarkets, which are full of New Zealand Braeburns, Australian Granny Smiths, and Washington state Red Delicious. Oversized, a waxed shine, $1.19 a pound. Even if the Massachusetts crop failed completely, the supermarkets would still have their apples. Except for a few brief fall weeks imports always take up most of the alloted display space in the produce sections even as the early regional varieties—Paula Reds, Jerseymacs—appear in late August. No matter the flavor, people buy with their eyes, and everyone here would tell you those Washington state Red Delicious are meant for eyes alone.

Most of the men are large, and the tables feel a little crowded. Many haven't seen each other all winter, and there's a rumble of catch-up talk: *Retire, no, Christ, that's when they plant ya . . . I was going to put out some oil next week, if it warms up a bit. Going to be a cold spring.* Many are talking about the cider situation. In recent years, several *E. coli* outbreaks have sent a scare through health officials and the public and now the rumors are all about requiring pasteurization for cider. No drops in the mix either, nothing that's touched the ground: *We might try running it through the milk pasteurizer. I tell you, regulations are what will kill us. It's ridiculous.*

The voices all feel familiar to me—this is the kind of talk I've overheard for much of my life. Still, I feel shy. I recognize a few old acquaintances of my dad. One or two step over to see me for a moment—"How's it going over there? How's your mother holding up?" "If I can be of some help . . ." We talk politely for a few minutes until the conversation trails off, and they amble away to join another group.

The orchard tour is first, and the large lot of us pile into the backs of some of the pickups and jostle along the cart path to higher ground to look at some Galas that had been planted half a dozen years ago. The top of the hill is crowned with a nineteenth-century white clapboard homestead, which is shaded by a few ancient specimen apple trees long past production now. I guess they're what's left of an old orchard meant to span years and years. They, with their rough gray bark, more bullish trunk than crown, the new growth of long slim branches weeping towards the ground, make us almost believe such trees always were, and that it's we who are the first to change. But the idea of beauty that they suggest and that we hold—*old romantic orchards*—of trees planted in rows across open land, began only when apple growing became a deliberate occupation, going back no more than a few lifetimes. The earliest New England farmers tucked in wild apple trees where they could, and the result was nothing like the blocks and rows we've all come to know and love. Theirs was a disorder Thoreau loved and saw passing in his time:

I fear that he who walks over these hills a century hence will not know the pleasure of knocking off wild apples. Ah, poor man! there are many pleasures which he will be debarred from! Notwithstanding the prevalence of the Baldwin and the Porter, I doubt if as extensive orchards are set out today in this town as there were a century ago when these vast straggling cider-orchards were planted. Men stuck in a tree then by every wallside and let it take its chance. I see nobody planting trees today in such out of the way places, along almost every road and lane and wall-side, and at the bottom of dells in the wood. Now that they have grafted trees and pay a price for them, they collect them into a plot by their houses and fence them in.

So we also are aftercomers of a kind and cannot guess the beauty been.

"It's good to be up high," someone says, and it is, as I look west towards the Connecticut River Valley and the long blue of the greater hills. The sky is already deepening in the east. The cold sun, low in the west. Frail branches of the small trees cast long shadows. The farm manager talks about his Galas, his successes and failures, and as we stand among the trees some hang on his words, others look for themselves at the pruning job, or the way the buds are breaking. Then we trundle back down the hill to the storage house to drink coffee, eat hot dogs, and listen to the Extension agent from the Department of Plant and Soil Sciences give a rundown on rootstock trials and foliar calcium sprays for apples. Another takes his turn and discusses the oil spray schedule. Specific, scientific, where does such knowledge go beyond this room and these few people? An undertow of small talk begins to break out. It's been a long day, and attention is drifting away.

The agent brings us round by raising his voice a bit: "OK. Listen up. One last thing. The Pesticide Disclosure Act is in front of the Ag Committee in the House. I thought we had the Chair's support, but it doesn't look that way. Give a call. If this goes through it will require you to notify every abutter a week before every application. We'll have a hard time doing our job." His sense of urgency quiets the group down. Some pull out paper and scribble a note down. He gives them some numbers to call. One thing you can count on to make this crowd feel like a group again—themselves alone—is the threat of new regulations and legislation.

There isn't one here, I don't imagine, even the successful ones, who isn't staring off into the unknown. Farms have always been lost, more lost than ever revived. Middlesex County has been going certainly away from this way of life for a century and a half. The farmland here, with its cleared, well-drained soils, has been sought out for development for a long time, and when it passes from one generation to the next, unless it qualifies for special valuation, unless the children commit themselves to ten years of farming, the land is assessed for highest and best use, which prices it far beyond any agricultural future.

However uncertain, it's also true that those who remain must love the work—from the countless hours of pruning in the lengthening days of late winter to the final harvest in the brief days of autumn—even if it doesn't seem they've had a choice except to carry on from the ones who've gone before them. And some, maybe many, who love it have not survived for a thousand reasons—family, finances, hard luck among them. But in a world of choices whoever doesn't love it has no chance at all. Next meeting, next month, in a longer twilight, in warmer air, the apples will be in pink, and it will be a busier time for work, though most who are here will attend that meeting also.

After the last talk dies down everyone drifts off to the parking lot. I stand at our truck waiting for David, who's stayed behind to talk to one of the Extension agents. I look back at the storage house where we all met, and the building suddenly seems overlit in the growing dark. I can hear bits of conversation in the distance. *They're saying frost in the lower valleys. Can you still take care of my cider apples this year? Let's talk about it before next meeting.* Yes, if, when it goes, all the particular talk will go, too. The words and their fullest meanings: *petal fall, fruit set, June drop. Stay with it.* The last truck doors slam shut.

It's after the evening commute so the road is quiet, and the birds are already silent. The silver-green tips of the apple trees, glinting. Lights go out in the storage house. One by one I can hear the engines turn over and the trucks drive off. Silence. A lingering sting of gasoline in the air, then the smell of the spring buds returns.

18

A Tour of the Farm Stands

Though we have many substantial houses of brick or stone, the prosperity of the farm is still measured by the degree to which the barn overshadows the house.

—THOREAU

THESE DAYS, IN the counties surrounding Boston, the prosperity of the farm is measured by the degree to which the farm stand overshadows the house. And the prominence of such farm stands is a true marker of the changes in agriculture here since Thoreau's time. Thoreau, of course, was being critical: *To what end, pray, is so much stone hammered? In Arcadia, when I was there, I did not see any hammering stone. Nations are possessed with an insane ambition to perpetuate the memories of themselves by the amount of hammered stone they leave. . . .* But when I gaze at the photographs of the linked buildings that were once this farm, turned in on themselves to keep out the wind, ending in an outsized white barn far larger than the house, I don't think of ambition so much as I think of a sturdy, self-contained world that spoke for hope.

With the barn gone, the remaining house and carriage shed form a modest L. When I knock at, then enter the side door to bring my aunt some of the first corn of the year or a cupful of peas, the two-hundred-year-old rooms—hushed except for the ticking clocks—feel separate and seemingly far from everything else on the farm today, the true workings of which—packing house, the greenhouse, and the storage buildings for machinery—are across the road from where she lives. There haven't been overcrowded noonday meals around the dining room table for years now that most of the workers, who eat their sandwiches in the packing shed or in back of the stand, are hired out of Lawrence or Lowell.

The farm stand itself appears solid and on its own some distance from the house. It's separate even from my memories of the quiet, three-sided shack we first called the farm stand, which I tended as a child, where we sold a few bushels of corn and a basket of tomatoes and beans on a summer day. Almost separate, too, from our practical old customers who are dwindling: the ones looking for green tomatoes every September and bushels of pickling cukes, the old Italians coming out from Lawrence for boxes of canning tomatoes, cheap in the late August glut, the Lebanese women looking for kousa, a dozen or two or three all the same size, so they'll cook evenly. Mixing with the old Italians and Lebanese are the newcomers in town who've settled in the houses on old fields, and the newcomers to the cities—Hispanic and Cambodian families—looking for hot peppers and shell beans.

They walk into a broader, beamier place than in the past, a place aproned by a paved parking space. In season the double doors are propped open, greeting and calling the passing world, and window boxes and half-barrels brim with impatiens or petunias or marigolds. After the punishing summer heat passes, pumpkins, apples, and bunched cornstalks are ranged across its front. We've always opened for business in mid-July with the first of the corn, and we've closed right after Halloween—a straggle of pumpkins unsold, some apples and squash to wholesale.

When the farm stand closes its doors to the winter, the place itself appears as strangely plain as the fields that are planted in winter rye. There's the last lingering scent of pears on the air. I've heard, in the waning days of October, our own customers say *I wish you were open all year.* What I think they really mean is that they'd rather the growing season was far less brief. Sixty days of tomatoes, seventy-five days of corn hardly seem enough when all is said and done, and when the fall stars are burning out of the summer haze, no other flavor is so greatly desired as what has just passed.

It makes economic sense now, though, to think about extending the season beyond the hundred days of corn and pumpkins and apples in order to have some cash flow early, and then late—something many farm stands in the region have already done, bringing in produce and bedding plants in early spring, then Easter lilies, and hanging plants for Mother's Day. They keep going all the way to Christmas trees and poinsettias, and are busier than ever, the demand for the aura of the rural life being all that much greater as the available farmland continues to shrink.

So on this raw, rainy afternoon in mid-May, David and I have set out on the road to look at some other farm stands—to see how they're set up, to see what they carry in the early season. I've charted a route north and west of us where I know there are half a dozen larger farms, and we've set out across the Merrimack into the low hill country of southern New Hampshire. Once across the river, through the city of Nashua and beyond the worst congestion of the Boston commute, the road starts to wind through open country—fields, mixed woods, white pine, some hemlock—and small towns, white with their churches set on high ground, old local mills and a small river running through, where the teasel still may be growing wild. Our own town, cramped and crammed with its spill out of both mill cities, with its proximity to Boston, hasn't looked like this for years. The farther we go, the steeper the terrain becomes: more woods, the country a little wilder, the growing season a little shorter. Though even here, on the roads between town centers, newer houses on broad bare lots are sprouting up on once-open land.

The apple trees are blossoming now in the late, wet spring. The whole spring has been so cool that every kind of blossom is lasting a long time, and the blooming seasons of different trees and shrubs are growing into one another rather than giving way—the forsythias, still bright yellow as the lilacs begin to open out. It makes for a fragrant air. Who doesn't love the work now, before the summer pests, the heat, the actuality of the crop—for now it is all possibility—will never love it. We pass a crew in oilers setting out lettuce on some early land. A field planted in peas. "Further along than ours," says Dave. I think of my dad and the way when he was driving anywhere he would notice the farms in the landscape more than anything. He seemed to be able to spot them anywhere, in places where no one else would think to look. He'd see the hint of an orchard behind a screen of pines, or a squash field between office buildings.

The first stand we stop at is as large as they come around here and I'm guessing it must be open most of the year. Barn rustic. Hanging plants are hooked along the eaves—geraniums and impatiens. Its front is full of bedding plants—pansies and perennials—set on low benches along the front. The pansies are the showiest with the bold purples and the yellows of their pie faces. The perennials are still a lock of promise, though the damp day has brought out the scent of lavender—a popular plant these days—spicy and clean.

Almost no native vegetables or fruits are ready yet. Asparagus and rhubarb are the only early New England choices, so the displays of fresh produce in the interior of the stand are full of imported fruits and vegetables: California artichokes, avocados, peaches, Florida citrus, Holland tomatoes. Even the apples are from away, and heavily waxed. They may be of higher quality than can be found in the local supermarkets, but they are also much more expensive. And the blueberries are five dollars a pint. David raises his eyebrows. I can imagine one of the white-haired women who come weekly to our stand, leaning on her cane to pick up the berries, looking hard at the price. She'd set them quickly back down again, and inspect everything more hesitantly afterwards.

More than a hundred days, and you're really just another kind of merchant. The shelves above the produce bins are stocked with jams, jellies, preserves, and conserves whose names always conjure the rural—*Weathervane Farms, Greenwood Farms, Molly's Kitchen*—and whose labels are adorned with etchings of workhorses and moldboard plows or specimen oaks. Candles, cookies, breads, candies. Lavender, statice, strawflower wreaths on the walls, and bundles of dried oregano, thyme, and bay hanging from the beams. Even as I'm wondering who buys all this, I jot down some brand names and addresses off the bottles of goods on the shelves. We look at the way the bins are constructed, and at the little greenhouse warts the stands have on them, talk about how to construct them, how to make the most of a small space, where to place the shelves. I see rusted farm implements resting on the rafters. Saw blades and ice tongs hang on the walls. We all make the most of the past. Even though most of the produce is from other places, the stand smells familiar, as always, with its scent of old wood and potting soil, and something green and astringent like parsley or celery.

As we continue on down the back roads from farm stand to farm stand, we can't help but separately wonder what our own place will become. Dave and I joke about a future of doughnuts, hayrides, ice cream, a petting zoo. He recalls how a neighboring farmer attested: *We made money on the pumpkins, sure, but we made more on the candy apples and popcorn and sausage sandwiches.* Today we've stopped at places that have llamas, sheep, a few ducks. They host birthday parties for children. The name of the occupation has changed: husbandryman, farmer, grower, retailer. Creeping in are words like *agro-entertainment* and *agro-tourism*.

The last farm stand I'd marked on our route was a little farther north on the other side of the river. There are few places to cross the Merrimack, and most bridges are in the heart of the old mill cities so we have to jog through the narrow streets of Nashua in order to cross, sensing our way to the river by looking for millstacks and low land until we catch a glimpse of the waters that were once described, at the height of industrial pollution in the sixties, as too thick to pour, too thin to plow.

"In Ecuador, sometimes you'd have to travel for hours just to cross the river." Dave had only been back here a month, and sometimes his mind seemed alive with two places at the same time. His world had grown thick with its crossings and accretions and comparisons. "People talk so much here. In my village, people would spend the whole rainy afternoon together, six men, two playing cards, four watching, and not a word between them." As David remembers—and I imagine—Ecuador, the far shore appears open and green.

We turn north and travel a road that runs parallel to the river's east bank along the crest of a small hill. To our left we can look across to river valley land that sweeps down to the Merrimack's edge. In places there are new Colonials standing square on their lots, without landscaping, and yet without furnishing or curtains so you can see clear through the twelve-light windows to the river beyond. They've been built on even, stoneless, rich soils, on what had been some of the best farmland left in the area, something like the Ondawa fine sandy loam of our part of the valley: *this land is easily plowed and cultivated, coming readily into good tilth. . . . The overflow in spring is depended on in a measure to keep up fertility. Comparatively small deposits are made by the streams of this region at flood time. Crops are rarely damaged by floods, which occur when little or nothing is growing on the land.*

We arrive at the last farm stand—a refurbished long barn with a forthright simplicity—a few minutes before six, just at closing time. The river runs straight and swift in back, the fields are planted right up to the parking lot. With our arrival comes a little flurry of activity as people stop on the way home from work. Most have come for asparagus, maybe enough potatoes to tide them over, and their purposeful errands remind me of the way our own stand sometimes feels in the waning days of the season. I imagine however much is ordered and sold from away and beyond, any farm can still at times be a world of its own, with its talk about the harvested frosted heads of cabbage, or the first peas of the season, talk entirely local and our own even as, above us, crossing and recrossing the sky, satellites map the earth. We walk through the bedding plants. I buy some asparagus.

As we begin our journey back on the same road I keep my eyes on the river running behind the new houses. The braided, swift waters must be the breakup of the last stubborn ice of winter—just now dissolving into water and flowing into daylight—that had lain under shadowed hemlock woods in the White Mountains. The road continues to hug the river, and I begin to realize we must be traveling on the road Thoreau and his brother could see in a moment's rest from their river journey at the end of summer in 1839: *The open and sunny interval still stretched away from the river, sometimes by two or more terraces, to the distant hill country, and when we climbed the bank we commonly found an irregular copse-wood skirting the river, the primitive having floated down stream long ago to—the "King's navy." Sometimes we saw the river road a quarter or half a mile distant, and the particolored Concord stage, with its cloud of dust, its van of earnest travelling faces, and its rear of dusty trunks, reminding us that the country had its places of rendezvous for restless Yankee men. There dwelt along at considerable distances on this interval a quiet agricultural and pastoral people, with every house its well, as we sometimes proved, and every household, though never so still and remote it appeared in the noontide, its dinner about these times. There they lived on those New England people, farmer lives, father and grand-father and great-grandfather, on and on without noise, keeping up tradition and expecting, beside fair weather and abundant harvest, we did not learn what.*

And as we travel along the road they saw from the river, I think of how impossible their journey would be now, our lives as far from theirs, surely, as theirs were from Passaconaway's. Thoreau and his brother camped where they docked and knocked at houses to ask about buying a loaf of bread, a melon from the field, a drink from the well. They slept in the open at any sheltered place along the bank. What would we make of them if they were to come to our doors now, asking for bread, travel-scuffed, out of nowhere?

For Thoreau, there were worlds to lament, the world of river freight was passing, the canal locks were falling into disuse as the railroads took trade off the rivers. A world we don't know to miss anymore, the river having changed so since, having become an industrial commodity for the mill cities whose bricks were shipped down on the water, and the mill cities having died away, and commerce and industry both leaving the Merrimack as surely as salmon and sturgeon.

When he sailed the river my ancestors lived in the hills of Italy and Lebanon, and spoke no English, and there wasn't even the rumor of work that would bring my grandparents to Lawrence. The crest stones were set on the Lawrence dam as he labored at Walden over drafts of *A Week on the Concord and Merrimack Rivers*, setting down what he desired—what anyone might desire—our lives to be: *The frontiers are not east or west, north or south, but wherever a man* fronts *a fact, though that fact be his neighbor, there is an unsettled wilderness between him and Canada, between him and the setting sun, or, further still, between him and* it. *Let him build himself a log-house with the bark on where he is,* fronting *IT, and wage there an Old French war for seven or seventy years, with Indians and Rangers, or whatever else may come between him and the reality, and save his scalp if he can.*

As his words cross with the immigrant accounts still swirling in my head, I can't help but feel the gap between what has been of so many lives along this river—*Keep on going and going till night, keep on doffing all the time, fast and fast. Come! The boss is to come. Come on, are you still there?*—and what we have hoped our lives to be as we build on what's crumbled to dust and changed beyond reckoning, the rusting mill gates and the collapsed barn, build on the ruins of those who have been forced away, those who have walked away from home and past homes and beyond the river murmuring, glimpsing, gone.

My own imaginings send tremors—like flawed glass, prismatic—through every solid thing I see. Smoke and snow distort; and petal fall, and voices tumbling through the years. It's the voices that leave as much as hammered stone, and all who've been voiceless, all whose tongues were uncomprehended, whose experience has been answered with silence. Always the hope is that somewhere in time they will finally be understood—all the declarations and questions, and what is still so much beyond speech, or too elusive for speech: *Got in three loads . . . Got in four loads . . . A hard wind blow . . . I have got the most of my wool spun and two webs wove. . . . Sometimes we did not have enough to buy bread one day or two days. . . . Come, the boss is to come. . . . Are you still there? . . . Resolved, that as the laborer is worthy of his hire, the price for labor shall be sufficient. . . . What did we know of the war. . . . We saw churches and gravestones smashed to dust. . . . We knew it was serious but we didn't know how serious . . . And I gave place to them, I that can make the dry leaf turn green again . . . Peace, peace is the last wish . . . It droppeth as the gentle rain upon the place beneath. . . . Who's going to be farming in this valley ten years from now?*

My father's gravelly, practical, last voice has joined all the others of a valley seen new each time, the way Thoreau saw his own trip once in the journey, twice in the words found to write of it, and yet again as he lay dying at the age of forty-four. His recollections labored over, layered over, to join with a present day. Forty-four years old—more than three of his lifetimes fit between then and now. His trip down the river had been the striking out of a young man, the retelling of the trip had been an elegy for his brother, dead at twenty-nine. When, in his own last days, his sister read aloud the passage he'd written of the long swift journey home past Nashua and into Lowell with the first autumn wind at their backs, the story goes he smiled at the remembrance of it, and whispered *now comes good sailing.*

I keep thinking of Thoreau and the lives of the valley as David and I, in order to avoid the last of the city traffic, veer away from the river and cut into the back way over the hill roads—Jeremy Hill, Bush Hill, Marsh Hill Road—that ride the contours of the valley. One moment we have a high clear view, the next we're nestled in wooded low places. No matter the names for them and their meanings—drumlins, eskers, terraces, hills—they are all the work of glaciers, and of time, are the soils that swirl in the survey maps—Coloma fine sandy soil, Hinckley, Merrimack—whose fugitive colors not so long ago charted each place here for its agricultural purpose.

This May evening is the coolest in a long time. We pass through most of the last of the farmland and open land, more fields and farms and woods than you'd believe from the main roads and the river roads. Is this an illusory way home or the real way home? Or one of many? Smoke rises out of the last fires of winter. Here and there windows are open to the fresh air. Even with the cold and gray—it has been gray since November—two or three times today I've overheard people say *everything's so green.* Beside me, Dave is running over with ideas about the early season. As he talks about putting up a greenhouse for some early tomatoes and maybe trying some bedding plants, a teeming rain falls soundlessly on the windshield. The roadside branches, stained dark and dripping; the last light, leaving the woods and the dense tangle of black alder, maple, and pine. We come up to the edge of the farm just at the old apple orchard—now comes good sailing—and see the blossoms lighting the dusk.

CLEARING LAND

Legacies of the American Farm

For my mother, Antoinette Brox

Our signals from the past are very weak, and our means for recovering their meaning still are most imperfect . . . The beginnings are much hazier than the endings, where at least the catastrophic action of external events can be determined . . . Now and in the past, most of the time the majority of people live by borrowed ideas and upon traditional accumulations, yet at every moment the fabric is being undone and a new one is woven to replace the old, while from time to time the whole pattern shakes and quivers, settling into new shapes and figures. These processes of change are all mysterious uncharted regions where the traveler soon loses direction and stumbles in darkness.

—GEORGE KUBLER, *The Shape of Time*

1
Inheritance

HORSEMAN, PASS BY, I used to whisper as the sirens made their long way down the road from the center of town. Now: one, two, three, four, five, six, seven gone out of a generation—my father, both his sisters, four of his six brothers. They had possessed a collective strength that gave definition not only to the family but to the farm itself, to the hundred acres of woods and streambeds, fields and orchards we have called ours that lie across the worn coastal hills north of Boston. I know time itself helped to establish a sense of security, time in place, time and a generation's fidelity to each other. Even as six of the brothers married and moved on and the care of the farm passed to my own father, there was a particular bond among all those siblings that lasted their entire lives. In their later, quieter years, though the farmhouse was no longer the gathering place for the extended family at holidays, all of my uncles were still in and out of it, often almost daily. In the years before his death my father lingered long at the table by the kitchen window, exchanging the news of the family, of the day. I can see him there still, bent over the local paper, talking with his sister Bertha.

That no one in the family lives in the old farmhouse anymore may be the strangest thing about all these passings. For almost a hundred years its nineteenth-century forthrightness had been at our center, and had so worked into our imaginations that, more frequently than I look at any of the pictures of my ancestors, I contemplate the 1901 photograph of the farmhouse with its linkage of buildings: summer kitchen, carriage house, barn. As I study the blades of the windmill above the roofline and imagine the silo just out of the frame, the world appears sturdy against a backdrop of sober daylight. A patient horse is swaddled in ropes and harnesses. Men and women look up from their work in the muddy yard. *However it is in some other world*, their uncomplaining gazes seem to say, *I know that this is the way in ours*. Farming in New England was already in decline, with woods growing up on long-cleared lands as the mill cities and prospects to the west pulled people away from this countryside, but it was still the common life, wide open under a big sky, and one farm's holdings adjoined those of the next and the next all down the road—pasture and field and orchard extending as far as the eye could see.

Scrawled across the back of my copy of the photograph in my father's hand is *Brox Farm 1901*, so for a long time I'd imagined it to depict a moment just after my grandparents took possession of the place, after their emigration from Lebanon, after peddling wares in upstate New York, and briefly enduring tenement life in the city of Lawrence six miles to the east of the farm. Even when I learned that it actually captures the last days of ownership of the family who'd lived there before my grandparents, it hardly seemed to matter. My family had simply taken over a way of life that had been accumulating for centuries, and there was little enough difference between last days and first.

In the years after my grandparents came into possession of the farm, the windmill blew down, the silo burned, but one after another child was born, the size of the milk herd increased, and their halting English became more certain. In time, teams of ordinary horses gave way to tractors as the farm steadied into the one I knew where irises and roses flourished at the fence, full-grown shade trees tossed high above the roof peak in the storms, and my entire extended family—thirty, forty of us—would gather at the farmhouse during the holidays. Aunts, uncles, cousins crowded the kitchen and dining room. We ate Lebanese *kibbeh* and stuffed grape leaves, we tore off pieces of Syrian bread to scoop up *hummous bi tahini*. The heat chuffed, the warm air was filled with voices, while beyond us, beyond the watery old glass of the farmhouse windows, the world was bright with the stark New England winter.

That the farmhouse remained for so long at the center of our world surely had to do with the fact that my aunts, Bertha and Del, who never married, lived there all their lives, along with my bachelor uncle Joe. My brothers and male cousins have little anecdotes even now about Joe. He taught them how to check a tire for leaks and how to change the oil in a truck. He stood over them as they planted the tulip bulbs in the farmhouse yard under his direction: "Is that hole four inches deep?" "Four inches," any boy would affirm. "Bulb on its side?" "Bulb on its side." But it's Bertha and Del whom I recall most often. When I was very young the farmhouse had been—like the meadow, the brook, the woods—one of the stations of my childhood. I would go there almost daily with one or three or four of my girl cousins who lived near me. We tracked in mud, we tracked in snow. We spoke of them—Auntie Bertha and Auntie Del—in the same breath, as if one word. It seemed all the noise there ever was we brought with us into the kitchen as we settled in the rockers and easy chairs or around the table by the east windows where we found ourselves bent over Scrabble with Bertha or searching through the puzzle pieces laid out and partially interlocked. The unentered rooms, made more still by our games in the kitchen, were sometimes peopled by the recollection of evenings when there were so many for supper they had to eat in shifts at the oak table—a silver cup full of silver spoons at its center—or of the living room thick with cigar smoke when the brothers had brought friends home after a late night out.

The apples blossomed, the grass dried in the August heat, my aunts' attention continued to turn to the sound of a car pulling into the drive, to the tenor of their brothers' voices hailing them as they walked up the porch stairs. In time my own world became crowded with life beyond the farm, with friends from high school and dreams of going away—it's hard to remember when I hadn't been intent on going away. I bent to my books during those late teenage nights, my imagining and hopes running on faster than the frequencies on the radio dial I flipped through when I was restless: static, voices, snatches of song, everything in the world there if you could just tune in, if you could just settle on the right place . . . My visits to my aunts became less spontaneous then, more brief and formal, made dutifully, prodded by my parents. I sat stiffly by the window, feeling the stillness of the house now that the games had fallen away. The family pride—"You don't act that way . . ." "Remember who you are . . ."—began to feel far too confining, and I tried not to stir it up as Del sat there rocking and Bertha knitted panels in patterns of cable and box stitches to make afghans for each departing niece to take away to college. I wondered about their single lives, their devotion to the family, the protection and regard of their brothers, the way you wonder about those flocks of swifts you see in late August, readying themselves, making their glinting turns in the turbulent air—are they helped by the wind or made helpless by it?

"I'd sell pies and ice cream," Bertha recalled, her needles clicking into the quiet as she told me about the tearoom her father had built for her to run just across the road from the farmhouse. She elaborated on the story during later visits: "We called it the Red Wing—everyone along the road would stop in . . ." I can still hear her mild, soft chuckle, and a final comment: "Anything to keep me home." Then she turned again to the work in her lap, which by then may even have been my own afghan in shades of blue that I was to drape over myself as I read poetry and studied ecosystems, Shakespeare, and Russian history in my cinder-block dorm room almost two hundred miles away.

Del was the first of that generation to die. When the cancer was diagnosed she was set and determined to begin with. "I want to know the news, however bad," they say she told the doctors. Then came the affront of treatment: "They made me drink barium." She'd always carried her large self with dignity, had carried out everything with deliberate care, even something as small as shaping Easter cookies, hundreds each the same size and intricately shaped. Her generation of the family did not record much, but everything worth noting is set down in her meticulous, flourished hand, every wedding notice and obituary of the extended family—third cousins, cousins twice removed—clipped and pasted by her into a hardbound album. I don't remember her ever registering fear, even as her voice grew faint and hoarse. "I've had a good life," she said to me squarely. In the later stages of her illness she began to shrink rapidly in her dresses. A stairlift appeared on the farmhouse stairs. She became a frail shape, unimaginable even to herself: "Who'd ever have thought I could feel my bones against the seat of the chair?" she said in a kind of wonder.

"You should come back for a visit now," my father told me one day, and then he went on to warn me I might not recognize her. I had been living away for years by then while my father and brother carried on the work of the farm. At the time I was on Nantucket Island, thirty miles off the coast of Massachusetts, so the return felt like a real journey. The ferry pulled away from the dock in the harbor dawn and glided past the last sandy stretches of wild, scoured land, then past cormorants drying themselves on the jetties. It took hours to cross the sound then, and after the quiet suspended world of us travelers—just our murmurs and the humming engines—I was always startled by the bright, jostling morning of the working world when I debarked on the Cape. However quiet the ferry had been, it couldn't compare to how tamped down the farm felt when I arrived back that time. My aunt's failing was all the concern. Small helpless gestures: her brothers would bring her boiled lobster—one of the few things she could keep down—and they'd urge her to eat just a little something. She might pick at the meat in the legs or at the fine white flakes in the cavity of the body before pushing it aside. The few deaths I'd endured until then had been sudden. I'd never seen anyone passing out of the world before.

When I returned a few weeks later for the funeral and stepped out of the car on a warm, clear summer morning—the sky: a white haze on blue—I was startled that the farmhouse appeared as serene as always. With the rows of tasseling corn and trellised tomatoes following the curve of the old hills, the placid geometry of the farm had never seemed so absolute before. Whether heightened by contrast to my life away or by death, or both, I can't say. I only know how set and ordered the farm appeared so early in June—always the most perfect time, when all is hoped for and full of energy, before drought sets in, and small failures—and I seemed to see for the first time the kind of care it took to keep up the place. I swear I remember it that way, though the last time I was at the family grave site I saw that Del had died in winter, and my sister insists she was then nearly nine month's pregnant with my nephew, who was born in January.

I've lost track of the time between that first death and my return to the farm to help out—was it fours years, or six, seven?—but I'm sure Del's passing and the way I remember the farm having appeared to me that day played a part in my decision. Perhaps it's only that the sense it aroused of a place steady, assured, and vulnerable had begun to meet my wish for some definition in my life. Already time had worn on my desire for freedom, or had made that desire seem ordinary. Nantucket had given way to other places, to western Massachusetts, to towns outside Boston where the wail of sirens was anonymous to me. I'd seen many of my friends marry and begin to establish families of their own. With Del's death, maybe I began to recognize the fragility of my one and only family, and the farm. My father and brother had always had their differences concerning the running of things, my brother wanting to expand for the future with a larger farm stand and greenhouses, wanting to experiment with methods and crops; my father, more cautious. As my brother fell further into the drugs he'd begun using as a teenager, his ideas were paired with an unreliability that worsened the contentions between them. In my years away I'd tried to ignore the arguments—I felt helpless in the face of them—but with the family growing old, they began to seem consequential.

WHATEVER ELSE MY return to the farm has meant for my life, it has made the later years of that older generation more clear to me than their earlier ones, none more so than Bertha's. I can't imagine her life beyond the farm. I know she traveled when she was younger and spent some summers working in a resort—in the Adirondacks, was it? But the life I knew was defined by her singular fidelity, and the fidelity it engendered in return.

After my father and four of my uncles had followed Del to their graves, Bertha sometimes forgot that they had gone, and I had to tell her again that Stanley had passed away, and again her eyes registered the grief as clearly as the first time it had been told to her. She was in her nineties by then, and a neighbor had moved in to help her with daily life. She had bristled at the thought of someone outside the family coming into the farmhouse, but she no longer had the strength to refuse. Joe's plaid coat and creased hat still hung on the coatrack. The dresses of Del's large, healthy self still hung in the closet. Bertha wouldn't have it any other way. "I like to have my familiar things around me," she said, standing stooped among the stove and sink and counters where she'd spent so much of her life working steadily at four things at once if need be: leveling off the measured flour with the blade of a knife, punching down dough in an earthenware bowl, saving the chicken fat, saving scraps. Always at the end of my brief visits, after I'd filled the air with patter about small things—the growing season, the weather—and had repeated myself just to keep the air stirring, after I'd said I should be going, she'd query me: "What's your hurry?" I could feel the silence I would leave behind me.

Her last summer was full of heat and drought. Amid the late terse calls of gathering birds, pine branches cracked overhead as squirrels collected nuts for their stores. The irrigation ponds had been pumped down to nothing and still there wasn't enough water for the corn. The apples reddened small, the honey darkened to a flavor deeper than I'd ever tasted, and by September the air smelled of ripe grapes on one wind, and on another of the dust that swirled in back of the pickups. We began to discern a new map of our world, one on which all the susceptible, stressed places in the land stood out: the younger maples and the ones by the road where the soil was laden with salt began to brown before they had a chance to flame with their fall colors. Each leaf dried up, from the edges in. The last green of the year lingered in the low marshy places. Bertha was always by the window when I stopped by, and every few minutes she'd gaze up at the unmitigated hard blue above and say, "Look, what a beautiful day, not a cloud in the sky."

Toward the end of that growing season she rarely had the strength to keep her place at the window and began to spend most of her time lying on a bed that had been brought downstairs and set up in the old summer kitchen. The shades were drawn against the sun; she: half sleeping, small. Still, no one else could touch the strength she got from her remaining brothers as they stood saddened and bewildered in front of her. She seemed less than half aware of me by then, as I chattered on about the world as it was, how the winter rye was coming up everywhere now. What was I talking for? To keep her in this world? To keep myself from wondering about her life, steeped in its certain fidelity, which led me to think of my own now lived for its own fidelities? Once I grew ashamed of my patter, and said simply, "Well, this is pretty tough." She focused her clouded eyes on me, shook her head with surprising strength, and uttered a deep, imperious "Ohhhhh," and I blanched and looked away, knowing I didn't know the half of it.

Her two surviving brothers spent a long painful time trying to sum up her life for the local paper, to conjure up something beyond the household. "Have them list all the nieces and nephews," one of my uncles said to me. "Remember she was a seamstress for a while, have them put that in." Her funeral was quiet, and maybe twenty-five of us—my cousins and siblings and their scattering of children and teenagers, a few old friends of the family, the priest—went back to the farmhouse afterward for a meal in the backyard. It was a beautiful June day to begin with, though a thunderstorm came in from the west and we all rushed into the close, bright kitchen. The house smelled of our dank hair and clothes and the rain brought out a little must. Someone opened a window for air, the curtains billowed in the wind, and we could hear the downpour and thunder as we dried ourselves off and poured ourselves coffee. We talked on and on about the house, the times, our own childhoods, the whole long lives of them all. "She would have liked this," one of us said. That turned out to be the last time we gathered as a family in the farmhouse. It took a good while for anyone to think of what to do with all the things, and the clothes of the three of them hung in the closets for the longest time. Afterward, when I would see a cousin of mine, one of us would say: "It's so strange not to have them there." And the other: "I know."

THE FARM, IF you were to drive by it now, would appear as measured and ordered as ever, even though no one in the family works the land anymore. I'd stopped helping out the year before my father died: the antagonism between my brother and him had just been too difficult for me. My brother didn't carry on for more than a year after my father's death. I think for all his grief—his was probably the most complex grief of all—he'd imagined a freedom for himself after my father died, though he was to find he was no more heir to the farm than anyone else. He had siblings to contend with, and uncles. His chances would never be clear, what with all the family history behind the farm. His own difficult past had made us all wary of trusting him with the running of the place, even though no one else knew enough or had the time or desire to see to it: a small farm had become an intricate business, and with the costs of owning land, the taxes and insurance, just letting it be wasn't an option. In today's world a farm isn't often abandoned to woods. Some other human scheme is waiting.

In hopes of keeping the farm going I eventually made a call to one of the countless boys who'd worked for my father over the years to ask if he would come back. David had worked on the farm all through high school and college, and he'd always had a particular love and affinity for it. After he finished college he joined the Peace Corps agricultural program, and he was working in Ecuador at the time of my father's death. I had little hope that he'd agree to come back, but it seemed, in those white days, that anything could be tried. We talked over crackling wires, a world away from each other; he still felt an allegiance to the place and to my father, and he said he'd take on the farm after his stint in the Peace Corps ended. His agreement felt like a great simplification. At the very least it would allow the family to get its bearings again, and to live on in the old image for a time. It helped me to believe that though our place here was passing, the world we'd established might continue.

Though Dave has been running the farm for years now, has hired his own crew, and has his own ideas for growth, there are practical things he and I talk over still, and sometimes he asks for advice. There's much to be anxious about, as there always has been, though I usually hold my tongue or choose my words to frame observations rather than suggestions: "Corn will never be the draw it once was . . ." I say, fading off, walking away. I try to resist the impulse to drag my past into his ideas for the future—"We always did it this way." It's a bit of a joke between us. When he started some early corn in the greenhouse one year, I laughed and said, "What's that I hear—the sound of my father turning in his grave?"

I miss the clarity you get from training an eye on the frosts and failures, from the exhaustion of working in the open air, and the way that exhaustion sharpens the wine smell of the apple cellar when it's packed to the ceiling in late fall. Over the past years I've begun to feel lighter on the land, until now I am almost a ghost when I walk across the fields to pick a few tomatoes for dinner, or pull up a corner of my shirt to collect some apples for sauce. Sometimes I watch the crew down at the far edge of a field calling to each other in Spanish as they bend to the plantings among bright yellow tubs loaded with tomatoes, squash, and cukes. A new man might stand and stare at me for a moment before narrowing his attention to the work again, but he doesn't hail me. Once when I went to find David to talk about some business and asked one of the men where he might be, he turned to another: "Where David? The lady is looking for David."

It's not the changes themselves that feel so strange—the larger scale, the different crops, the new workers. There have always been changes, and had I been part of them, I would have been swept along with them. But standing apart makes even the small changes feel significant. It can be a relief to slip into the woods at the edge of the farm, so closed in and intimate come summer. Up the dry hill beyond the brook, the moss and old nurse logs breathe their punky smell of dankness and decay. You might imagine you're walking on a moderate sea, the way the unearthed roots of old pines and the hollows they've left behind have moldered into swells and troughs beneath the duff, and fallen trunks with their whirls of limbs lie like the staves of old wrecks, half buried and fretted with the work of insects. But these woods are the abandoned far pastures from a time when farming was the common life, so you can't walk for long before your thoughts are turned inward by remnants of past labors gone to seed. A little south of the brook, just beyond an old cattle bridge, you'll come upon a forgotten orchard where trees still carry a trace of the mind and hand that shaped them, and everywhere through the woods are runs of stone walls and barbed wire, heaps of rusting metal and broken bottles, a corroding milk can with a lady fern unfurling from a lacy hole in its side. The rust and ruin, the riot of green—mutually incongruous—have created a society of their own. The rain and sun and shadow that decay one feed the other.

IF YOU TRAVEL a few hours west, beyond the Worcester hills and the Connecticut Valley, you come to quieter, steeper country, a later frontier where cultivation has always been more difficult and tenuous than on our milder coastal terrain. Still, since there's less residential and industrial development there, the patterns of agriculture are more obvious: ancient sugar maples line the roads, and you can count an unbroken string of farmhouses one after another. As I drive through I find myself reciting the child's song: *big house, little house, back house, barn.* A few sheep graze on a slope, a herd of dairy cows lie in a pasture. Modest farm stands dot the roadsides, little more than tables stacked with tomatoes and corn, a jar for money, maybe a sign: EVERYTHING WE SELL WE GROW HERE. My family's farm looked the same fifty years ago.

In some places the woods closing in over the mantle of the hills are nearing the backs of the buildings. You can peer into the trees and see collapsed barns. When this world is released from cultivation it returns to a more northerly aspect than our own. Hemlocks and sugar maples climb the rough ledgy heights where bobcats can hide themselves. Summer people like it. There are driveways leading to new places hidden away in the woods. Young families have begun to buy up the old places for their weekends, though they have trouble keeping the fields. "Even in the last twenty years, we've lost so much open land," I heard one longtime resident complain. "The new people don't keep it clear. They start out thinking they can find some old farmer who'll mow it for free . . ."

I was there last in late August—even after years away from the demands of the farm, traveling in August still feels freeing to me—and the first apples of the year were out. I pulled into a place with no name, just a hand-lettered sign that said: APPLES. PEACHES. I couldn't see any outdoor work being done. Everything was still. You could tell the weaker maples would begin to turn in another week or two. I stepped down into a long, low wooden room with old white walls, cave cool, cave dark, full of benches and tables stacked with crates of apples, and the air was dense with that familiar smell. A woman alone, maybe a little older than me, was grading apples, turning each one, looking for signs of insect damage, sizing them. I knew right away by her unhurried pace that she was just helping the place out to its end. There's often a lone woman at the end.

"Smells great in here," I said.

"That's what people say. I'm here all the time, so I don't notice it anymore."

"Have you got any Gravensteins? Half a dozen?"

"Sure." We talked a little bit about them, how they make such good pies.

"You don't make cider anymore, do you?"

"No, no." Which led to talk about the crisis of a few years back when a handful of people were sickened by *E. coli* in some unpasteurized apple juice in Colorado. The word of those illnesses had been so insistent on the nightly news that everyone shied away from unpasteurized cider. Sales plummeted, the state debated putting new regulations in place to require pasteurization—untenable for small orchards that squeeze their own drops and seconds in a mill, something that a farm might depend on to make a marginal crop worthwhile.

"We'd take people down to show them how we washed the apples before we pressed them into cider," she said, "and still they were afraid to buy it. Fifty years we never had a problem and then—"

"I know, I know. I haven't had any cider since. It just doesn't taste the same."

It seemed we wanted to talk about everything, our families getting older, the fate of the farms. "Two dairy farms in town went out just last year . . . And you go over toward the river, and you see all that beautiful level, loamy farmland where you don't have to dig up the rocks, where the tractors don't tip, and you see them building houses on it."

After a while I grew a little self-conscious about interfering with her work, though she didn't seem to mind. Or maybe the discomfort I felt had to do with her devotion to the place, which seemed so assured to me, and the way she read some devotion in me. She went on, "We lost 95 percent of our crop last year."

"Oh, the frost in May—you're colder here—it must have been worse." We ran on for a while again about the apples.

"It's nice to talk to someone who knows."

"Yes, yes, it is," I said back, and I left her there, sizing and grading the Paula Reds and the Gravensteins in the deep, rich smell of the place. When I walked out I saw the brushy, unkempt edge of the fields and it was clear to me that the woods were creeping into the upper reaches of the farm, that the deer were already coming down, and the thrushes were calling out from the low places.

Often when it comes, the end of cultivation is no louder than the tumbling of apples into crates in a cave-cool, cave-dark room, but the life lived in the wake of its disappearance is a break with a long history, and the days that follow—as the worked and tended country disappears, along with its bales and stacks, rows and grids, the men and women moving among them—are different in intent and kind. It may take awhile for the idea of it to die away—there may be a romanticized echo, in which farming's rewards are imagined more vividly than its costs—but its end is one of those times the whole pattern shakes and quivers and settles into new shapes and figures, perhaps all the more so in a country dependent on the idea of agriculture for its identity. In America, not only do individual dreams have their origins in farming, the notion of the Republic is stowed there as well. *Cultivators of the earth are the most valuable citizens,* insisted Thomas Jefferson. *They are the most vigorous, the most independent, the most virtuous, & they are tied to their country & wedded to its liberty & interests by the most lasting bonds.*

On the worked-over lands of the Northeast, where the new has settled on top of the old again and again, the boundary between the smallest field and woodlands remains a kind of frontier. Or perhaps it's more accurate to think of it as a contour of human time—it's always a question of time: how much to acquiesce and try to inhabit its flow, how much to try to make a mark against it, how much heed to give to the past attempts everywhere evident on the land. If you have maintained such a boundary all your life and you see a little brush growing into the clearing, you can get a creeping feeling at the back of your neck. You understand how small your own enterprise is, and how temporary. *That's what I despise—brush and trees growing in hay fields,* remarked one farmer early in the twentieth century. I've been startled to hear another insist: "I'd rather see a hundred houses in my fields than have the woods come back on it."

The woods *have* come back on it even where agriculture hasn't been entirely lost. As New England dairy herds were sold and farmers turned more frequently to market garden cultivation, the rougher pastures were let go. My family's farm, in the years I've known it, contains much less cleared land than it had in 1901. Today the row crops, the corn and tomatoes and peppers and beans, are concentrated on the most productive land. Without adjoining farms, and only a few others scattered among suburban tracts, ours is smaller within the community as well, though in some strange way it is also more prominent now that farming is no longer the common life.

Though the margin between our cleared land and woods remains certain for now, I can't help feeling the farm may be swallowed up soon, and a whole world will go with it, one that has been both particular to ourselves and representative of the many who have contended with the land and brought their histories to bear upon it. As if understanding can alleviate loss, I am trying to place our own time within the larger story of cultivation. I hope constraint amplifies, that by giving shape to the stories, to the persistent half-lives of the vanished who roam even on stinging winter days, I can see more clearly where we belong in the accumulation of beliefs, ideas, violence, necessities, and desires that have determined this country. But the beginnings *are* hazy, and the people there seem so different. The moment I try to articulate something of the story, which is its own attempt at defining a frontier once and for all, the boundaries seem to change again. Even as I write this down now my understanding clears and grows back in, clears and grows back in, and I know if I let go, if I turn my back on it even for a little bit, I'll have to remake it out of the rank that has grown back. The stories flux like fall migrations, forming and re-forming, part of them gaining as another part recedes, the labor of it continuous, and still the lines dream their own way, on the wind.

2
Agricultural Time

THE EARLIEST DEPICTION I've seen of fields on these shores are the eight stands of ripening corn sketched on Samuel de Champlain's 1605 map of what he called Port St. Louis. The place was later established as the settlement of Plymouth, but on Champlain's map smoke rises from wigwams in the Wampanoag village of Patuxet, leafy trees are scattered along the shore, and dense forest crowds spits of land at either end of the harbor. Depth soundings—1, 2, 8, 9, 10—are plotted on the water. How long Patuxet had existed before that map I can't know, but it wasn't to last much longer. An epidemic in 1617 killed two-thirds of the Indian inhabitants of the region: thousands lay dead in and around their wigwams, so many that the living were unable to bury them, and their bones were bleached by the sun. The disease wiped out the village almost entirely. Squanto, the one survivor, had been sold into slavery in Europe and avoided death by being absent. After his eventual escape and improbable journey back to his own shore, he found his home deserted. By the time the *Mayflower* anchored in the harbor in December 1620, the fields Champlain had drawn were rank with weeds. William Bradford's account of the settlement at Plymouth makes no mention of the disease, or of wigwams. *They sounded the harbor and found it fit for shipping, and marched into the land and found divers cornfields and little running brooks, a place (as they supposed) fit for situation,* he wrote. *At least it was the best they could find, and the season and their present necessity made them glad to accept of it.*

Within days the Pilgrims began building rudimentary shelters; within months they'd made a treaty with Massasoit, chief of the Wampanoag, who'd agreed:

1. *That neither he nor any of his should injure or do hurt to any of their people.*
2. *That if any of his did hurt to any of theirs, he should send the offender, that they might punish him.*
3. *That if anything were taken away from any of theirs, he should cause it to be restored; and they should do the like to his.*
4. *If any did unjustly war against him, they would aid him; if any did war against them, he should aid them.*
5. *He should send to his neighbours confederates to certify them of this, that they might not wrong them, but might be likewise comprised in the conditions of peace.*
6. *That when their men came to them, they should leave their bows and arrows behind them.*

However detailed the words, the treaty never did allay the apprehensions of the colonists. William Bradford continued to see the Algonquin as skulkers at the borders: in the pages after the recording of the treaty he writes of the killing of some settlers, and for him it stood as proof of *how far these people were from peace, and with what danger this plantation was begun, save as the powerful hand of the Lord did protect them.*

All the while the winter took its toll on the colony. Only one of the Pilgrims had perished during the crossing, yet in those first months at Plymouth sometimes two or three a day would die of hunger, cold, or disease: . . . *of 100 and odd persons, scarce fifty remained,* wrote Bradford. *And of these, in the time of most distress, there was but six or seven sound persons who . . . fetched them wood, made them fires, dressed them meat, made their beds, washed their loathsome clothes, clothed and unclothed them.* By the time the frost was out of the ground only twenty-one men and six grown boys were strong enough to prepare the fields for planting, and they could hardly be called farmers. Most had been involved in the clothing industry in Holland. William Brewster had been a village postmaster; John Carver, a businessman; Edward Winslow, a printer; Myles Standish, a soldier. They had few adequate tools. No oxen or plows are listed on the *Mayflower* bill of lading.

They would not have survived without those fields at Patuxet, which saved them the labor of clearing land for planting, and the ten bushels of stored Indian flint corn the Pilgrims had unearthed during their explorations of

Cape Cod prior to settlement at Plymouth, and Squanto, who *stood them in great stead, showing them both the manner how to set [corn], and after how to dress and tend it. Also he told them, except they got fish and set with it in these old grounds it would come to nothing . . . and taught them how to take it, and where to get other provisions necessary for them. All which they found true by trial and experience. Some English seed they sowed, as wheat and pease, but it came not to good, either by the badness of the seed or lateness of the season, or both, or some other defect.*

The land they'd cultivated in Plymouth was not only the best they could find, it was about as good as it would get for farming. *The soil is for general a warm kind of earth,* wrote William Wood in his *New England's Prospect* in 1634, *there being little cold spewing land, no moorish fens, no quagmires . . . Such is the rankness of the ground that it must be sown the first year with Indian corn, which is a soaking grain, before it will be fit to receive English seed. In a word, as there is no ground so purely good as the long forced and improved grounds of England, so is there none so extremely bad as in many places of England that as yet have not been manured and improved . . . Wherefore it is neither impossible, nor much improbable, that upon improvements the soil may be as good in time as England.*

But New England soil has always had its own demands. It's largely an inconsistent swirl of glacial till—often rock-strewn, acidic, and quick to lose its fertility. *We found after five or six years that it grows barren beyond belief,* one colonist would comment early on. The Wampanoag, like other Algonquin, had cleared and planted such lands without oxen, iron, or knowledge of the wheel. Their tools were made of shell and wood, stone and bone, and were useful only on light, well-drained soils that could be turned up with the shoulder bones of deer. Of necessity, then, they left the heavy lands and clay soils untouched, keeping their cultivations to the gentle hilltops, the south-facing slopes that warm up early, the richer soils in the intervales, the clearings along the coast. To make new fields they burned stands of white pine, which thrive on lighter soils. The softwood could be felled with a stone ax, and fire—helped by resin and dry needles—would work fast. Fire once to burn the brush, fire twice to burn the trunks, and again to make a cindery bed. They grew beans and squash, but their mainstay was flint corn, which they beat into meal and sifted through a basket. Sometimes they parched it, or wrapped the cornmeal in leaves and baked it in ashes, or made small cakes and boiled them.

Since the Algonquin supplemented their cultivated stores with wild meat, berries, plants, and acorns from the surrounding woodlands, they burned the underbrush in some of the forested areas for easier hunting and gathering. In winter they moved their villages inland, away from the bitter ocean winds, and returned to the coast in summer to fish, and to plant in the milder season there. They kept no livestock, so they had no need to raise extensive fields of grasses for feed and fodder. Their compact and movable life, shoaling between woodlands and cultivations, created a pattern on the land close to the drift of the soils with their runs and swirls and settlings. It bewildered English eyes, which were long accustomed to determined villages and an ordered, patchwork landscape. *The Indians are not able to make use of one fourth part of the land,* one observer insisted, *neither have they any settled places, as towns, to dwell in; nor any ground as they challenge for their own possession, but change their habitation from place to place.*

In much of England and central Europe you'd have had to go back to the ninth and tenth centuries to find a world that depended on only lighter soils for cultivation. Few in the seventeenth century would have had any memory of the incremental clearing of the land, or the initial plowing of fields, of the slow evolution of cart tracks into strong roads as villages became established and trade routes and market days flourished. Drainage systems had, by then, obscured old bounds. Horses plowed established fields quickly while oxen reclaimed heavier lands from the wild, clearing more and more efficiently as collars and yokes became sturdier, and plows stronger. Populations had grown and had become more mobile in the increasing stability. More horses, cattle, oxen, and workers required more acreage in wheat and oats, barley and rye.

By the seventeenth century there had long existed written instruction on how to keep oxen and how to inspect cattle. Documents dating back to the Middle Ages listed expected yields of grain, and instructed farmers on how to feed the dung heap, how to fertilize and rotate crops. The character of those who work the soil had also been set down: *The ploughmen ought to be men of intelligence, and ought to know how to sow, and how to repair and mend broken ploughs and harrows, and to till the land well, and crop it rightly; and they ought to know also how to yoke and drive the oxen, without beating or hurting them . . . And they and the keepers must make ditches, and build and remove the earth . . .* Such an idea of the cultivator isn't so far from Plato's description of the farmers of prehistoric Athens, *a class of genuine full-time agriculturists with good natural talents and high standards.*

Plato compared that legendary Athens to the one of his own time. The fertile lands and careful practices of former days lived large in memory, and seemed to chasten his present: . . . *the rich, soft soil has all run away leaving the land nothing but skin and bone. But in those days the damage had not taken place, the hills had high crests, the rocky plain of Phelleus was covered with rich soil, and the mountains were covered by thick woods, of which there are some traces today. For some mountains which today will only support bees produced not so long ago trees which when cut provided roof beams for huge buildings whose roofs are still standing. And there were a lot of tall cultivated trees which bore unlimited quantities of fodder for beasts. The soil benefited from an annual rainfall which did not run to waste off the bare earth as it does today . . .*

The past, with its own power and life, imbues all kinds of land. Sometimes an idea is born of erosions. René Dubos would speculate on the damaged lands of Greece, the white rocks and the sun-loving aromatic plants—the idea of Greece that has grown in our own minds: *I have wondered whether the dark and ferocious divinities of the pre-classical Greek period did not become more serene and more playful precisely because they had emerged from the dark forest into the open landscape. Would logic have flourished if Greece had remained covered with an opaque tangle of trees?*

Would our sense of a patchwork agricultural world as a proper country have flourished if seventeenth-century Europe hadn't long lost its tangle of trees? Woodlands had grown scarce by then because of the increased clearing of fields and the constant demand for wood. Wood was needed for housing and fuel, for masts and hulls of ships, to make casks, vats, and vine props for vineyards, to produce resin for torches, bark for rope makers, ash, and charcoal. Strict rules concerning the felling of trees had been set down. *Every one of our commune shall be bound to plant each year ten domestic trees . . . Let no one dare to cut larchpoles to make firewood on this mountain . . . Let no one dare to cut wood from Xomo up to the rock of Slamer and from the peak of the mountains towards Xomo, save to construct a house or for the fire, on pain of five sous per tree.*

I've wondered where the idea of wilderness goes when the woods disappear, the woods that had sheltered beggars and hid salt smugglers and thieves, that—however frightening—had also been useful to men and women bound by a feudal system which, no matter how much they worked, left them begging at the gates of the rich come winter. They could turn their animals out to feed on beech mast, they could gather mushrooms, berries, and herbs to appease their hunger. The mystery of the indifferent and necessary woodlands could not be distinguished from its terrors, the many terrors could not be distinguished one from the other.

As cultivation progressed and forests diminished, and as both European villagers and farmers came to rely more on cultivation and less and less on foraging in the dwindling woodlands for their survival, I imagine that what wilderness remained became more daunting—less of a refuge, less complex in people's understanding, so that by the seventeenth century, when Europe was largely a cultivated landscape, the woodlands had become incomprehensible. When the first white settlers were confronted with the extensive eastern forests of the New World, however much they may have been burned and cultivated by the Wampanoag, Nipmuck, and others, they couldn't see a way through, and marveled at the way the Indians could: *A man may travel many days and never find [a path. The Indian, however,] sees it instantly . . . and will mark his courses as he runs more readily than most Travellers who steer by the Compass.*

In the first years of settlement the colonists at Plymouth kept their plantings tight within an impaled world—in part for the protection it offered from the Indian tribes and from everything they understood as wilderness, but also out of necessity. The work had to be done communally—they hadn't the strength to cultivate large swaths individually. As well, they were obliged to work together to send quantities of clapboard, fur, and sassafras to their backers in England. Only after several years of settlement, after the constraints imposed by their backers lessened, and their desires for their own lives had room to grow, was each colonist granted a modest plot. William Bradford remembers: *And to every person was given only one acre of land, to them and theirs, as near the town as might be . . . The reason was that they might be kept close together, both for more safety and defense, and the better improvement of the general employments. Which condition of theirs did make me often think of what I had read in Pliny of the Romans' first beginnings in Romulus's time. How every man contented himself with two acres of land, and had no more assigned them . . . And long after, the greatest present given to a Captain that had got a victory over their enemies, was as much ground as they could till in one day. And he was not counted a good, but a dangerous man, that would not content himself with seven acres of land.*

As the colonists became more secure in the New World, the varieties and amounts of crops they grew increased. They planted root crops of English origin such as turnips, parsnips, and onions alongside the wheat and barley they had been cultivating from the start, and *they began now highly to prize corn as more precious than silver.* In addition, they were raising livestock, which required extensive fields and pasture. Each family necessarily required more land for planting and grazing, and Bradford details the way in which the bounds of the community were loosened: *For now as their stocks increased, and the increase vendible, there was no longer anything holding them together, but now they must of necessity go to their own great lots. They could not otherwise keep their cattle, and having oxen grown they must have land for plowing and tillage. And no man now thought he could live except he had cattle and a great deal of ground to keep them, all striving to increase their stocks. By which means they were scattered all over the Bay quickly and the town, in which they had lived compactly till now was left very thin and in a short time almost desolate.*

More expansion rapidly followed. By 1638, less than twenty years after the settlement began, Plymouth Colony consisted of 356 fields covering five square miles, and a considerable number of other towns had established themselves along the coast and up the rivers, natural places for settlement since they afforded travel by water—few roads reached into the interior—and colonists could avail themselves of nearby wild grasslands.

EUROPEANS HAD LONG depended on cultivated grasses such as timothy and redtop to feed their livestock. Such grasses weren't native to the New World, and weren't to become established until early in the 1700s, after timothy was discovered growing in New Hampshire, having sown itself there, it is supposed, from seeds carried over in the bedding and ballast of ships. But in the first years of settlement, fresh meadows along the rivers and, more important, the vast salt grass marshlands along the sandy shores of the coast—vaster than anything imaginable now that so many of them have been ditched and drained—stood in for the calm and order of English fields, and were essential to colonial survival. They were time given, since wild grasses served as a ready crop on land farmers didn't need to clear or tend. Men could harvest feed for their livestock and turn to their other concerns. *There is so much hay ground in the country as the richest voyagers that shall venture thither need not fear want of fodder, though his herd increase into thousands, there being thousands of acres that yet was never meddled with,* William Wood was to comment. *True it is that it is not so fine to the eye as English grass, but it is not sour, though it grow thus rank, but being made into hay the cattle eat it as well as it were lea hay and like as well with it.*

The salt marshes really are no more than built-up detritus: season after season of thatch and stubble and the wash and drift that gets caught in roots and stalks, earth made of soft spots, little mires, marsh pools, peat and pan, smelling of low tide, of dead sea life and brine. *Spartina alterniflora*, a crude grass that's heavy enough to be used for the thatch on houses, defines the open edges of the marshlands. If you nose in with your kayak on a clear summer day, you can see the salt glinting on its stalks. Come winter the ice cuts the withered grasses, and as they mat and decay they build up the earth into solid enough ground for *Spartina patens*—a finer grass, most prized for salt hay—to establish itself. The grasses possess a complex set of adaptations that enable them to exclude and excrete much of the salt in seawater in order to survive on the tidal marshes. It's not that they necessarily thrive there—they'd grow larger and contain more nutritive value on higher, fresher ground—but spartina can't successfully compete with other plants in an inland meadow, and so it keeps its place on the sea-laden edge of land where little else can grow. No, salt hay could never be confused with English hay. Sea pickle and worts and sea lavender grow among it, dead crabs and wrack get caught in the stalks. It brought the smell of the sea into colonial barns and bedding and households. Heavy and crude, it didn't scatter on the wind like cut timothy. It served as fodder but wouldn't fatten the cattle. An experienced colonial farmer, less optimistic than Wood, complained: *Our beasts grow lousy with feeding upon it and are much out of heart and liking.*

The salt hay harvest began in late summer, during the spring tide, just when the waters had ebbed and the stalks of grass were still moist for cutting. On farms bordering the marshes, the men might work together to get in the harvest. They'd cut the spartina and the adjacent black grass on the higher grounds through September until the frost, and sometimes after the frost, though what was harvested then was used for banking houses rather than for fodder, since once the hay froze it lost its value as feed. *[They] took scythes out to mow the marsh in August when the grass was tall but not yet mature . . . The men moved along, three abreast, and cut the high marsh grasses. After the hay was cured the men gathered it by pitchfork and piled it into stacks . . . On the lower reaches sure to be flooded, the salt hay was cut and carried to the upland, perhaps to be put in the barn. On the high marsh, the hay was piled on small poles driven into the surface of the soil in clusters about two feet high and two feet apart. On these groups of small pilings, called staddles, the hay was safe from high tides. The hay was collected in the fall if necessary, but usually it was left until winter came and the frozen marsh surface could support an ox team and sledge . . . Sometimes an unusually high tide occurred when a regular high was made higher by strong winds in an autumn storm. Then the carefully piled haystacks floated off the staddles intact and finally lodged up against the land somewhere . . .*

The dependency on those marshes was to diminish once colonists established fields of English hay on inland grounds, but the distinction between wild grasses and cultivated ones lived on in the imagination. *Rise free from care before dawn, and seek adventures,* instructed Thoreau in the nineteenth century. *Let the noon find thee by other lakes, and night overtake thee every where at home. There are no larger fields than these, no worthier games than may here be played. Grow wild according to thy nature, like these sedges and brakes, which will never become English hay.* When Thoreau made his comparison between wild and tame lives, agriculture had already begun its decline in New England. The cultivated bordered the abandoned; the ceaseless effort of farming was thrown up against itself. No wonder he would free the human spirit from the demands of the field: *Let the thunder rumble; what if it threaten ruin to farmers crops? that is not its errand to thee. Take shelter under the cloud, while they flee to carts and sheds . . . Men come tamely home at night only from the next field or street, where their household echoes haunt and their life pines because it breathes its own breath over and over again; their shadows morning and evening reach farther than their daily steps.*

When I look at the salt marshes now I see in the planes of field and sea and sky a place apparently washed of human effort—you would never imagine a workday as you gaze at them, or how they once defined a world and allowed it to continue. Sometimes kayaks and other small craft course through the meanders, but otherwise few humans enter the marshes, and to do so feels almost forbidden. Spare and quiet, soothing, fetid, contemplative, they now feel like the most solitary part of our crowded, overbuilt section of New England. An egret stalks its prey, the grasses cowl and whisper in a storyless world, seabirds cry overhead—always closer to a keening than a song.

FENCED-IN FIELDS DISTINGUISHED the settled world more clearly from disordered nature, and also from the less distinctive clearings and movable villages of the Algonquin. The colonists were never to regard Indian villages as land possessed—*As for the Natiues in New England, they inclose no Land*, noted Roger Williams, *neither haue any setled habytation, nor any tame Cattle to proue the Land by*—and they persistently expanded onto Indian lands. The Wampanoag, Nipmuck, and other eastern Massachusetts tribes were vexed by cattle trampling their cornfields and their clam beds, and they were pushed farther and farther back from the coast and the river valleys, from the light soils they had cleared and planted. *You have driven us out of our own Countrie and then pursued us to our Great Miserie, and Your own, and we are Forced to live upon you,* one Indian was to claim. The colonists were not only arriving in unimagined numbers and claiming extensive lands, they were following the Algonquin patterns of settlement along the coast and up the rivers. On Cape Cod and the south coast of Massachusetts: Plymouth, Sandwich, Eastham, Harwich, Yarmouth, Barnstable, Dartmouth, Taunton, Rehoboth; along what is now the Rhode Island coast: Providence and Warwick; the north shore of Massachusetts and up the Merrimack River: Gloucester, Ipswich, Newbury, Amesbury, Salisbury, Haverhill . . . Like Plymouth itself, all these towns—almost every early New England town—had at their hearts an Algonquin field.

With each colonial push for new settlement, the fragile peace that had been established between Massasoit and Plymouth Colony in those first months of the first year frayed more. In 1675 Massasoit's son, Philip, embarked on what is known as King Philip's War, which still stands as the bloodiest per capita in the history of the country. Indian attacks on established colonial villages destroyed lives, cattle, crops, and houses. *Though English man hath provoked us to anger & wrath & we care not though we have war with you this 21 years for there are many of us 300 hundred of which hath fought with you at this town[. . .] we hauve nothing but our lives to loose but thou has many fair houses cattell & much good things,* reads the note a retreating Indian tacked to a tree as the town of Medfield burned. And another elsewhere: *You know, and we know, you have great sorrowful with crying; for you lost many, many hundred men, and all your house, all your land, and woman, child, and cattle, and all your things that you have lost.*

As Jill Lepore observes in *The Name of War*, by destroying colonial possessions the Algonquin undermined the idea of the world the colonists carried with them, and had striven to apply to new land: *Colonial writers understood the destruction of houses as a blow not only to their property but also to the very Englishness of the landscape . . . Nearly all of the damage to the English during King Philip's War—the burning of houses, the spilling of blood . . . —was understood as attacks on bounded systems. While disorder threatened to rule New England, military strategists sought means to draw a line to keep Indians—and chaos—out. In Massachusetts, alarmed colonists even debated building an eight-foot-high wall of stone or wood all the way from the Charles River to the bay, "by which meanes that whole tract will [be] environed for the security & safety (under God) of the people, their houses, goods & cattel; from the rage & fury of the enimy."*

After a year of war, the bounded world of the colonists was almost completely destroyed: *In Narraganset, not one House left standing. At Warwick, but one. At Providence, not above three . . . Marlborough, wholy laid in Ashes, except two or three Houses . . . Many Houses burnt at Springfield, Scituate, Lancaster, Brookefield, and Northampton. The greatest Part of Rehoboth and Taunton destroyed. Great spoil made at Hadley, Hatfield, and Chelmsford. Deerfield wholly, and Westfield much ruined . . . Besides particular Farms and Plantations, a great Number not to be reckoned up, wholly laid waste, or much damnified.*

If the Algonquin had only their lives to lose, it was their lives they lost. Nearly all the Christian Indians who had been captured and imprisoned on Deer Island in Boston Harbor starved to death during the winter of the war. Most of those along the coast who weren't killed were sold into slavery; a remnant were squeezed west against their longtime adversaries, the Mohawks.

Savage on both sides, exhausting and debilitating, the war never had an official end—skirmishes were to continue for years—and a vehemence regarding the Indians remained stubbornly coiled in the colonial mind, evidenced in the way Philip's remains were treated. In August 1676 he was cornered in a swamp near his home territory of Mount Hope. After he was shot dead he was beheaded and quartered. His attackers took his head to Plymouth and mounted it on a pole—*meat to the people inhabiting the wilderness*, Increase Mather was to say—where it was to remain even as Philip's cornfields were sold to a Boston businessman and the new owner began to raise and export onions to England, as English settlement pushed inland far beyond the light soils of the Algonquin fields, as new settlers made cropland out of rich soils beneath the sugar maple and beech forests, and made pasture out of the fair soils beneath the chestnut and oak in the uplands, as colonists cut enormous quantities of wood for fences, houses, fuel for fires, boxes, barrels, as wood was piled and burned, or sent to the sawmills until only the wet and rocky ground of hemlock and red maple remained forested. Years after Philip's death, Increase's son, Cotton Mather, journeyed to Plymouth to gaze upon what remained of the head and he *took off the Jaw from the Blasphemous exposed Skull of that Leviathan.*

AS COLONIAL SETTLERS inched westward and northward, their fields, though they may have been more marked and orderly than those of the Algonquin, were still fundamentally different from the long-established European fields of their memories, imaginations, and initial expectations. Where they'd ventured beyond the already-tilled light soils of the Algonquin and the natural grasslands along the waterways to make farmland out of the wild, inevitably the initial clearing of such land was disorderly and rough, with uprootings and slash, exposed soils and downed trunks. Colonists might spend the fall of one year cutting and burning a woodland, and the following spring plant their first crop—squash, beans, barley, corn, onions—among the stumps and char. In most places the soil would have been scattered with stones, the glacial debris which was always to define and limit New England agriculture. To visiting Europeans familiar only with cultivation on long-established fields, this process of making land out of the wild was bewildering. Upon seeing America for the first time, one visitor wrote: *The scene is truly savage, immense trees stripped of their foliage, and half consumed by fire extend their sprawling limbs, the parts of which untouched by the fire, now bleached by the weather, form a stronger contrast with the charring of the remainder; the ground is strewn with immense stones, many of them a size far too large to be movable, interspersed with the stumps of the lesser trees which have been cut off about a yard from the ground.*

Colonists continued to make new fields out of the wild in part because of increased settlement, but also because the sheer abundance of available land seemed to discourage prudent agricultural practices. European cultivation, having depended on a gradual progression of technologies, had grown slowly and methodically. In its early stages the rudimentary nature of the tools and the scarcity of work animals was to limit how much land could be cleared, and how quickly such clearing could be accomplished. Difficulty promoted care. Not for nothing was the plowman deemed exemplary who would work *to till the land well, and crop it rightly* so as to maintain fertility. Even later, after much of the countryside was cleared, European cultivation continued to be one of limits. Prior to the discovery of the New World, with its seemingly endless possibilities for expansion, Europeans understood that to support the growing population of their continent they necessarily had to maintain the fertility of their soils.

Once the initial fertility of the soil is lost, it is difficult to regain, but agricultural practices that maintain fertility are extremely labor-intensive—especially without the aid of modern fertilizers and machines—and in the colonial world would have involved the care of the dung heap, the spreading of manure, the planning that goes into crop rotation. In the New World, with so much available land to carve into—and to be ordered—if after a few years' time the vitality of the soil diminished, sometimes it was easier for a settler to cut a new field than to take time to manage the old one, to worry about fertilizing and rotating crops. The abandoned field, without trees or cultivation to hold it, would be subject to erosion. Once topsoil, the product of millions of years of weather and decay and buildup, was lost, it could not be recovered. Land no longer productive would be abandoned to the wild once again.

As the distinction between the old ideas of Europe and new realities of America grew more pronounced, the distinctiveness of the American farmer became something increasingly articulated, and increasingly complex, an argument with itself, with the wild, with the Europe left behind. In the later eighteenth century in J. Hector St. John de Crèvecœur's *Letters from an American Farmer*, his Farmer James would write to a French correspondent suggesting the first clearers of land were one particular kind of American: *He, who would wish to see America in its proper light, and to have a true idea of its feeble beginnings and barbarous rudiments, must visit our extended line of frontiers, where the last settlers dwell, and where he may see the first labours of settlement, the mode of clearing the earth, in all their different appearances . . . By living in or near the woods, their actions are regulated by the wildness of the neighbourhood. The deer often come to eat their grain, the wolves to destroy their sheep, the bears to kill their hogs, the foxes to catch their poultry. This surrounding hostility immediately puts the gun into their hands . . . Once hunters, farewel to the plough. The chase renders them ferocious, gloomy, and unsocial . . . These new manners, being grafted on old stock, produce a strange sort of lawless profligacy, the impressions of which are indelible. The manners of the Indian natives are respectable compared with this European medley . . . Thus are our first steps trodden, thus are our first trees felled, in general, by the most vicious of our people.*

Those who farmed in the settlers' wake were another kind of American, Crèvecœur believed, at once closer in kind to the cultivators of the squarely cleared fields of Europe, but distinguished from the European farmer by dint of possession: *and thus the path is opened for the arrival of a second and better class, the true American freeholders; the most respectable set of people in this part of the world: respectable for their industry, their happy independence, the great share of freedom they possess . . . Here nature opens her broad lap to receive the perpetual accession of new comers, and to supply them with food . . . Here we have, in some measure, regained the ancient dignity of our species; our laws are simple and just; we are a race of cultivators; our cultivation is unrestrained, and therefore everything is prosperous and flourishing . . . The instant I enter on my own land, the bright idea of property, of exclusive right of independence, exalt my mind. Precious soil, I say to myself . . . What should we American farmers be without the distinct possession of that soil?*

But cultivation's hold is always tenuous. The sense of order and safety it imparts will change if you turn your back on it: the brush grows in, the night comes on, old fears crowd you. It's a skittering truth—if it is a truth at all—and in danger of unimagined outside forces taking it away. Even the seeming certain world of Crèvecœur's true freeholder could be turned on its head by a change of fortune. By his twelfth and final letter, the American Revolution had begun, and Farmer James was writing out of a war-torn world. *I wish for a change of place*, his last letter begins, *the hour is come at last that I must fly from my house and abandon my farm! But what course shall I steer, inclosed as I am? . . . The property of farmers is not like that of merchants . . . my former respect, my former attachment vanishes with my safety . . . since I have ceased to consider myself as a member of the ancient state, now convulsed, I willingly descend into an inferior one. I will revert into a state approaching nearer to that of nature . . .* In the midst of war's destruction and its necessary realignments, he allied himself with a life lifted out of an old order, perhaps as only one without experience of the wild could dream and mythologize it: *Thus shall we metamorphose ourselves, from neat, decent, opulent, planters, surrounded with every conveniency which our external labour and internal industry could give, into a still simpler people divested of every thing beside hope, food, and the raiment of the woods: abandoning the large framed house, to dwell under the wigwham; and the feather-bed, to lie on the mat or bear's skin. There shall we sleep undisturbed by frightful dreams and apprehensions; rest and peace of mind will make us the most ample amends for what we shall leave behind.*

PERHAPS NO ONE more than Thomas Jefferson was to keep turning the idea of the cultivator over in his mind, refining it and redefining it as the nation changed and expanded. Jefferson understood the farmer to be the true citizen of the Republic: *Those who labour in the earth are the chosen people of God, if ever he had a chosen people, whose breasts he has made his peculiar deposit for substantial and genuine virtue. It is the focus in which he keeps alive that sacred fire, which otherwise might escape from the face of the earth.* And possession of the land was essential to the character of the farmer. He would exhort the Cherokee: *You are becoming farmers, learning the use of the plough and the hoe, enclosing your grounds and employing that labor in their cultivation which you formerly employed in hunting and in war . . . When a man has property, earned by his own labor, he will not like another to come and take it from him . . .*

As Jefferson had imagined cultivators, they would not be subsistence farmers but a liberally educated people free of excessive needs and wants who would make their clothing from their own wool and flax, who'd produce their own food and raise a surplus for export, so that there would be no need to depend on manufacturing for wealth. By encouraging a society of farmers, he had hopes of avoiding what he saw as the fate of Europe: nations living down through time, and hampered by decay, whose countrysides had suffered from farmers striking their hoes into rented fields, and others turning away from the land to work in factories. He writes from Nice, France: *I have been pleased to find among the people a less degree of physical misery than I had expected. They are generally well clothed, and have plenty of food, not animal indeed, but vegetable, which is as wholesome. Perhaps they are over worked, the excess of the rent required by the landlord, obliging them to too many hours of labor, in order to produce that, and wherewith to feed and clothe themselves. The soil of Champagne and Burgundy I have found more universally good than I had expected, and as I could not help making a comparison with England, I found that comparison more unfavorable to the latter than is generally admitted. The soil, the climate, and the production are superior to those of England and the husbandry as good, except one point; that of manure . . . Here the leases are either during pleasure, or for three, six, or nine years, which does not give the farmer time to repay himself for the expensive operation of well manuring, and therefore, he manures ill, or not at all.*

Jefferson himself may have struggled with debt, with the erosion of his Virginia soils, and failing fertility. Like all farmers, he was at the mercy of the soil and weather, but also like all farmers nearly in the same breath he would voice satisfaction with the first ripening crops:

May 4. *the blue ridge of mountains covered with snow.*

5. *a frost which destroyed almost every thing, it killed the wheat, rye, corn, many tobacco plants, and even large saplings, the leaves of the trees were entirely killed, all the shoots of vines, at Monticello near half the fruit of every kind was killed; and before this no instance had ever occurred of any fruit killed here by the frost . . .*

14. *cherries ripe.*

16. *first dish of pease from earliest patch.*

26. *a second patch of peas come to table.*

Still, his was a philosophy of farming laid out by a statesman who had traveled the world searching for varieties of seed that he could breed and crossbreed, who had slaves to till his soil and international acquaintances to correspond with about the idea of the cultivator. His desire was sometimes freed from necessity. I can't read his *Farm and Garden Books* with one after another entry of exotics—vines from Burgundy and the Cape of Good Hope, African early peas, pumpkins from South America and Malta, Italian peaches from Mazzei, sweet almonds from Cadiz, pear cuttings from Gallipolis—without thinking of the spare accountings my father left behind, which I'd found in the days after he'd died. My father's farm book advises: *An annual Farm Inventory will . . . show your Net Worth above all debts . . . show whether or not you are getting ahead financially and by how much.* In his first year of running the farm everything was assiduously recorded: the 1928 Dodge and the 1926 Dodge, the manure spreaders and harrows and cultivators, the outbuildings. Every cow, horse, and hen was accounted for. In his earnest accountings farming seems to stand apart from all philosophies and ideas. It is the names of the herd that strike me—the plainness of the names, the absence of flourish: *Cattle: Brown Swiss, Little White, Big White, Mule Ears, Long Legs, Old Jersey, Little Pure Bred, Big Pure Bred, ½ sister Hybred, Wisconsin Pure Bred, Horned Gurnsey . . . ; Calves: 3; Bulls: Alec; Horses, work: Big Dan, Little Dan . . .*

You can't always explain how or why an aperture opens, how you fall into a larger seeing, but reading those names, in his hand, after he had gone, was the first time I believe I truly understood how far he'd traveled in his life, and how much I had taken everything—most of all a feeling of security—for granted. I felt a little shame: all of a sudden I saw the deep hard work of it, the aspiration to be gotten bit by bit. I saw the luxury of my own dreaming mind. Maybe I'd been so late in coming to that understanding because the father I'd always known had gained some leeway—he'd been able to extend the farm by buying some of the surrounding land—and with that leeway came a sense of security, and of dominion. He had tried growing a fig tree, walnuts, champagne grapes. The same land plowed for so long out of necessity may have been working its way toward desire. Still, if you were to ask him why he farmed, he'd simply say he liked to see things grow.

For Jefferson, if the cultivators of the earth were to be eternal and the country was to grow, there would always have to be a frontier, new lands where new farmers could stream beyond original settlements and keep pushing across space, up against the Appalachians, then beyond into the backlands. If there was always a frontier, then the country could always be young with fencing and cropland and cattle defining a western border again and again, each place made by one man and one woman, again pulling up stumps and plowing earth, now with their forty acres—for forty acres was by then considered the necessary amount of land to support a family—their cultivated rows and their fenced-in cattle, the geese, the ducks, the hens, their bleaching fields, and spinning wheels. *I think our governments will remain virtuous for many centuries; as long as they are chiefly agricultural; and this will be as long as there shall be vacant lands in any part of America,* Jefferson wrote. *When they get piled upon one another in large cities, as in Europe, they will become corrupt as in Europe.* A nation of farmers would keep manufacturing at bay, but the new territory they moved into would contain the troubles of the nation: Jefferson would not oppose the expansion of slavery into the new territories, or the injustices of the acquisition of new territories through skirmishes and war. *There are but two means of acquiring the native title,* he wrote. *First, war; for even war may, sometimes, give a just title. Second, contracts or treaty.*

The heart of the New World had its own name: *prairie.* The long open center of the word conjures an endless, calm, undefined expanse, which would have stood in contrast to Jefferson's Monticello, bounded by far mountains and framed by the ascending wild: *Of prospect I have a rich profusion and offering itself at every point of the compass,* Jefferson wrote of his home. *Mountains distant & near, smooth & shaggy, single & in ridges, a little river hiding itself among the hills so as to shew in lagoons only, cultivated grounds under the eye and two small villages. To prevent a satiety of this is the principal difficulty. It may be successively offered, & in different proportions through vistas, or which will be better, between thickets so disposed as to serve as vistas, with the advantage of shifting the scenes as you advance on your way.* But to play out his dream for a whole country, to democratize it, what could be more perfect than one enormous field to be divided into individual great lots?

The first prairie the settlers encountered was the tallgrass prairie, and what it contained can seem countless: cordgrass, switchgrass, Junegrass, wild rye, milkweed, musk thistle, blue false indigo, sagewort, speedwell, sedge, sumac, watercress, yarrow, wild parsley, plum, wild strawberry, white prairie clover, wild onion. But it was defined by big bluestem, *Andropogon gerardii*, which in favorable conditions grows to almost twice human height, and can endure fire, drought, and extensive grazing since most of its mass is below ground: a mat of roots sometimes twelve feet deep, thickening into a deep and rich sod so dense other species have a difficult time thriving among it. It spreads by rhizomes, and needn't depend on scattering its seed to reproduce. After the dust bowl, after the rains finally came, it was big bluestem that recovered first. When you were in it, it was said that the only direction you could see was straight up. To find grazing cattle you had to stand up in the saddle and look for movement in the grass. When weather came over the mountains from the west it brought lightning-stoked fires, and once a fire started on the prairie, the carpetlike growth ensured it would burn for a long time. One pioneer noted: *The last 12 miles we travelled after sundown and by fire light over the Prairie, it being on fire. This was the grandest scene I ever saw . . . We had a view at one time from one to 5 miles of fire in a streak, burning from 2 to 6 feet high. In high grass it sometimes burns 30 feet high, if driven by fierce winds. By the light of this fire we could read fine print for a ½ mile or more . . . Till I saw this, I could never understand one part of the scripture. The cloud which overspread the camp of Israel and kept off the rays of the sun by day, was a* pillar of fire by night. *It was literally so with the smoke which rose from these fires.*

The space the prairie made, and the wide sky above it, was its own wilderness and it contained its own fear. It has been said: *The children of the American Revolution hesitated forty years on the western edges of the forest because they didn't trust the grasslands.* Once you entered it, how were you to mark your place? How were you to get across it? *When I saw a settler's child tripping out of home bounds, I had a feeling that it would never get back again*, one woman on the prairie was to say. *It looked like putting out to Lake Michigan in a canoe.* The extent of its complexity was something Herman Melville understood: in *Moby-Dick*, in a chapter he calls "The Prairie," the inscrutable brow of his leviathan turned in his mind to resemble that inland sea . . . *for you see no one point precisely; not one distinct feature is revealed . . . nothing but that one broad firmament of a forehead, pleated with riddles; dumbly lowering with the doom of boats, and ships, and men . . . his great genius is declared in his doing nothing particular to prove it. It is moreover declared in his pyramidical silence . . . I but put that brow before you. Read it if you can.*

In very early ways of measuring land, length and width were attached to its agricultural value: *a perch of poor soil was longer than one of fertile soil—but in the course of the sixteenth century it became standardized at 16½ feet. This inconvenient length was derived from the area of agricultural land that could be worked by one person in a day—hence the variability. The area was reckoned to be 2 perches by 2 perches (33 feet by 33 feet). Thus a daywork amounted to 4 square perches. Conveniently, there were 40 dayworks in an acre, the area that could be worked by a team of oxen in a day . . .*

By the time the East was settled, the area of an acre was standardized, but as one town sprouted another, natural advantages and obstructions often decided the limits of a farm. Holdings were not necessarily of any uniform shape or size. The boundaries of some Massachusetts plots laid out in the seventeenth century fray into uncertainties: *Northward lys the lott of Thomas Woodford beinge twelve [rods] broade and all the marish before it to ye uplande. Next the lott of Thomas Woodford lys the lott of Thomas Ufford beinge fourteene rod broade and all the marish before it to ye uplande.* Boundaries that still come down in the deeds—by the pin in the white oak, to the stake in the ground, to the corner of a wall—confound surveyors to this day. The oak is likely gone, the stone wall tumbled.

The prairie's enormity had been arrived at in an age of reason. It was mapped and marked as a grid—an idea promulgated by Jefferson—that took no exception to swales and hilltops, dry spots and stony places, different soil types and contours, which were everywhere on the prairie, however even it appears now that it is largely gone except in the imagination. Axmen and chainmen methodically laid out the territory: *At the end of 22 yards, a tally peg was inserted, the rear chainman came up, and the process was repeated. Ten chains made a furlong, 80 chains made a mile; 480 chains made one side of a township. At each mile they put in a marker post.* They were instructed not to veer from the grid, no matter what they encountered, and so divided the backlands and the prairie, the world of the buffalo herds and the People of the Wind:

> *6 Chains, 60 links, brook running South 20 degrees West.*
>
> *14 Chains, 40 links, steep narrow ridge nearly 170 feet high, perpendicular. Covered on east side with many bushes and weeds . . . Golden rod, the latter when timely used, and properly applied has been found efficacious in curing the bite of the most venomous Snake. Soil on ridge equal parts of sand and black mould.*
>
> *13 Chains, gradual descent, thicket with trees, the whole of the distance was cut through for the Chain carriers to pass.*
>
> *5 Chains, 63 links, makes two miles.*

One forty-acre parcel might be boggy, the next might lack a ready source of water—settlers were advised to see their land before purchase, and the smart ones did. It was not the best way to create one farm and then another, divorced as it was from any assessment of the agricultural value of the land. It was not the best way to create one community and then another—the grid created no natural settlement center. But it was an idea with a heretofore unknown clarity of demarcation, which still exists: *"When you see how easy it is to use the land survey," declares Lance Bishop, chief of the BLM's geographic services in California, "you have to admire Thomas Jefferson's foresight in choosing a grid. Every parcel of land has an identity. As an example, I've just bought a five acre parcel, and I can go back to the original records and see the shape of the original property, where it was first platted, where the original markers were set, and all subsequent records. It's very clear; there's no ambiguity about what you own."* Such a grid, as one commentator noted, had a purpose far beyond that of providing land for individual farmers and their families: *A uniform, invariable shape that took no account of springs or hills or swamps was an obstacle to efficient agriculture, but to a financier tracking the rise and fall in land values, it was a great convenience. The grid, designed by Thomas Jefferson to create republican farmers, also turned out to be ideal for buying, trading, and speculating.*

At the start settlement still proceeded as it had back east, little by little, along the braided rivers. The prairie might have given the settlers a deep rich sod to cultivate, and saved them the labor of felling trees to create fields, but nearness to watercourses was their only assurance of transport in a world of scarce roads, and the woodlands that ribboned the waterways provided the lumber for houses, barns, fencing, fuel for cooking fires and heat in winter. The soil they encountered, built up by countless years of grasses growing and dying, could be measured in feet rather than inches, and was thick with organic matter. It clung to the cast-iron plows that had served the eastern farmers, and had to be scraped off—a frustrating and inefficient process until John Deere invented the self-scouring steel plow. Even then, the heft of the work was enormous. A settler might not cut into the big bluestem himself but employ professional "prairie breakers" with their oversized plows. As the settlers along the rivers fought fires and plowed firebreaks and mowed fields, as the bluestem disappeared, the natural fuel sources for prairie fires disappeared. Once the fires stopped—fires that had always kept woodlands from establishing themselves on the same ground—saplings reappeared on lands that weren't plowed or used for pasture. Farmers let them grow, and the tallgrass prairie began to settle into a pattern of cultivated fields, pasture, and small woodlots.

EVEN JEFFERSON UNDERSTOOD that his dream had *peopled the Western States by silently breaking up those on the Atlantic.* The rumor of richer prairie sod to the west did spur the abandonment of New England farms, but additional factors contributed to the depopulation of the countryside. By the late eighteenth century not only had careless farming practices drawn down the fertility on formerly rich land, Massachusetts by then lacked enough tillable soil for agriculture to support its population: farming had overextended itself into rougher country with lower natural fertility, which produced marginal crop yields and only a meager living for its inhabitants. These poorer outlying farms would be among the first to be given up on, yet even those brief clearings retain a presence on the land to this day. Climb any of the hills on the low ranges or the scattered monadnocks of central New England rising just a few thousand feet above sea level, and after you've walked through beech forest and hemlock and lifted yourself up ledges, you just might walk into a scrubby clearing where walls mark the boundaries of old sheep meadows. The footpath through the pasture will be trodden down to bedrock, and the soil alongside the path supports only the cold, hardy life of mosses and lichen. Sheep haven't grazed there for over a century, but the weather is harsh enough so that large trees haven't been able to reestablish themselves. Every once in a while you can detect a certain amount of care in the way the stones have been laid for a wall, and you get an idea of what had been invested in such wild country. As you look out on the vast forested hills of your own century, it's hard to believe anyone would think to farm there, or that brief, quiet efforts could have rung such a persistent change.

Those who left the farms not only went west, they went to the industrial cities that were quickly rising up along the larger rivers in the Northeast as manufacturing became an increasing necessity. Industry was a means to prevent the further exodus of the population from the eastern states, but it was also deemed essential for the further development of the country. In the wake of the War of 1812, as cheap British goods threatened to flood the American markets and undermine fledgling industries, even Thomas Jefferson capitulated to the larger necessity of manufacturing. *We must now place the manufacturer by the side of the agriculturist,* he writes in 1816. *Shall we make our own comforts, or go without them, at the will of a foreign nation? He, therefore, who is now against domestic manufacture, must be for reducing us either to dependence on that foreign nation, or to be clothed in skins, and to live like wild beasts in dens and caverns. I am not one of these; experience has taught me that manufactures are now as necessary to our independence as to our comfort . . .*

In our own lower Merrimack Valley, one of the earliest areas of the country to be industrialized, textile cities—Lowell, Manchester, Haverhill, and Lawrence—sprouted along every substantial drop in the river. Alongside those cities, some places in the valley where the soils are rich enough—or the farmers tenacious enough—have held on to agriculture; a few farms even now have been in the same family for almost four hundred years. But on other farms, the years had been discouraging enough for families to consider change, and they looked to the cities for those changes. As the cities grew, not only by pulling people away from the farms but by attracting new immigrants from abroad, they also created new markets for farmers, and some of the outlying farms, rather than being abandoned, were sold to recent immigrants from southern and eastern Europe and the Middle East.

To take on farmland where others had already come and

gone, to gain your ground there, is its own kind of succession. The rain and sun and shadow that decay one world feed the other. Before my grandparents bought our farm in 1901, it saw several owners. Richardson sold to Packard, Packard sold to my grandfather, who bought not only the buildings and land but the horses and hens and chickens, the carts and feed, the hoes and harrows. I can imagine one life stood in for another in the nature of the works and days, that the plowing and planting required to keep the place going didn't change much from owner to owner.

For my grandparents, possession of a farm meant they would have greater freedom and control over their lives than the city offered, but it also meant they hadn't the ease of living among other Middle Eastern immigrants. How insular they must have felt, and how peculiar their lives must have seemed to the neighboring farmers, the eggplant and chickpeas, spiced lamb, pine nuts, dried figs, and pistachios at their table, the guttural tongue spoken among the white clapboard barns, the pastures, and orchards. My grandparents would have been compelled to speed up their efforts at assimilation: their children were never to speak more than a halting Arabic. No matter their effort to blend in, the natural suspicion of their foreignness would have been exacerbated by the knowledge that their presence underscored the fading away of a particular world, one that had been the sole province of the Yankee farmer. That suspicion is hinted at in a few veiled stories of the family past, something about a neighbor continually trying to expose my grandfather's hard cider enterprise.

I have no doubt the insularity contributed to the great

sense of fidelity among them, but my father and his brothers and sisters always chose to remember to us how one world grew into the other, how they sat down with their neighbors at box parties and danced with them at the Grange Hall, how out of the blue, a Yankee neighbor was to offer a loan to one of my uncles to attend college. Sometimes I like to imagine that within this mixture of stability and instability there is a strange turning of Jefferson's dream. In his travels in Europe he had once imagined how the southern European farmers would be of advantage to the American soil: *Emigrants too from the Mediterranean would be of much more value to our country in particular . . . They bring with them a skill in agriculture and other arts better adapted to our climate. I believe that had our country been peopled thence we should now have been farther advanced in rearing the several things which our country is capable of producing.*

"JUST LIKE YOUR father and his father," people would say once my brother had grown old enough to have his own ideas about things to be done. As long as he remained young—the boy on the tractor, the kind you see all the time—a twelve- or thirteen-year-old—puffed out with the responsibility in one moment, worn with the weight of it in another, older than his peers by virtue of work—for as long as he remained that boy, I imagine there had been more dream than difficulty. As he grew into his teenage years, when my father would call to my brother repeatedly to get up for work on early summer mornings, my brother willfully slept on. Some days they'd sit beside each other at lunch and again at dinner, speaking to one another only to disagree. It may be a trick of memory but their arguments seemed to dominate the household then. The rest of us receded in their wake.

As my brother became a grown man and expressed more fully his own desires about building a modern greenhouse, and then a larger farm stand, the tensions between them grew even more pronounced. The drugs he used withered everyone's trust, so that even his modest, sensible ideas—"It's not worth keeping the orchard, it's been losing money for years," he'd often say—were greeted with averted eyes and skepticism. Sometimes when he entered a room he'd walk into silence.

Whatever the situation of the land, there is always the lay of the family as well, which has as much to do with the success of farming as anything else. Farm life intensifies the demands on a family—I know it did in ours. Living with those arguments had to color my concept of the farm, and I don't wonder that in the conversations among farmers I overheard at trade shows and yearly meetings, amid the talk of sales and row covers, I seemed always to pick up on an undercurrent of conflict. When I heard an old farmer ponder whether it was time to give the farm over to a son who by then was in his sixties, I could recognize something darkly humorous about it—I retell it, and it always raises a chuckle—but I also felt sorrow. I could sense the long years of forbearance within the family.

Though I haven't been to a trade show in a long time,

old voices still swirl in my head as if I'd heard them just yesterday: "He was the oldest, he drank the wine, tasted the food, then they all ate. He's got Parkinson's—advanced. He'd just go out into the field and stand there until it got too bad, and then not at all. The kid comes back and wants to do it the new way. He just wanted to be consulted." "The boys will never keep the place together. You'll never have the nicer things in life by farming and there's always one who wants more." "I worked beside him for forty years, I changed the diapers on him, and then he just leases it out without even talking to me."

Upon my return to the farm, having witnessed for years those disagreements only from a distance, I suppose I thought that my helping out could alleviate some of the tensions that had made our world feel insular and fragile. I didn't fully comprehend how intertwined the relationship between my father and brother truly was.

"I don't know what to do," my father said to me one day in a moment when he was bewildered by my brother's intransigence.

"Maybe he should go his own way," I inched, thinking how it would release them both, release everyone.

"Cruel, aren't you?" he said quietly, and I half believed him. I could see then that there was no changing the direction of things, and perhaps that understanding was part, though never all, of the reason I would only allow myself to invest so much in that summer work. Even if I had wanted the farm to be my life, I knew that no matter how much I worked, a future—a real future—was unimaginable. It is a strange thing now to realize however much their relationship kept the rest of us at bay, it is likely also true that had my brother not stayed on my father could not have run the farm as a business into his eighties, and it would not exist at all.

I'd been back working for a few growing seasons when I saw the Midwest for the first time. I'd decided as the year wound down to its stillness in the spare world after the frost, and my exhaustion gradually lifted, that I'd take advantage of the winter to get away. During those first years back, in addition to being confounded by the stressful relationship between my father and brother, I was more self-conscious about my return than I am now—as if I was going against the grain—and I was haunted by the thought that I might be cutting off possibilities or making myself smaller, so to strike out for friends in Montana and then Utah on a crisp January dawn seemed freeing. I remember how much I anticipated the trip, was buoyed by the idea of the West, of the far distances between my friends, even as I worried about the snows I might encounter on my way there. "They won't let you through the passes without chains," friends would say to me. "Make sure you have blankets in the car in case you break down." "You always need to be watching the weather."

I imagined the Midwest as something to simply get through quickly as I slipped south onto I-70 in hopes of avoiding the worst of the snows for travel. As the pines gave way to deciduous patches here and there and the open spaces increased, I began to see more and more farms, large ones, enormous. The original forty-acre holdings that had fed a family and a few others had, over time, become farms that fed thirty others, then fifty, a hundred, one hundred and twenty-nine at last count. For the first time I began to understand ourselves in relation to something larger. Even in their dormancy I could sense the extent of the dream—the size of one field could be larger than our entire place—but what struck me even more was a feeling that there was no room for intimacy in that landscape. The day would glitter on over wide, wintry expanses where the attempts at intimacy were stark and obvious: the farmhouses were obscured behind tall conifers planted as windbreaks. Everything existed in relation to the horizontal. Nothing radiated out from a town—the innate loneliness of the configuration of the grid.

Still, I could sense a community of farming in that world. Every day I would start out in the dark, drive a few hours, then pull off the highway for breakfast. Inevitably I walked into a world of farms. Amid the aroma of coffee and bacon frying, the clattering dishes and steamed-up plate glass, men in peaked caps and overalls sat on stools and in booths, along with grain salesmen and Agway reps, and talked to each other about farming, farming, farming. The crops may have been different from ours, and the scale much larger, but the kind of talk felt assuring and familiar, and made that world seem whole: "The price of soybeans . . ." "The new cultivator does the work in half the time . . ." I'd drive on for another several hundred miles and find them at lunch, one day and the next.

On the winter morning I began to drive through Kansas I'd started out in the dark as usual, expecting nothing more than a long day across a flat state and meals among farmers. But as the dawn lightened the sky I could sense how the land was rising up on either side of me, and I found myself growing attentive to some shouldering presence. In a little while I saw rime everywhere on hills of dried, stunted winter bluestem. There was no height to hold me—the grass had gone to its reserves—but the swell and the space held me, and the color of the grasses, which was as tawny as the summer pelt of a fawn. I'd come upon a vestige of the tallgrass prairie—and a vestige is all that remains of it, along railbeds, in old cemeteries, on substantial hills. Later on I was to understand that I had arrived in the Flint Hills, which had always been too steep to plow and, with their shallow, rocky soils, had been too poor to cultivate. Cattle may have replaced buffalo there as the tallgrass prairie became tallgrass pasture, but the dream of the grid meant little to it, and while elsewhere the grasses were plowed under, and the soil was planted in row crops, and the prairie became known as the corn belt, the big bluestem on those hills continued to weave its mat of roots. Now the Flint Hills stand as the last large expanse of tallgrass prairie, and are preserved as a presettlement ecosystem where research scientists measure biotic production, trace the fate of the soils, and examine the effects of fire and grazing on the vegetation. Sometimes they burn grasses methodically in an attempt to replicate and understand the old nature of the land.

As the day came on and the prairie hills rose to their

true immensity, I found myself under a sky that was even wider than the Nantucket sky I'd known years ago, wider than the salt marsh sky. I was on the highway going sixty, sixty-five, pulled forward by my aim but watching sideward. Maybe I was susceptible to it—something so immense and full of forgetting—because I'd been traveling through all that sequestered farmland in the days before. Even as Kansas settled into cultivation again, the prairie wouldn't leave me, and now my memory of those hills has grown outsized, living on in my mind more than the Colorado Rockies and the red rock canyons of Utah I was to see afterward. I'm sure the Flint Hills are the one place I would recognize again, and not only for their sheer singularity—the last wild is not a sea anymore, only an island; I think more and more of what I imagine to be their resilience and patience. I have dreamt of walking into those grasses, of the feeling I would have of open wildness, of my own smallness among something so endless, a world clear to the horizon with no goal other than to be its stark, intricate, immeasurable self.

I have never been able to think of anything so large as the brow of a whale or the doom of ships when I've looked out on our land. But if, within the summer stillness, you stand in the right place among the worn hills and the leafed-out hampered branches of our orchard trees, the earth can seem to hold as much coiled energy as a cresting wave or a van Gogh landscape. And sometimes within the confines of an old meadow full of short grasses and rocky outcrops, I have imagined the emerging granites to be the backs of small mammals breaching the swells of a watery surface. Our houses are small craft upon it, but there's no place for the *Pequod* running under full sail, or the *Rachel* appearing on the horizon of a miraculous abeyance: *The unharming sharks, they glided by as if with padlocks on their mouths; the savage sea-hawks sailed with sheathed beaks . . .*

3
The New City

IT COULD CHANGE—the way memory is always at risk of changing, being one insight away from a recollecting of the self—but I feel a resolution when I remember my father. An ease had grown between us, one we never would have had were it not for having worked together in his late years. That ease stemmed in part from mutually dependent work but also from a lightness of expectations. We had no history: our dealings were simpler and freer than the longstanding ones my father had with my brother, caught up as they were with the expectations of inheritance, from which I'd always been separate.

When I worked on the farm as a child my chores were partial, ones I shared with my sister, my girl cousins, the neighbor who'd crashed his motorcycle at the bend in the road and had a plate in his head: picking strawberries, picking beans, three- or four-hour stints selling corn and tomatoes. The farm was at the heart of all the talk during summer lunches then. My father and Pete, his right-hand man—and later, my brothers—filled the air with discussions about what needed to be done: "You might as well plow down the Salem Road piece." "The rye will be coming tomorrow." "The Macs are reddening up pretty nice." "I should put some water on the corn." I couldn't place my voice within the conversation, and my mother's voice was absent, too.

After I returned, as I ran the farm store and saw to the wholesale orders and kept accounts—work I came to enjoy for its straightforwardness—I thought of the time as an interim. I carried with me the notion that my future ran underground all through the long summers, that it resided in the wintertime when I'd return to writing again. Even so, I took to the work wholeheartedly, and when I'd sometimes join my parents for lunch on a summer day, my father and I would naturally talk about what had to be done: "We'll need more lettuce and radishes by morning." "The corn sales are starting to quiet down." "The Huntleys want some tomatoes. I could drive them over tomorrow afternoon if there's a free truck." Necessity made a world of its own, running on at its certain pace. There was always so much to do, and time, time, time, you were always working against time, and the heat, and the rain. Each day was one hurry after another, and I always felt I didn't have a moment to think about anything else. My mother would be as quiet as she'd always been, setting down a platter of ham, dishes of steaming corn or baked tomatoes, olives, cheese, bread. One day I was helping to clear the dishes—I'm sure I was abstracted, already thinking ahead to the work of the afternoon—and as I set them in the sink and began to head on my way, she turned to me and said: "It's always about the farm, isn't it?"

It was, and I have never thought of my mother as having any attachment to the work of it. She'd been raised in the city of Lawrence, six miles to the east, and had lived on the third floor of a tenement until she married my father in the early 1950s. The world he made for her was almost as distinct from the farm as that of a suburban housewife. The Cape Cod Colonial he had built is down the road from the original farmhouse and is distinguished from the fields and orchards by a trimmed lawn in front and back. There are no barns, outbuildings, or add-ons anywhere near it. Every once in a while there'd be a tractor parked in the drive when my father or brother had left off plowing and come right in to lunch, but no one passing would think a farmer lived there. Some farmwives up and down the road helped with haying and planting—it wasn't unusual—but my father had never asked my mother to be part of that world, and I think it was a mark of pride with him that he could keep her apart from it. He brought the farm to the door. Eggplant, tight and gleaming, sat in baskets on the porch. Squash, tomatoes, peppers. Late in August he left her corn by the bushel, which she would freeze for the winter. In September, butternut squash, Macs for applesauce, Cortlands for pies, and after Cortlands, Northern Spies.

Even the old farmhouse, I can see, had separated itself from the working farm by then. When my aunts and uncles were growing up, work moved easily in and out of the house, everyone was involved. As at every farm along the road, men and boys chopped wood and hitched up the teams just beyond the kitchen doorway. Women cut seed potatoes on the stoop, or plucked chickens. Children milked cows, fed chickens, gathered eggs, hauled water from the well. Bertha and Del, the oldest, also saw to their brothers. I've heard it said of Bertha more than once: "She was like a mother to them . . ."

The summer kitchen of the farmhouse eventually was turned into a sitting room. The barn, silo, and windmill: gone. These days a tractor or two will be stored in the carriage house, but mostly it's a place filled with disused tools—scythes, rusted collars, broken hoes, ropes. The chain of buildings attached to the house is like a vestigial limb, a phantom of old use. Part of the separation of house from farmland was inevitable, a consequence of modernization. As wood fuel was replaced by oil, and kerosene by electricity, as salt and ice gave way to refrigeration, the farm became more integrated with the world at large, and more dependent on it for its energy and food. The household required less of the immediate world.

When the farm became my father's to work, as his brothers moved on to their own lives and occupations, as chickens and cows were sold, the division between house and land became even more distinct. My uncle Joe continued to tinker in the shed and to keep his own garden plot to the side—there's still some straggling asparagus and rhubarb that sprout up spring after spring—and Bertha continued to walk through the fields and woods, but the farm was fully a business in their later years, and had grown into a definition beyond them. The work of it was done by men and boys outside the family, while the farmhouse remained the repository of family life and memory. Late in their lives Bertha, Del, and Joe sat on the porch in their green rocking chairs and watched three seasons pass. Their yard was swept neat, and the irises along the old hitching post bloomed and faded in late spring. The farm they knew continued in their imagination: "Pop would tell me to go get a chicken that had stopped laying. 'Are you sure?' he'd ask. 'I'm sure,' I said. Boy, did I catch hell when he pulled out that string of eggs . . ." "I used to carry the milk over to the well. Imagine that—trusting me with the milk—I was nine years old!"

There is a small oil painting I love, *Florentine Villas* by Paul Klee. On first seeing it from a distance, I could only make out muted blurry patches of color, reds and earth tones that appeared sun-bleached and washed so you'd think they had been stitched together out of old cloth. As you peer closer you see that the patches are washes of paint patterned by Klee's spidery, scratchy marks, so simply delicate and certain they seem to hark back to the drawings on the caves at Lascaux. There are no people in the picture, but the human imprint is patterned everywhere on those washed squares. Some of the patches represent homes, others mark row crops, vines, stairwells, roads. Only in looking more closely can you distinguish the cultivated from the built. The vines and crops are softer human marks than the stairs, walls, and roads, but there isn't much distinction in the order between outdoors and in. The wild with its different idea is somewhere beyond. Where is it we live? I remember thinking as I looked at the households and the care spreading out from them, and how integral they seemed, each to each.

While many of the things of the old work life still hang in the carriage house—moldering husks disintegrating into themselves—after Bertha died we carefully went through the household, all of us, and settled out things and carried them away to our own houses for remembrance: hats and coats, letters, books, lamps, plates, sifters, china pitchers, mixing bowls. It took so much time. We'd unearth one thing and then another, and each had memories and stories attached to it. What I remember most clearly, though, is the late-February evening half a year after Bertha's death when all of us nieces met to divide our aunts' jewelry. More than seventy pieces had been found in the boxes in the upper drawers of the dressers, in the safe set in the old fireplace, in all the lightless sequester of the house. "Christ, where did it all come from?" someone said. Time had worn away a little of the luster of the metals, a little of the shine of the stones. The cardboard boxes were brown, and the plump velvet cushions within the velvet boxes were threadbare.

After a dinner of fish stew and a salad of baby greens we cleared the table and sat down again to draw our lots out of a ceramic bowl. One of my cousins had assigned a number to all seventy pieces, the gold bracelet and gold cross, the diamond rings as well as the imitation pearls and the plastic necklaces. Which was Bertha's, which Del's, we didn't always know. None of us remembered them having worn much more than a brooch on a lamb's-wool sweater, or a necklace and matching earrings to a formal gathering. A few of the pieces were fine things cherished and deliberately handed down from their own mother: a wedding ring, a watch, and several hammered-gold bracelets. Most had been gifts from their brothers, given at Christmas and birthdays, or as souvenirs from trips. There were cameos from Italy, one who loved the sea gave scrimshaw, the one who went to Hawaii brought back shell beads.

We talked on about the trips we'd taken—"Provence! You're so lucky"—and Boston restaurants, careers, children. "We can open the floor to trades afterward," one of my cousins sang out, and we joked about a gaudy coral piece: "You could get away with it on a black coat." We said we'd fix the watches, adjust the rings, and polish up some of the metals. "Oh, you got a good one," my cousin in charge said as she set down a gold cross in front of me. "Oh yes," another affirmed. I stared down at it, not knowing what to do, or how to feel. It was one of the oldest pieces, which everyone had seen in an old photograph hanging around a teenage Del's neck on a fine chain that brought it to rest in the center of her breast. She was wearing a bowed and flounced white dress and had one firm arm around a shy, inward Bertha—dressed the same—and another around two of her baby brothers, as protective and determined as I'd ever seen her. I couldn't imagine ever wearing the cross, not so anyone could see it anyway, not with the force of its convictions. "And a not so good one," my cousin continued as she set down a choker made of false blue pearls.

Everyone laughed and I let myself be drawn back into the laughter. We held the pins up to our jerseys and lifted the earrings to the sides of our faces until all the numbers had been drawn and all pieces distributed, and we each had before us a little glittering mound built of paste and pearls and gold and plate.

I THINK I can trace a faint line, scratchy yet certain as Klee's, that separates farmhouses from farmland. It extends back at least to the nineteenth century, when New England girls followed the roads down from rocky, marginal Vermont and New Hampshire farms to meet industry in the first textile cities along the rivers. These young women were the first to lose their place within the farm families. During the first centuries of settlement they'd been a steady part of the household economy, bringing in added income as they'd spun and woven wool and flax into cloth for the family not only to wear but to sell for additional income. They'd also made cheese, butter, and brooms, tended the family garden, helped their mothers with the demands of cooking and cleaning, raising children, caring for the sick. With the advent of the textile mills, even rural families found it was cheaper to buy cloth than to rely on homespun, and in the stores their farm-made goods could no longer compete. The textile mills offered a small income to the girls, and work there meant one less mouth to feed at home. The mill wasn't to provide a constant place for them, and most never envisioned it as their entire future: they worked to build up a dowry, or while awaiting marriage. Some worked through the winter and returned to the farm in summer.

Their early letters home bespeak a readiness for experience, a determination and acceptance, often an appealing openness. *I get along very well with my work. I can doff as fast as any girl in our room,* one young woman wrote home. *I think I shall have frames before long. The usual time allowed for learning is six months but I think I shall have frames before I have been in three as I get along so fast. I think that the factory is the best place for me . . .* And in a subsequent letter: *You wanted to know what I am doing. I am at work in a spinning room and tending four sides of a warp which is one girls work. The overseer tells me that he never had a girl get along better than I do and that he will do the best he can by me. I stand it well, though they tell me I am growing very poor. I was paid nine shillings a week last payment and am to have more this one though we have been out considerable for backwater which will take off a good deal . . .*

They might see the city in a way that was not imagined in the larger discourses, with a softer division in the world: *I arrived here safe and sound, after being well jolted over the rocks and hills of New Hampshire; and when (it was then evening) a gentleman in the stage first pointed out Lowell to me, with its lights twinkling through the gloom, I could think of nothing but Passampscot swamp when brilliantly illuminated by "lightning bugs." You, I know, will excuse all my "up-country" phrases, for I have not yet got the rust off . . . You will ask what I have already seen . . . There are stores filled with beautiful goods upon either side, and some handsome public buildings . . . I waited, one day, to see the cars come in from Boston. They moved, as you know, very swiftly but not so much like "a streak of lightning" as I had anticipated. If all country girls are like me their first impressions of a city are far below their previous conceptions, and they think there is more difference than there really is. Little as I know of it now I see that the difference is more apparent than real. There are the same passions at work beneath another surface.*

But as the textile industry matured, the speed of the machines increased, production lines grew more frantic, and working conditions worsened. The young woman who'd believed the factory was the best place for her reported a far more sobering story just three years later: *It is* very *hard indeed and sometimes I think I shall not be able to endure it. I never worked so hard in my life but perhaps I shall get used to it. I shall try hard to do so for there is no other work that I can do unless I spin and that I shall not undertake on any account. I presume you have heard before this that the wages are to be reduced on the 20th of this month . . . The companies pretend they are losing immense sums every* day *and therefore they are obliged to lessen the wages, but this seems perfectly absurd to me for they are constantly making* repairs *and it seems to me that this would not be if there were really any danger of their being obliged to* stop *the mills.*

Eventually most of the young women were to lose their place again as worsening conditions in the factory made the work untenable for them and immigrant laborers took over, but the experience had changed their options. However much more softly they'd seen the divisions in the world, however great their fidelity to home and family, when they married many chose not to marry farmers. Perhaps they'd become accustomed to the busier, more varied world of the city and didn't wish to return to the remote life of their childhoods. But they also knew the demands that world would bring with it: *They contemn the calling of their father,* one nineteenth-century critic wrote, *and will nine times out of ten, marry a mechanic in preference to a farmer. They know that marrying a farmer is a serious business. They remember their worn-out mothers.*

WITH THE COMING of industry the physical change in the countryside was rapid and extreme. Thoreau was witness to one such transformation: *At the time of our voyage Manchester was a village of about two thousand inhabitants, where we landed for a moment to get some cool water, and where an inhabitant told us that he was accustomed to go across the river into Goffstown for his water. But now, after nine years, as I have been told and indeed have witnessed, it contains sixteen thousand inhabitants. From a hill on the road between Goffstown and Hooksett, four miles distant, I have since seen a thunder shower pass over, and the sun break out and shine on a city there, where I had landed nine years before in the fields to get a draught of water; and there was waving the flag of its museum,—where "the only perfect skeleton of a Greenland or river whale in the United States" was to be seen . . .*

Where my mother's native city of Lawrence now stands, there had been scarcely a house on the land in the early eighteenth century, until the skirmishes with the Indian tribes died away and some of the settlers moved from the villages on the outlying hard rock hills to homesteads nearer their plantings. A crossroads settlement was the nearest thing to a village on that plain; there was even less then to mark any kind of centrality than there had been when the Agawam gathered near the drop in the river, which would come to be known as Bodwell's Falls, to fish for shad and salmon. There'd been alewives in the tributaries. The floodplain soil was suitable for corn and beans.

The river's course there was so wide it had been considered an obstacle in the colonial world. Not only was it difficult to cross—in places near Lawrence it was over seven hundred feet wide—it was impossible to make use of, unlike the smaller rivers and streams where a few men could construct a dam for modest power with wooden cribs or branches and stumps. By the early nineteenth century there were still fewer than several hundred settlers in the area, which was kept largely in cornfields, rye fields, orchards, woodlands. The place was sometimes known as Moose Country. History wasn't much more than anecdote, little legends. In exchange for a roll of cloth, it's said, a settler purchased of the Indians the rights for all the lands he could surround in a day's travel through the forest. And there remain some things the explanation of which is lost, if ever there was one. When the town on the south side of the river came to be known as the Plain of Sodom, the town to the north called itself Gomorrah in return.

During the emerging years of the Industrial Revolution, Bodwell's Falls was still seen as a quandary: the river was so wide that if a dam were to be built, it would have to be the largest in the Western world. So, while other places on the river—narrower expanses with greater drops—were dammed, as industry followed the old patterns of settlement north, as the machines sped up and output increased, the world around Bodwell's Falls remained agricultural.

The dam that was eventually designed for the city of Lawrence there represented the long accumulation of architect Charles Storrow's education in Europe and America. At the time of its building in the three years between 1845 and 1848, such a span still had never been successfully crossed by a dam. For the project Storrow studied the structure and defects of timber dams—their tendency to rot, and their inability to withstand severe overtopping. He inspected dams on the Croton River, the Schuylkill, the Kennebec, and studied the designs of English dams. He pored over treatises on retaining walls and mapped equations concerning the effects of the force of water on stone. He found no strong examples to draw on, though his studies helped him to eliminate timber and earthen dams from consideration, and also helped him to structure the masonry dam he eventually designed. The project was so large and expensive he knew he had to get it right on the first try, since it would be impossible to repair. Any broach would spell disaster—financial and otherwise—for the city. It would not only destroy the mills in Lawrence but wreak havoc on the cities downstream.

Storrow designed the city as well, he laid out its streets, factories, and houses. It was built in concert with the dam and seemed to rise up almost overnight. A nineteenth-century eye saw it this way: *For two centuries the river and the land literally ran to waste; but sparsely settled, in productiveness meagrely requiting the tiller's industry, it seemed destined . . . to a career of barrenness and comparative worthlessness, until the splendid water-power caught the eye of the sagacious manufacturer, when a change, rapid as wonderful, came over the scene: the desert waste grew green, active, busy life dispelled the unpleasant silence, and the solitary place forthwith resounded with the cheerful rattle of machinery, the ring of the anvil, the vigorous strokes of the artisan and mechanic, the whirl and bustle of trade, and the constant rush of steadily augmenting throngs where once the few hardy fishermen, along the falls, at the mouth of the Shawsheen and Spicket, captured the magnificent and delicious salmon, the bony shad, and the slimy, squirming eel, a change is wrought sudden and complete.*

The effort was exhaustive: *In all the region in which the city now stands there was no spot where one could escape the din and dust . . . Beginning at the gneiss ledge, situated nearly two miles south, the stone from which composes the river wall and mill foundations—or at North Andover, three miles east, then the depository of bricks and lumber by railway—or at Pelham, some eight miles west, from whence came the granite for the dam—there was an almost endless string of slow plodding teams, loaded to the utmost, all centering from the dam to the Spicket River to deposit their loads. But here were not the only signs of activity. All over the place buildings were rising with most astonishing celerity. For twelve hours a day, the heavy teams, here removing hills, there filling valleys, or loaded with building materials, plodded through the suffocating dust of dry weather, or the almost bottomless mud of the rainy season . . . In 1846 and early 1847 there was a large accession to the population. Mechanics, merchants, physicians and lawyers began to locate here, and order began to rise out of chaos. In 1846 the first religious service was held, and by the following year most of the leading sects were established here. In October 1846, the first newspaper was issued . . .*

The factories built there were larger than any previously constructed, production faster. Bodwell's Falls turned out to be the perfect place to meet the future, as it made possible a city more compact and efficient than any other: *The site had advantages not even enjoyed by Lowell. The banks of the river were high enough to accommodate mill islands parallel to the direction of the Merrimack, making short straight power canals and short tailraces possible. These canals could waste their excess water into small streams which flowed into the Merrimack on either side of the river . . . Where the designers of Lowell found it necessary to build over 8 miles of canals to obtain 10,000 horsepower, at Bodwell's Falls a single canal 1 mile in length delivered approximately the same power.*

The city, known first as the New City, was a mere seven square miles and almost evenly laid out on both sides of the river. Even today, when you look at a map of northeastern Massachusetts you can see how its small round shape sets it apart, how it seems to be a concentrated drop applied atop the broad, irregular towns soldiering up and down the river.

WE TAKE THE nineteenth-century cities of New England to be brick cities. Their resolute stacks and factory walls of red clay are what we see before us, and what artists render when they want to evoke those places. But the scale of those textile cities would not have been possible had men not discovered an efficient way to quarry large blocks of granite, which builders relied on for the infrastructure of the cities. Charles Storrow drew on quarries in Pelham, New Hampshire, and Cohasset, Quincy, and Rockport, Massachusetts, for the stone to build his dam. He used granite for the canal walls and the foundations of factories, and the lintels on row-house windows. The nineteenth century seemed to depend on granite as much as early farmers had depended on salt marsh hay, and a good measure of New England granite came from coastal quarries, from the headlands where the sand shore and salt marshes stand in abeyance.

In the colonial years those headlands remained untouched, primal, like something out of Noah's flood—*far and near we have an image of the pristine earth, the planet in its nakedness*—while colonists used so much fieldstone for cellar walls, well holes, ice houses, and fences that they began to fear they'd run out. The town of Braintree, Massachusetts, in 1715 decreed that *no person shall dig or carry off any stone on the said commons, or undivided lands upon any account whatever without license . . .* What farmer wouldn't laugh mildly at that now, since granite has continued to surface with every frost and thaw, with every turn of the plow and spade, and the rough and rounded stones seem not only endless but integral to the idea old New Englanders have of their stubborn, resistant selves. *I see him there*, wrote Frost of his wall mender:

Bringing a stone grasped firmly by the top
In each hand, like an old-stone savage armed.
He moves in darkness as it seems to me,
Not of woods only and the shade of trees . . .

Yet even in the eighteenth century, men dreamed of the hidden opportunity beneath soil and fieldstone. *As it is from the surface of the ground, which we till, that we have gathered the wealth we possess, the surface of that ground is therefore the only thing that has hitherto been known*, wrote J. Hector St. John de Crèvecœur. *It will require the industry of subsequent ages, the energy of future generations, ere mankind here will have leisure and abilities to penetrate deep, and, in the bowels of this continent, search for the subterranean riches it no doubt contains.*

It's not that the rock could not be split. Granite, though extremely hard, is full of stress, and will give along seams, and crack into stratified beds: you can see that easily along Cape Ann on the north shore of Massachusetts where the sea has worn the rock so smooth it almost flows, and the seams have eroded into something like the loose folds of a mammal's skin. The job is to find the strains and apply the right touch. Ancient Egyptians jammed wood into natural cracks in granite, wet it, and let it expand. In higher latitudes the Finns poured water into fissures and waited for it to freeze. Others burned brushfires on the surface of the stone, and when the heat began to crack it they'd clear the ashes off and strike the weakened place with an iron ball. But such methods were inefficient and rarely yielded even-sized building blocks.

What finally made large-scale quarrying viable was a precise and patient method of splitting stone called hammering, which was first practiced at the beginning of the nineteenth century by German immigrants in the Cape Ann quarries. One man would hold the chisel while two others took turns striking it; a turn of the chisel, and they'd strike again, and again strike and turn, and stop a moment to spoon out the stone dust, then strike and turn again. In this way they tapped straight rows of holes six inches apart, about four inches deep into the stone. They then set shims and wedges, sometimes called feathers and pins, into the holes, and a man would tap into hole after hole along the line. I saw a demonstration once, lifted out of the din of a quarry. The stonecutter seemed to be playing an instrument, hammering one and then another of the wedges up and down the row and back again. Tap, tap, tapping. Then he'd pause a bit and cock his ear to the stone. "You have to listen to your stone," he said. "Let the stone breathe." His solitary effort chimed with the sound of a collateral world, the one that has always existed—particular, domestic, devoted, intent on underpinnings and doorsteps and graves. Tap, tap, tapping. Then the granite cracked quietly, no louder than mild surf falling back over cobbles.

I imagine quarried stone to be its own substance entirely, wrought out of the rising dust of the pits—deep, deeper, deepest—amid a cacophony, granite not just for home truths but perfect for the ambitions of the nineteenth century: hewn and hammered, worked straight and smooth toward an endless desire for dams, columns, bridges, banks, post offices, and dry docks, the eagles of the Boston Customs House, a monument to Myles Standish, paving blocks for the streets of Boston and the streets of Havana. *Quarry*, from the Middle French *quarre*, meaning "cut stone," from the Latin *quadrum*, square. And it did make square a world that had once satisfied itself with cobbles and fieldstone. Cape Ann granite turned out to be the strongest in New England, the same as that which had been hauled out of the quarries of Aswan: mottled green and mottled gray, glinting with quartz, the sapstone turning a pinkish red in time. When it was finished and polished it gleamed the way fieldstone never had, and it seems Emerson was right: every new relation *is* a new word.

Granite, *grano*, *granum*, grainy old photos of the Finns and Swedes and Italians at their workstations. They have the intent eyes of birds, and barrel chests. Behind them stretch canyons of stone where everything is happening at once: men preparing to blast rock, men dwarfed by the derricks they are operating and the dimension stone being hauled up out of the pits. Dust is rising and settling on the shoulders and caps and in the lungs of quarry workers bent to the center of the noise. Scattered around the rims of the pits are wooden sheds and rigged canvas shelters where cutters are shaping blocks or carving decorative pieces and monuments. *You can look through those shed windows when the sun is shining and that old dust is everywhere . . . It used to get so dusty in there you couldn't see the other men.*

Out beyond, between the larger quarries that produced dimension stone, were scattered small one- or two-man quarries where men hammered out paving blocks. Their hands may have been burned, broken, and callused; still, it's said, they could sense in their fingertips where the rock would give most easily. They'd cut and chip one hundred and fifty to two hundred blocks a day, at five cents a block, thirty-two different sizes of pavers before they were through: Manhattan Specials, Washingtons, Philadelphia blocks, Belgian blocks, Cuba blocks, crosswalk stones. *You have to take advantage of your stone . . . You can't be afraid of the stone. If you are, you won't hit it hard enough to get anywhere, but if you hit it too hard, you'll spoil it.*

When the quarries were in full production, pumps ran day and night to keep the flooding at bay as the men worked down and down to get out the stone. The deeper they cut, the fewer seams they encountered in the rock, and the larger the blocks they could haul out, so they might have felt that if they cut deep enough they'd encounter a fractureless absolute. Maybe that's what they were working toward at the height of the quarrying days, when millions of tons of stone were blasted from the headlands, shaped, and shipped—they sold everything but the sound, it was said, even the stone chips and the dust, which gardeners spread around the roots of their fruit trees. *People couldn't imagine life any other way . . .*

BY THE TIME my mother's parents arrived in Lawrence the city had been increasing in density for more than fifty years. The place was never to grow beyond its seven square miles, but as its industrial ambitions continued to increase during the nineteenth century its influence reached even into the North Country. To regulate the level of the river so as to prevent shutdowns and idleness especially in summer, the Essex Company, members of which had founded both Lawrence and Lowell, gained control of water near the source of the Merrimack River, at Newfound Lake, Squam, and Winnipesaukee. They built dams to control the flow. In the low-water season, as they raised the river they flooded the hay fields of farmers a hundred miles north of the city.

Within the city, as the size of the factories and the speed of the machines increased, the labor strife increased, with brief walkouts during the late nineteenth and early twentieth centuries, and major strikes in 1882 and 1912. The scale of the tragedies increased as well. It had been a mark of pride that only one man had lost his life during construction of what was to become known as the Great Stone Dam. The early mill deaths had occurred in ones and twos as in the first years at Lowell, when a young girl would write to her father: *My life and health are spared while others are cut off. Last Thursday one girl fell down and broke her neck which caused instant death . . . The same day a man was killed by the cars. Another had nearly all of his ribs broken. Another was nearly killed by falling down and having a bale of cotton fall on him.* When the poorly cast iron pillars of the Pemberton Mill in Lawrence gave way on January 10, 1860, the building collapsed onto 670 workers. A few hours later, a lantern carried by a rescue worker broke, and the ensuing fire raged out of control. Eighty-eight people died, 275 were injured. The majority of those killed were women, many from overseas. The majority in the mills at that time were still women.

The tenements low on the plain grew more crowded as new tenements were built on back lots. In 1911 it was reckoned that one-third of the population lived on one-thirteenth of the city's land: over 33,000 persons on 300 acres. *It is not a home but a tool-box*, a critic of the city's tenements wrote. But women would make what they could of the life. They watched their neighbors' children and shared cooking utensils which they hung on the outsides of the houses. *I used to hear my mother call me from the kitchen and the smell of food told me it was time to eat*, one remembered. *But when I'd get there sometimes, it wasn't my mother, but a voice from next door, through the window! I could reach out and eat off my neighbor's table we were so close*... And another: *You know how we workers lived then, like slaves, but we all helped each other out, that's just the way it was back then, you did for others.*

The first house my mother remembers was on the outskirts of the city, but my grandmother insisted they move to the city's center, where she could walk everywhere. Maybe she felt more at home shouldered in by her own language. She never knew more than a little broken English even into her old age, after she'd been in this country more than seventy years. She kept largely to her house and her Italian neighborhood while my grandfather became conversant in English, even took night classes while he worked in the Wood Mill. She raised my mother and her sisters to understand Italian but to speak only English in a neighborhood which was no less a village, really, than the one she'd left behind in Italy. She never learned to drive, the iceman came to the door, the milkman left bottles at the bottom of the stairs, she bought her bread, cheese, meat, fruit, and salt cod from the corner stores.

Of prospect she would have had very little: wooden tenements near crowded out a view of wooden tenements distant, and if she could have seen beyond her Italian neighborhood she would have seen the plain thick with Syrian and Belgian and French Canadian neighborhoods. There was no way to prevent a satiety of it, to have it successively offered and in different proportions, no advantage of shifting scenes as you advanced on your way, not until the land began to rise from the plain to the west and north, and on those heights the mill owners' Victorians, with their turrets, fish-scale shingles, and finials, would keep their graceful distinctive air well into the twentieth century.

To the east and south, on the same plain as the tenements, factories lined the canals. The Wood Mill, built in 1906, was set on a sandy spot where the Agawam had once had a factory for projectile points. At the time of its construction it was the largest worsted mill in the world. Nine hundred men spent eight months building its two wings, which were each a half mile long. You'd think they'd have had to create their own word to measure it, but the way we comprehend it is to say: its four stories contained thirty acres of floor space. *Acre*: that measurement once dependent on the speed of oxen. In the thirty acres of the Wood Mill men and women processed as much as a million pounds of wool in a week. Work was measured by so much output per hour. The Poles were in the spinning rooms, the Greeks in the dyehouse, the English in the sorting room, the Italians in the weaving room. Each room contained its separate duty, dependent on another. If work in one room stopped, work in the other rooms stopped. Some workers were known for strength, others skill, everyone endurance. The acres were stacked upon acres, the machines were parceled out in rows and in columns, the workers were spaced like apple trees: so many feet apart, so many to the row, so many rows to the orchard.

HOW, OUT OF the din of the city, would my mother have had the opportunity to marry a farmer? She had been working as a secretary at the New England Milk Producers Association during the years my father served on its board. "It's such a funny thing," she remembers, "my mother bought a television. We must have been the first in the neighborhood with a television. Your father would come visit on Saturday nights, and we would watch the shows." I imagine that to my grandparents, to have a son-in-law with a farm was an opportunity to return to some of the light and air of the village they'd known in Italy. Those fig trees my father tried to grow were for them. When my grandmother would come out from Lawrence to stay with us at the farm for two weeks in early summer, she would sit on our porch and shell peas. The husks fell light as chaff into a basket, the peas rolled in the milk-glass bowl, the summer was not yet fraught. She'd doze and rouse herself, her mind would roam, she'd mutter something in Italian and then look out on the fields of young corn, a fresh summer breeze coming in—it was not yet the stillest of afternoons, though the trees had arrived at their deepest green. A long sigh: *beezina.* Runs of unintelligible Italian. Sometimes a pidgin of Italian and English.

But for my mother, to have moved six miles east to the farm in the 1950s—it took fifteen minutes to drive between the two places—was to move to an utterly unfamiliar world. I'm not sure it was any less dramatic than the journey experienced by those young women coming down from the hill farms to work in the factories, entering a world that hadn't existed before, with its different understanding of work and time, which they let change them. Emerson said they were born with knives in their brains. My mother will laugh today at the memory of the woman she'd been back then, buying corn for my father at the grocery store. "I thought I'd surprise him . . ." I can't begin to imagine what he himself thought, with the first block of the thirty acres of it just ripening outside the door, as she set the ears down on their lone dinner table.

When she tells that story I know she is stunned by time: it seems to her it happened yesterday. Yet she has long since grown into the world around her. She may never have participated in the farmwork, but the life she built within the house maintains its certainties, even in widowhood, and she loves it still, the look of the fields beyond her, the quiet and peace of it. She is living in another kind of time from me: just these days, one after another, and I wonder what it is like for her to wake day after day into the drift of unspecified time. Sometimes her contentment makes me restless only because I can't imagine the same for myself: ambitionless time, time at its most elastic, almost shapeless. I depend on the hours at my desk, forming and re-forming experience and time, and I don't know what life would be without them. But my mother and I have our similarities. She has grown shy about the farm, and if I don't drop vegetables off on her porch, sometimes she'll buy apples and tomatoes at the supermarket even though Dave is growing them outside her door. It seems we are both ghosts.

4
Island

AND I HAVE *sent them a shell taken out of my well thirty-nine feet below the face of the earth*, wrote Zaccheus Macy of Nantucket in 1792, *and I have taken many sorts of shells out of wells near forty feet down. And one time when the old men were digging a well at the stage called Siasconset, it is said, they found a whale's bone near thirty feet below the face of the earth, which things are past our accounting for.* Geologists have long since proven those shells were the accretions of ice ages—layers, in turn, of the vestiges of warm-water life that had burgeoned in mild times, and of cold-water whelks and barnacles that made their way down from the north in advance of the ice. After the last glacier stopped and melted, it left a rough scatter of debris—the moraine—most of which was eventually drowned by a rising sea so that beneath the southernmost stretch of the Gulf of Maine lie the shoals ships must always navigate, and the land that has not yet drowned remains as the spine and low heights and sand plains of a chain of islands—Long Island, Martha's Vineyard, Nantucket—Nantucket, nearly thirty miles off the New England coast, being the one that reaches farthest into the ocean, its tangled life accruing at the incremental pace insisted upon by salt and wind, its cliffs, spits, and reaches changing shape with every driven wave and storm.

When I think of the island, twenty years after having lived there, I often see myself walking through the heart of it. What was—and still is—the rough till deposited by the glacier had once been known as hunting grounds, then grazing grounds, and is now called the Moors: a low rolling landscape of scrub oak and pitch pine, bayberry, blueberry, pasture rose, grasses, mosses, false heather. Plants and animals abundant there are often rare beyond it, and sometimes strange. An owl hunts by day. Lady's slippers, those woodland orchids, thrive in sunlight. Always, there's the unmitigated sky. During my time summer houses had not yet been built on high places on the Moors, so often as I walked on the rutted dirt roads that crossed them I couldn't see a human structure. I had no sense of scale, and could believe, at the center of an island three miles wide and fifteen miles long, that I was in the midst of a vast heath. The roads themselves seemed to have no larger destination than a cranberry bog or one of the ponds—the kettle holes, where ice had once lingered after most of the glacier had melted back—and often they led only onto other roads, unpurposeful, and perfect for a dreaming mind. I wasn't the only one to see it as such: when artists paint and photograph the Moors these days they hardly ever represent humans in the works save for an occasional solitary gazer, as if it were entirely a place for onlookers, an immense consolation, an answer to desire. Though perhaps not for everyone: "It's like roaming an empty jail," I once heard someone say.

The landscape stood in absolute contrast to the inland world I had long known, and during my years in it I felt there was nothing attached to my own history, that I'd found a world I was utterly free to carve into, full of possibilities, a place to find my intended life. I remember feeling, upon leaving college, that I could go anywhere, though without some restriction on the possibilities—no other, no professional calling—I had no real idea of where to go. I'd settled on Nantucket because some friends lived on the island. I can see me there still, waiting at the ferry dock for my first crossing on a cold March morning, just a few boats at anchor in the harbor at Woods Hole. The land was gray, as was the sea, but you just knew the closedown was about to lift and a bit of green would be breaking out soon. My books, clothes, and little else were packed in a few soft bags slung over a bicycle. That's all I carried with me, that and a wish to become a writer, though I'd no idea at all of how to accomplish such a thing.

Very soon after settling into life on the island I discovered the Moors for what I hope they'll always be to me. Sometimes when I walked there under that sky I'd feel vulnerable to the exposure, especially when I saw another person in the distance, but more often I had a feeling of being out of reach and exposed only to immensity, as aloof and surveying as Henry Moore's *King and Queen*, that stoic sculpture he planted on the Scottish uplands. His idea for the work had begun as he was idly playing with a piece of modeling wax, which came first to resemble the head of Pan. *Then it grew a crown, and I recognized it immediately as the head of a king,* Moore had said. By the time of the sculpture's completion he had created two figures side by side whose delicate, complex heads appear graceful and intent on the vastness in front of them. *The landscape is so bleak and impressive, so lacking in frilly bits of detail, so monumental,* Moore had said of that heath. *The light is always changing too . . . The open air dwarfs everything because you relate it to the sky which is fathomless, endless, and to distance, which can be enormous.* For all the refinement of their heads, the rooted, weighty lower bodies of his sculpture seem to grow into the earth—or to emerge from it. The airy heads, the stolid feet somehow hold equal sway so you don't really know where power resides. Walking on the Moors was like that, making you feel both lifted and earthbound.

When the Wampanoag alone inhabited the island, the Moors had been their hunting grounds—after a good rain I sometimes found arrowheads that had washed onto the roads. The Indians had prospered longer on Nantucket than on the mainland coast: early epidemics hadn't reached them, and white settlement came relatively late to the island. Nantucket's Wampanoag population even grew a bit as other Indians, in flight from colonial settlement, arrived. They had cultivated cornfields on parts of the island, and as for the Moors, it's likely they burned them to keep the growth down for choice hunting—they had been so successful at the hunt that deer had been just about decimated by the seventeenth century. Such burning would have given the low-growing heathland and grassland vegetation breathing space to develop free of competition from larger plants for sunlight and for nutrients.

With the arrival in 1659 of the first white settler, Thomas Macy, the different understanding of land use and possession that the colonial world always brought with it came to Nantucket. Macy had left his established life at the mouth of the Merrimack River on the coastal plain above Boston after having provoked the ire of the Puritan governors the year before by giving shelter to some Quakers during a rainstorm, though even prior to the incident he had been looking for a way to establish a settlement beyond Puritan control, and the island presented a distinct possibility for resettlement since it wasn't under any colonial jurisdiction. It may be true that settling an island thirty miles out to sea was not, comparably, such an enormous undertaking in the middle of the seventeenth century, when roads scarcely penetrated the densely wooded interior of Massachusetts and settlement still hugged the coast, when water was a common means of travel. Although Thomas Macy needed a man who knew the local waters to help him navigate the shoals around the island, it would have been less difficult for him to reach Nantucket than to breach the inland frontier. Nantucket may, at that time, have felt closer to the north coast of Boston than it would centuries later, when the inland world was joined to the coast by road and then rail, though you still had to travel to the island by wooden ship and canvas sail.

Still, it unsettles me to think of him leaving all that he had established for such an unknown—after all, Macy's life on the mainland could hardly have been called insufferable. He'd owned significant fertile, well-drained acreage in an agricultural world. One historian, in looking back at the settlement of Nantucket, speculated that Macy and his fellow founders were dreaming of re-creating the landscape they'd left behind in England: *The open talk that boomed Nantucket would not have been: "Macy and Starbuck and others are anxious to get away to some outland where a man can think as he chooses . . ." Rather it was: ". . . There are tidy little fortunes, folks say, to be made there in cattle and sheep. Land much like the Devonshire commons, and the Wiltshire grazing runs."* To dream such a thing in a new world is something other than a longing to re-create the past. It's also a longing for a past free from the past, for the appearance of grazing runs and commons without the confinements of a long-settled world. Maybe what D. H. Lawrence had claimed for the earliest settlers, whose desires for possession were marked by cattle and fencing and fields, was also true for Nantucket's first colonists: *Liberty in America has meant so far the breaking away from all dominion. The true liberty will only begin when Americans discover IT, and proceed possibly to fulfil IT. IT being the deepest whole self of man, the self in its wholeness, not idealistic halfness . . . We cannot see that invisible winds carry us, as they carry swarms of locusts, that invisible magnetism brings us as it brings the migrating birds to their unforeknown goal. But it is so. We are not the marvellous choosers and deciders we think we are. IT chooses for us, and decides for us . . . [I]f we are living people, in touch with the source, IT drives us and decides us.*

Macy's division of the island into an agricultural world—pasture, tillage, home—established the old Indian hunting grounds in the midst of the island as the Sheep Commons: the twenty original settlers each received a share, and the fourteen artisans—the coopers and wheelwrights and carpenters who came with them—each received a half share. A share consisted of 720 sheep commons; one sheep common was equal to an acre and a half, which would furnish pasture for one sheep or, alternatively, four geese. Each shareholder was allowed to keep the number of animals his share could support. A shareholder could keep one cow for every eight sheep commons held, and one horse for every sixteen. With no wolves on the island, there was no need for fencing. Each sheep was earmarked and roamed freely. On occasion animals wandered into town, and through the open doors of houses.

In the ensuing centuries, as the island moved from an agricultural world to one that followed the whaling trade farther and farther out to sea, the sheep continued to nibble away at the heathlands. The herd might diminish—during the meager times of the American Revolution it fell to no more than 3,000 sheep—but it was not to disappear until the nineteenth century. At times there were 15,000 sheep scattered across the moraine and the outwash plain. Their grazing, in addition to the constant salt spray and the poor quality of the soil, helped to keep the shrubs and small trees down, and so encouraged the heathland plants, just as fire may have in Indian times.

Cultivation changes the world and the light that falls on the world. René Dubos suggests something of how complex is the relationship between the human and wild when he considers the exploited landscape of Greece: *The humanization of the Greek wilderness has been achieved at great ecological loss.* At the same time he speculates: *While visiting the Moni Kaisarianis monastery, I noticed a dark opaque zone on the slopes of Mount Hymettus; this area had been reforested with pines. To me, it looked like an inkblot on the luminous landscape, especially at sunset, when the subtle violet atmosphere suffuses the bare rocks throughout the mountain range. The "divine illumination" lost much of its magic where it was absorbed by the pine trees . . . Not only did removal of the trees permit the growth of sun-loving aromatic plants and favor the spread of honeybees, as Plato had recognized; more importantly, it revealed the underlying architecture of the area and perhaps helped the soaring of the human mind.*

Nantucket, more northerly, surrounded by the cold green Atlantic, the snow-and-starred winters, produces a more brooding light on its once-grazed lands.

As the sheep were cropping the middle of the island, keeping open a place for the heathland plants, elsewhere under the plow the quality of the soils quickly declined. Early accounts had attested to the worth of the soil: *When first settled by the English, the soil was good and produced equal to any part of the country . . . Ebenezer Barnard, a man of strict veracity, in the year 1729, tilled five acres in the general cornfield, at that time on the north side of the island, between the Long Pond, so called, and the west end of the town, a tract of land below the medium quality. From these five acres he gathered 250 bushels of good corn, and this quantity was considered rather less than average for that year's growth. This may be accounted an uncommon growth for any country . . . In the year 1773, the cornfield was at Madaket and Smith's Point, at the northwesterly part of the island. The land then produced 20 bushels, on average, to the acre, which was considered a remarkably good crop. Since that time the crops have gradually lessened, and within a few years they would not average more than 10 or 15 bushels to the acre.*

I think of Nantucket, in a way, as the intensification of New England's agricultural fate. All through the Northeast in the eighteenth century fertility and topsoil were being lost to erosion and poor agricultural practices, but on Nantucket, with the winds off the water, with the exposure and salt, the effects were exacerbated: *At the time of the settlement of the island it was covered with wood, which protected the crops from the raw easterly winds, and by a continued supply of falling leaves and other decaying vegetation preserved the richness of the soil. The frequent ploughing of the land, since it was cleared of trees, has exposed the soil to the action of bleak winds, to which the island is very subject, and by which it is blown into the sea.* The soil had undergone so much degradation by the time of the American Revolution that Nantucketers—who numbered five thousand then—could not support themselves on the land: *The soil will not produce a subsistence for one third part of the people.*

Wholly destitute of fire wood and but a little clothing, such being their situation and circumstances . . . there will many people perish for want, before the end of another winter.

The degradation of the soil had been so exacerbated by the loss of woodlands that the island was to become, in Herman Melville's words, *a mere hillock, and elbow of sand; all beach, without a background. There is more sand there than you would use in twenty years as a substitute for blotting paper. Some gamesome wights will tell you that they have to plant weeds there, they don't grow naturally; that they import Canada thistles; that they have to send beyond seas for a spile to stop a leak in an oil cask; that pieces of wood in Nantucket are carried about like bits of the true cross in Rome.* At one point in the nineteenth century, during a particularly cold winter, wood grew so scarce that islanders dismantled and burned two houses for warmth.

If with its salt and wind-stunted vegetation the island had pulled itself even farther away from the mainland, and from what the New England eye had grown used to, Nantucketers themselves seemed to be pulling away from the deemed character of New England, even in the way they kept their sheep: *The island being owned and improved in common, the sheep have not had that attention in the winter, which it is the general practice of farmers in the country to give to them. They are suffered to run at large throughout the year, exposed in winter to the bleak winds and cold storms, with no place of shelter provided for them. The forest has disappeared, and the greatest part of the island is left a naked plain, where the gale meets with no obstruction and animals find no refuge. It sometimes happens that many sheep are covered in heavy falls of snow, and perish before relief can be afforded, though large numbers of men are employed to release them. This mode of keeping sheep may to some appear wrong and even cruel; but it may be observed that the proprietors have always been in that practice, and, by long custom, have become so reconciled to the measure, that the thought of doing wrong has almost become extinct.*

Later writers would imagine that the separation from the mainland had been a profound one from the start. J. Hector St. John de Crèvecœur, upon seeing Nantucket in the last half of the eighteenth century, saw a place that in no way resembled the cultivated farmland of the mainland, and could not imagine that those who first settled the island were dreaming of grazing runs, or that its barrenness had been something exacerbated by settlement. He understood Thomas Macy's settlement of the island in a particular light: *Who would have imagined that any people should have abandoned a fruitful and extensive continent, filled with the riches which the most ample vegetation affords, replete with good soil, enamelled meadows, rich pastures, every kind of timber, and with all other materials necessary to render life happy and comfortable, to come and inhabit a little sand-bank, to which nature had refused those advantages; to dwell on a spot where there scarcely grew a shrub to announce, by the budding of its leaves, the arrival of spring . . .* And he ascribed to the first founders an intent they may not have had to begin with: *They found it so universally barren, and so unfit for cultivation, that they mutually agreed not to divide it, as each could neither live on, nor improve, that lot which might fall to his share. They then cast their eyes on the sea . . .*

❦

The Whale Fishery had begun as a source of supplementary income, with the stripping of pilot whales that had washed up on the beaches. The pilot whales' element, like the mariner's, is the deep, but sometimes they follow the squid near shore. They navigate by sounding, and the shifting, sandy bottoms of the old moraine can confuse them. They become like birds caught in a cage. Pilot whales move in pods, and when they become disoriented dozens can become stranded at one time. It was this tendency to beach in numbers that made shore whaling attractive to the early colonists. By the late seventeenth century it was so widely practiced that appointed inspectors decided whose property the beached whales were. The discoverer received a third share, the colonial government a third, the town a third.

Eventually men began to pursue whales in the local waters: *When they come within our harbors, boats surround them. They are as easily driven to the shore as cattle or sheep are driven on land. The tide leaves them, and they are easily killed. They are a fish of the whale kind, and will average a barrel of oil each. I have seen nearly four hundred at one time lying dead on the shore.* Nantucketers understood the rich potential of the harvest early on. Historian Obed Macy recounts, *In the year 1690 . . . some persons were on a high hill, afterwards called Folly House Hill, observing the whales spouting and sporting with each other, when one observed "there," pointing to the sea, "is a green pasture where our children's grandchildren will go for bread."*

The sea can be understood as prairie, the human adventure on it taken up in the manner of any other cultivation.

Like medieval villagers moving beyond their bounds, carving a bit more from the wilderness as their plows and animals improved and the population grew, over time men pursued different kinds of whales in more distant waters. After a while, there was no returning from the sea at the end of the day. The hours on the hunt became days, then seasons, until whalers were spending years sailing to the South Pacific in search of the sperm whale. The sailors became more and more separate from life onshore. The Nantucketer became entirely peculiar. Herman Melville was to write, . . . *these sea hermits issuing from their ant-hill in the sea, overrun and conquered the watery world like so many Alexanders; parcelling out among them the Atlantic, Pacific, and Indian oceans . . . [T]wo thirds of this terraqueous globe are the Nantucketer's. . . . For years he knows not the land; so that when he comes to it at last it smells like another world, more strangely than the moon would to an Earthsman. With the landless gull, that at sunset folds her wings and is rocked to sleep between billows; so at nightfall, the Nantucketer, out of sight of land, furls his sails, and lays him to his rest, while under his very pillow rush herds of walruses and whales.*

By the nineteenth century whaling ships were sharing the seas with countless other craft: wooden ships under canvas sail carried immigrants, travelers, cargoes of seeds, cattle, sheep, paving stones, furniture. Most captains navigated near land by coasting—shadowing the shore from one landmark to the next. Sometimes the only passage from one place to another was over shoals. In a stretch of quiet summer weather the changes in them might be almost imperceptible, but sandy shoals—embryonic lands that have not yet broken through the sea's surface, old lands dead, dying, onetime islands long since become rumor—may change shape faster than the navigation charts can be revised. In winter, in rough weather, spits can wash away overnight. Points become islands, islands become spits. Nineteenth-century ships lost steering, there were fires on board. They had no long-range forecasts. The lighthouses, with their oil lamps, could fail in the worst weather. A lightship might run off course. When the lights failed there was the foghorn, but sound bends on the wind, in the water and fog. You can't get your bearings by it. It can only let you know that you are near. If you are lost for a moment you could be lost for good, even in our sea, so sounded and certain, read from the air and scanned by radar, its currents tracked and mapped. A storm surge puts every mariner back into the world Melville charted: *In that gale, the port, the land, is that ship's direst jeopardy; she must fly all hospitality; one touch of land, though it but graze the keel, would make her shudder through and through. With all her might she crowds all sail off shore; in so doing, fights 'gainst the very winds that fain would blow her homeward; seeks all the lashed sea's landlessness again.*

Thoreau, on Cape Cod, observed, . . . *and sometimes more than a dozen wrecks are visible from this point at one time. The inhabitants hear the crash of vessels going to pieces as they sit round their hearths, and they commonly date from some memorable shipwreck.* Each wreck had its society: there would be those who worked to aid the dead and the near dead, a few would go on collecting nearby seaweed that the storm had washed in, others scavenged what remained of the ship, or were on the watch for whatever valuables the tide brought in: turnip seed, barrels of apples, scrap iron, wood, books, ties . . . The coastal world, both the cultivated and the wild, could seem to be half built out of flotsam. *Another . . . showed me, growing in his garden, many pear and plum trees which washed ashore . . . all nicely tied up and labeled,* Thoreau remarked during his journey on the Cape. He also heard of a man who saw *something green growing in the pure sand of the beach, just at high-water mark, and on approaching found it to be a bed of beets flourishing vigourously, probably from seed washed out of the Franklin.* Whatever the profit, there was always a loss: *When I remarked to an old wrecker partially blind, who was sitting on the edge of the bank smoking a pipe, which he had just lit with a match of dried beach-grass, that I supposed he liked to hear the sound of the surf, he answered: "No I do not like to hear the sound of the surf." He had lost at least one son in "the memorable gale," and could tell many a tale of the shipwrecks which he had witnessed there.*

There had been, by the mid-nineteenth century, so many wrecks and deaths in the coastal waters that eventually, especially along the sandy shores—the granite shoals, after all, proved more constant—the Humane Society placed huts on the desolate reaches: *The traveler stands for example on the southern shore of the island of Nantucket, and after looking off over the boundless ocean which stretches in that direction without limit or shore for thousands of miles, and upon the surf rolling in incessantly on the beach, whose smooth expanse is dotted here and there with the skeleton remains of ships that were lost in former storms, and are now half buried in the sand, he sees, at length, a hut, standing upon the shore just above the reach of the water—the only human structure to be seen. He enters the hut. The surf boat is there, resting upon its rollers, all ready to be launched, and with its oars and all its furniture and appliances complete, and ready for the sea. The fireplace is there, with the wood laid, and matches ready for the kindling. Supplies of food and clothing are also at hand—and a compass: and on a placard, conspicuously posted, are the words:* SHIPWRECKED MARINERS REACHING THIS HUT, IN FOG OR SNOW, WILL FIND THE TOWN OF NANTUCKET TWO MILES DISTANT, DUE WEST. I imagine it open-mouthed facing the sea, not a harbor or village in sight, just the solitary house that will save ten or one or seven survivors.

The edge of the sea attracts people looking for a definite limit, and freedom. Twenty-some years ago, anyway, the island was a place you could tuck into for a while, not too obvious among the descendants of the first founders—the Coffins and Macys—and the Portuguese from the Azores descended from the whalers, the summer people turned permanent, artists and writers—the complex of communities that make up many shifting coastal towns now that tourism rather than fishing is the mainstay. The human community there has as many small and competing interests as the heath, and it has more in common with the sandy shoals than it knows.

Nantucket felt more apart from anything I had ever known, and that apartness sometimes helped with definition. Others were "from away." I was "from away" to others on the island, I know, though I told myself my life was tucked up enough so I wasn't really noticed. I loved its closeness, and it would be years before I found it confining. There was still a place for sidetracked life, for people who didn't mind being the last one, though even then you could see in the increasing affluence of the town how those who had once been ordinary people in the community were becoming characters, and might soon become nuisances to the idea of the place. I was grateful for the way the island allowed me to travel my own hidden path while I nursed my hopes of becoming a writer, and also afforded me a bit of protection from the expectant eyes of my family.

During my first years on the island my attempts at writing were sporadic, and I told myself the distractions were the stuff of life—I was busy with friends and work in restaurants and bookstores, work that paid my rent—but my failure to make headway weighed on me so much that I began to look for a way to impose greater discipline on myself. One winter three years after having arrived there, I agreed to house-sit on the Squam Road, a rutted dirt track that ran between the swamp and the sea on the eastern edge of Nantucket. I thought by putting myself farther away on the island I could find enough solitude and silence to begin writing in earnest. The winter landscape of Squam felt utterly wild: bayberry, swamp oak, black alder—gray, grayer, grayest—were all stunted by poor soil and salt and gnarled by the wind. Everything had long been growing into everything else, and had become impenetrable to humans, crossed over by deer paths, eyed by marsh hawks on their wavering, tipping wings. I don't think there was anyone else living on the Squam Road that winter—at least I never saw a light except the one I left on for myself. Most of the gray-shingled houses that faced the open ocean, their windows washed and stained with salt spray, had been built for summer.

Even now I think I should have named that wind, and added it to the list of legendary winds of the world—the foehn and the mistral, the chinook and the Santa Ana—though it was a private wind that didn't seem to reach anyone but me as it blew steadily in from the Atlantic. All winter long it blew, and it came up most fiercely after the sudden twilights that ushered in the long nights in a world going down into winter. It defined my walls and windows and made the house creak and bray as it rushed across my unformed life and then across the tangled scrub of the swamp. I had never feared the wind the way I did then, on nights so dark I couldn't see my hand in front of my face, and the bright, polished stars were so many they weighed down on me, much as I wanted to be able to stand and stare, to remark with matter-of-fact wonder: *one could not have put a finger in between them.* The darkness seemed to compound the fury of the wind, and all of it would have driven me crazy if I hadn't lit a large fire every night, stacking log upon log in the brick fireplace. When the creaking and braying cut especially deep, I'd crash more logs onto the flames and sleep right beside it, curled in my blankets, solitary daughter of those who were never to ask for help, of sturdy inland souls whose voices I imagined hearing during those nights. *Keep at it,* they'd said. *Use your head.*

Maybe it's true that the greater part of my fear was simply of being alone on that road, and a woman alone having to face her own life, but I don't think that could have been everything. For all its beauty, and however much I loved it by day—the inlets and salt creeks, the beach rose, the red berries of the black alder shining out of the swamp oak—I was also bewildered by a terrain that was so different from those water-worn hills north of Boston, where wilderness was kept at bay by cultivated rows and furrows. I think now I was most afraid of an absolutely unbounded world, and me silent and singular within it. As I remember the fires, those stars, that wind, it seems as if it was all a part of being tested, and having failed.

Still, when I sat down to work in the mornings as the sky lightened over the Atlantic and the wind calmed for a while, I remember having an unusual amount of energy, not only from the sheer relief at having gotten through the night but also from attempting to begin what I'd longed for. I loved the daytime there in unusual proportion, loved the feeling of being at absolute sea level, the density in the sea air, the dampness and thickness of it. Absent the night wind, a stunning, palpable quiet surrounded me as I worked, while beyond the window, as daylight slowly silvered the swamp, the uncultivated landscape of Squam grew more and more distinct.

If I think of Squam long enough, I remember the fires of that April, too, when the days were more insistent than the nights, and the wind had a warmth to it, and a fragrance, and lost much of its power over me. I'd sit in the front doorway and bask in the mild dusk and the first red energy of spring while behind me—half in my view, half forgotten—the last split oak of the year burned gently in the fireplace. Those nights I fell asleep easily as the foghorns sounded one after another across shrouded water. Though the mildness of the new season was strong enough to allay my fears, my winter in Squam turned out to mark my last year on the island. I had created for myself a greater discipline and focus, and the accomplishment seemed to create a new restlessness in me. I found myself wanting more of the world than the island, even with its vast sky, could provide. My arrival four years earlier may have been a questioning of what I was aiming for; to leave was a questioning again—all leavings are a questioning of what is left behind, the eternal questions of immigration. To leave is to want change, however much you may want what is left behind somehow to remain the same.

I wanted a correction for the remoteness of those years, a place more in the thick of things, so I settled into an apartment outside Boston. For the longest time I felt shy of the world at large, and I couldn't quite believe I wouldn't be returning to the island. When fog settled around my apartment, I was disoriented without the low warnings of foghorns. I remember walking down the streets thinking I couldn't get enough air. The air wasn't dense enough: Squam air had an almost solid weight. Outside the double-decker where I lived, human energy was everywhere palpable, and the streets were always busy, even after a snow. Before the last flakes had fallen the shovelers were out in full force, and the sand trucks and plows moved through in formation, clearing every last bit away. The cars going by sounded as if they were driving through rain. I was overwhelmed, even as I knew I had left Squam partly for fear of growing too far away from all the noise of life. Or that's how I put it to myself, anyway. And still sometimes I dreamt of the world I had had to myself, and what I remembered as the immense quiet of it all. Sometimes I chastised myself for not having loved those solitary nights when I could.

I must have been after that same sense of silence and solitude I'd experienced in Squam when, nearly fifteen years later, I decided to build a working space separate from my home on the farm. My small house was feeling noisy by then: it sits on a slight rise and is oriented towards what these days is a busy commuter road, and away from the unperturbed life of the apple orchard. There had been only a few small windows giving onto that back view, just large enough to let some sun through, enough to peer out at the winter shadows stretching early and long across the snow, the house shadow blending with the shadows of those cragged apple trees, which blend with those of the high white pines of the woods beyond as the last light lingers in the treetops a hundred feet above.

The house has an old shed attached to it which had been built seventy-five years before, from salvaged barn timbers. All kinds of nails and rusted hooks hang from the walls; the wood has long gone brown with age. The rear of it slopes towards the orchard, and I began to imagine a room carved out of the back which would look out on those trees. It would have to be a room turned around from the practical, with half the windows facing north. I thought I could keep much of the shed intact—just shore up some timbers and raise a few walls—but the carpenter said squaring it would be too costly, and the most he could do was save the old beams. He cut a full third of the shed away and built me an entirely new room: full of windows, white, spare, square, true, and how I love that back view.

In winter when I lift my eyes from work and look straight into the measured rows of the orchard—without blossoms or leaves, without the spring haze of incipience—I see thickened trunks and branches trained to resilient strength and spareness by years of pruning. You'll know them if you've seen Mondrian's *The Grey Tree* painted an ocean away, nearly a century ago, just before he moved into pure abstraction. His gray limbs sprout branches that catch the turbulent and dense gray air and finally thin into the merest slips that are caught by the air, even as they cut it. They seem to stand for all the contending forces in the world. Mondrian, who in later life, when asked why he was reworking his earlier canvases, said, *I don't want pictures. I just want to find things out.*

When my work goes well I'm surrounded by the same palpable silence I remember from Squam. In my study—larger than any room in my house, lower, brighter—days go by and papers and books pile up around the edges of my desk, crowding me into a world of its own accretion, with its own absorption and fidelity. It's not far from Squam in its hopes still, though the days are accompanied by greater discipline, and the pencil and paper I used on that table twenty years ago have given way to the whirs and grunts of a Macintosh. *There has to be,* writes Adrienne Rich, *an imaginative transformation of reality which is in no way passive. And a certain freedom of the mind is needed—freedom to press on, to enter the currents of your thought like a glider pilot, knowing that your motion can be sustained, that the buoyancy of your attention will not be suddenly snatched away. Moreover, if the imagination is to transcend and transform experience it has to question, to challenge, to conceive of alternatives, perhaps to the very life you are living at that moment. You have to be free to play around with the notion that day might be night, love might be hate; nothing can be too sacred for the imagination to turn into its opposite or to call experimentally by another name. For writing is re-naming.*

Such a world is inexplicable in its richness and its costs, and creates in you a more solitary creature than you might otherwise have been. Or it may meet your own solitary nature. I know when I walk out into the day, that interior I've left behind can seem matchless and inarticulate even to myself, never mind to others working on the outside. And yet the reality of the open day seems to depend so much on those stowed-away hours. *I am here alone for the first time in weeks, to take up my "real" life again at last,* wrote May Sarton. *That is what is strange—that friends, even passionate love, are not my real life unless there is time alone in which to explore and to discover what is happening or has happened. Without the interruptions, nourishing and maddening, this life would become arid. Yet I taste it fully only when I am alone here and "the house and I resume old conversations."*

Every once in a while my work is unnerving—flat or obdurate—and I have the same old feeling of being tested and having failed that I endured during those wind-compounded nights on the island. Then I go back into the inch-deep solace of my house and sit down at the dining room table to work. Easier, I tell myself, with the swish of cars going by on the road and the juncos pecking at seeds in the snow, the kettle whistling and the splash of last summer's geraniums brought in, their sharp scent come to life by the heat of the wood burning in the stove.

I still have some old friends living on the island, and I like to go down for a fall or winter visit—never in summer with the crowds, not if I can help it. In recent years, even in winter the island has come to feel much less remote to me. Maybe that has something to do with the greater number of houses and people, or with how quickly you can get there now on the ferry, but I think it might have as much to do with my mere dipping into that world. You need to dwell in a place for a good while before its true remoteness sinks in and moves in your imagination. All I know is that as the boat nears the island I still habitually move to its port side so as to look out at Coatue, the barrier beach that protects the harbor from the sound. It's just a spit of sand with grasses and some scrub, so low you can believe it might sometimes be washed over by tides. It seems as vulnerable and beautiful as always, but it used to mark more for me, as I came and went during the years I lived there, as the winter ferry slipped past so early in the morning, the sun rising behind Monomoy, the engine humming, the boat quiet with passengers dozing on their coats or sipping coffee. Our voices never seemed to gain strength until we passed all land and were well into the sound. Nantucket had mattered so much to me for a handful of years. It had been difficult to leave. Now to see Coatue vulnerable and beautiful as ever, and not to be stung by longing, is its own sadness.

The island itself seems conscious of its own rapid change, and unable to slow it. Sometimes when I stop to chat with old acquaintances in the street they ask if I've been out to Squam.

"No, no, not yet."

"Don't go. You won't recognize it." The summer houses have gotten bigger, they say, and there are so many more of them. It might be true that in summer or on an unseasonably mild off-season day I wouldn't recognize the place, but I'd like to believe that even if there are three or four lights on the Squam Road now as the January sun goes down, I'd know it for what it always was on those winter nights.

When Nantucket remembers itself it remembers itself as a whaling town thrumming with the work of coopers and smiths and stitchers at work in the sail lofts, a town smelling of the tryworks, like *the left wing of the day of judgement*, though that world had burned to the ground in the nineteenth century, and sand sifted into the harbor mouth and silted it up and made the passage of larger ships untenable. The whale fishery became concentrated in New Bedford even as the gold rush was sending desiring imaginations elsewhere and the advent of kerosene had begun to swamp the whale oil industry. Still, the more time passes, the more established becomes the remembrance of that world. The historical association, long a healthy presence in town, is planning to expand its museum to include a whale oil and candle factory as well as the 1850 Fresnel lens from the Sankaty Lighthouse and the full skeleton of a sperm whale that washed up on the beach a few years ago.

These days when the pilot whales beach themselves on the Cape and islands there is an outpouring of concern along with enduring curiosity. When more than fifty of them drifted onto the shores of the outer Cape a few summers ago, the event made national news for a week. Vacationers came out by the hundreds in a frantic effort to lead them back to sea. Once the pilot whales are out of their element, though, their own weight begins to crush their internal organs. Their skin burns and blisters in the sun. While people waited for the tide to haul them back out, they covered them with wet towels and bedsheets, and schoolchildren carried buckets of water to pour on them. A photo from the air makes the scene look almost like a carnival, with all the bright colors of their clothes and paraphernalia. Eventually, several men guided each whale back out beyond the surf, but getting beyond the surf was the least of it. Volunteers stood watch all through the evening and again at first light to see if the whales would return, and they inevitably did, being too exhausted to find their way to open water. There were more volunteers and less help for them with the second effort to get them back to sea. A third return. Even more died, and they died more quickly, until finally veterinarians euthanized the rest.

The sheep have been gone from the heart of the island for more than a century. They had grazed on until the island economy declined in the 1860s, when not only had whaling declined, cheaper wool from the West decreased local demand. After the grazing herds diminished, the heath began to lose its foothold, and it might have disappeared entirely under the growth of scrub oak and pitch pine had not residents, sometime later in the nineteenth century, begun to burn what they now called the Moors. The burning was meant to attract large flocks of American golden plovers and curlews, which were killed for market. As Fred Bosworth notes in *Last of the Curlews*: *And sometimes, during northeast storms, tremendous numbers of the curlews would be carried in from the Atlantic Ocean to the beaches of New England, where at times they would land in a state of great exhaustion, and they could be chased and easily knocked down with clubs when they attempted to fly. Often they alighted on Nantucket in such numbers that the shot supply of the island would become exhausted and the slaughter would have to stop until more shot could be secured from the mainland . . . The gunner's name for them was "dough-bird," for it was so fat when it reached us in the fall that its breast would often burst open when it fell to the ground, and the thick layer of fat was so soft that it felt like a ball of dough. It is no wonder that it was so popular as a game bird, for it must have made a delicious morsel for the table. It was so tame and unsuspicious and it flew in such dense flocks that it was easily* *killed in large numbers . . . Two Massachusetts market gunners sold $300 worth from one flight . . . Boys offered the birds for sale at 6 cents apiece . . . In 1882 two hunters on Nantucket shot 87 Eskimo curlew in one morning . . . By 1894 there was only one dough-bird offered for sale on the Boston market.*

The burnings stopped when the birds disappeared, opening the heath once more to competition from scrubby oaks and pines. In recent years conservationists, in order to encourage the heathland habitat, have attempted to replicate the effects of grazing and old fires by performing controlled burns. Sometimes you can detect in the effort a desire for purity, for a world before our own design. *We have built roads, which stop the spread of fires, and extinguished fires that would have burned prior to European settlement,* one scientist commented. We *have built houses and planted trees that interrupt the salt spray. We are not letting nature take its course.*

I had taken for granted, when I walked on the Moors, that I was walking on free space, and there was nothing stopping me from striking out long and far. It was one of the few unmarked places where I'd never been shadowed by a feeling of trespass. I had no idea that shares of the Commons, as originally set out by Thomas Macy, were still owned and had been divided and handed down through generations, diminished by successive estates so that now bits of them were the property of people spread out over the world. In modern times, if one wants to buy or sell a parcel, one has to track down the scattered inheritors of the Commons, or prove the rightful owners cannot be found. Developers do, as do the conservation organizations, which acquire land on the Moors just as aggressively as they acquire shoreline in an effort to keep the boundaries of the island as it has been known, and the growth of the Moors as they have been known. NO MOOR HOUSES, the bumper stickers said when I was there.

There were Moor houses, of course. Nantucket became more prosperous through the 1980s and 1990s, and executives built large summer homes even on the part of the moraine that had once been known as hunting grounds, then grazing grounds. The town itself seems to work towards that prosperity the way the old Nantucket town had worked towards whaling. More and more of the island people sell houses, or work for the moneyed, keeping and building houses in support of that trade. There aren't enough ordinary people on the island to fill certain demands now that the cost of housing has risen beyond the reach of the island's children and the laborers. Carpentry crews fly in daily from the mainland.

The five-and-dime store and the modest department store on Main Street have gone out of business. Many storefronts display beautiful decorative wares from central Italy and southern France, but you have to search off-island to buy things for daily life. Sometimes wedding guests lining the church steps open envelopes stuffed with live butterflies as the bride and groom depart after their vows. Controversies flare over four-wheel-drive rights to beach access, and traffic ties up Main Street all summer long. Still, the Moors possess a scent all their own, nothing like the sea, nothing like *the left wing of the day of judgement.* When I catch the scent I sometimes imagine that, in supporting a rare habitat, the nibbling mouths of sheep may have had a more lasting impact than the leviathan, and that the bayberry, with its heady oils, having grown on long after the grazing has stopped, will make a candle to light a smaller, more intimate world than whale oil ever illuminated.

I can't imagine my factual understanding of Nantucket ever being clearer than it is now, the way it has been measured and mapped and studied under glass, so much more certain than when Zaccheus Macy puzzled out the bones and shells he took from wells. Even so, it is a place that can't be fixed by the clearest photo taken from space, in which the Atlantic shows itself as an even expanse of oceanic blue while long-established lands stand resolutely green against the drab gray and beige of human settlements. The ground grains of fossil and stone that make up the shoreline show white and appear cauterized to a permanence, as if the edge of land could ever be so crisp and stilled. The photo suggests nothing of the way the sands shift and change with every tide and gale, now losing ground to the sea as the sands absorb the power of a storm surge, now gaining ground as a quiet tide seeps in. On vast stretches of sandy coast there's nothing for shore life to cling to, and what survives survives by burrowing. *I was always aware that I was treading on the thin rooftops of an underground city*, Rachel Carson once wrote of walking along just such a shore. *Of the inhabitants themselves little or nothing was visible. There were the chimneys and stacks and ventilating pipes of underground dwellings, and various passages and runways leading down into darkness. There were little heaps of refuse that had been brought up to the surface as though in an attempt at some sort of civic sanitation. But the inhabitants remained hidden, dwelling silently in their dark, incomprehensible world.*

It is thought, from all the erosion and the inexorable rising of the sea, that the island may disappear completely in another three centuries, but for now, if I look at that photo long enough, my imagination puts Nantucket-as-seen-from-space in motion. Most everything else feels as if it comes under the pull of the mainland, the way Cape Cod curves inward to Massachusetts Bay, and the Vineyard leans into the Elizabeth Islands, while Nantucket appears intent on putting even more distance between itself and the mainland, the way Tom Nevers—its southeast headland—juts out. If Melville had seen it from space he would have thought of the brow of the whale itself, for it is every bit as blunt as the summit of Mount Greylock as seen from his hill-shrouded study window. I think first of the *Winged Victory of Samothrace* leading herself towards the open sea with the thrust of her torso. The northward points of Tuckernuck and Muskeget, and Great Point—those fragile spits of land—stream back like her wings, as if the island itself were in flight now that time is collapsing the distance between it and everything else. You can jet in from Boston in scant minutes, and the air above is full of engine noise, and the ferry across the sound goes twice as fast as it used to, getting you there in a perfect hour, hardly enough time to feel a sense of transgression or any singularity about the trip, however much I want to know it the way Melville imagined it—*away off shore, more lonely than the Eddystone lighthouse.* Surely its apartness must be the one constant in all its history. Surely something about Nantucket remains past accounting for.

5
Grange

PERHAPS IT'S ONLY that my father had no time for the sea and my mother was always dreaming of it, but sometimes I say *salt hay* to myself again and again for nothing beyond the pleasure of the slight discord I feel between the two words, discord of both sound and sense. *Salt*: the one curt syllable glints sharp and stinging. To stand at the headlands and smell it, however much you love the ocean, is to know there's no sheltered world. Steadily roaring breakers are constant in your ears, you gaze into an infinity, whatever weather moves out over the water and curls back contains a fierce and renewed strength. But *hay*, with its open *a*, is assuaging, and suggests an infinite calm: it belongs to an inland world, and lies at the heart of agriculture, for hay must always be made, and does not exist without human effort. Cultivation itself is a dream of shelter against the unpredictable, and essential to our dream of it is the sweet and redolent smell of cut grasses blooming and deepening on the air, then diminishing as the harvest dries and burnishes. In those marshes alone, on the margin between land and sea, salt and hay shoal one into the other, each word claiming the shifting boundaries of both worlds. A scythe sweeps the grass, somewhere out beyond the surf a buoy bell clangs. *Have salt in yourselves*, says Mark, *and have peace one with another*.

I would never have known a thing of the salt hay harvest had I not seen Martin Johnson Heade's paintings where, under a nineteenth-century sky, looming haystacks—now gold and gleaming, now in shadow—dominate the marshes. The stacks might be twelve feet high—outsized, the way he's painted them—and thatched so sturdily, without a waver or a list, they appear every bit as stable as the land itself, as if they'd never give in to gravity or wind, or drift onto the uplands in a stormy tide. They're larger than anything human. Men move among them, men and boys dwarfed by what they've made. Smaller, much smaller, among the stacks are a few cattle scattered on the marsh, a cart of hay being taken to the barn or up to the spreading grounds, some men raking in the distance. Skies change—storms wash over the marsh, sunrise, sunset, twilight, a glaring orange sun, or diffused light settling on everything—skies change, and the light that falls on the land changes, but the world remains steady: the cattle do not sink and need to be shot, the horses don't misstep and break a leg, a man does not misjudge the tide and get caught on the open water without a way back home.

It never existed, of course, a world of such constancy and peace. Cattle did sink, and horses misstep, and in a rare instance a man would lose his life while haying, but the marshes, and the human lives in their proximity, were inextricably tied to each other in the world Heade painted steadily for more than forty years, mainly in the permutations of its harvest season, leaving others to render the chasms and canyons of the frontier. What Heade painted was the last strong time of salt hay harvesting. The cutting of spartina, which had receded as the country expanded westward across the ranges and into the prairie, enjoyed a nineteenth-century revival when Northeast farms began to grow as the textile cities crowded in and the factories grew larger and farmers began to increase their herds to satisfy the demand for milk. They lacked adequate tillage land for their cows since many of the old fields had grown up into impenetrable thickets, then woods, but the salt marshes—the ones that hadn't been cut off from the sea by roads, or drained for development, or turned into cropland for corn, onions, and strawberries—remained open. Extremes, after all, have their own kind of stability: sometimes a different grass might have moved onto the marsh, but because of the salt, nothing of any stature—no cedar or pine, oak or hickory—had taken hold.

At the end of the nineteenth century those open marshes seemed more valuable than ever. *They used to tell you if you weren't sure of your bounds, you had better find out, for if you should cut over the line . . . you would be liable for damages*, one of the last to remember the salt hay harvest says. And: *They thought so much of that salt marsh hay, they would fight for the last spear.* Farmers brought down all they needed from higher ground and camped by the shore during harvesttime. Along with scythes and rakes, they carried bog shoes for the horses so their animals could better maneuver the spongy earth. They used flat-bottom boats, called gundalows, to haul the cut grass up the creeks. As before, they built staddles out of old logs to stack the hay on so that it would stand above the wash of floods and tides. Tools born of a marginal, particular world, belonging to a limited place in time and to no other. *On the south side of the path was Frank Pevear mowing with one horse . . . I can see him now. He was walking behind the machine. Warren H. Batchelder doing the mowing with a two-horse mower . . . John N. Sanborn mowing his 23 acres with his own red roan horse. He walked back of the machine. Then, Stephen Brown mowing his own 11 acres. We could count around 30 horses and 60 men sometimes. Then in about two days or more when the mowing was finished, there would be just a few horses but around 100 men and boys, raking and stacking. There were lots more over towards the Hampton River and up west, back of the trees towards the Hampton Falls dock. In the September season it was the same thing but not so many boys.*

It must have seemed, even in the last long August days of it, with the switchel cooling in the ditch, and the horses' bog shoes white with salt, and stacks growing higher and higher, that it was a world certain of its footing: *I would like to tell you how we used to haul the hay to the stack . . . We made that drag. It worked so well there was no way to improve on it. You had to drive with long reins, standing on the short plank. The hay would slide up the side of that plank and keep going forward all the time and by the time it had gotten to the horse's heels, it would be piled up in front of you and out the sides against the ropes. There would be enough to call it a load . . .*

They knew how many days in the month the tide ran in the marshes. They knew the greenheads wouldn't bite in the shade, and that the mosquitoes were worse going down than on the marsh itself. They knew not to lay a scythe on the ground where a horse or a man might step on it. That it was a bad thing to be caught on the marsh without fresh water. They watched the moons and the tides, knowing you had to live with your judgments, and with whatever was gathering on the gray-blue sea strewn with whitecaps: *One afternoon he was going down to get a stack of hay . . . It was on a low run of tide but it seems that there was a bad storm at sea that no one knew about and the tide coming up. He was on the marsh when the tide started to come up but knowing that the tide was a low run he didn't think much of it. As the tide kept coming he decided to turn around and go back but before he got near shore the tide was so deep he could not see the path. He stopped. The hired man thought he knew the way and went on but became lost in the river and drowned. John Thayer stayed, standing up in the pung and holding his son all night.*

They could not improve on it, but neither could they save a passing way of life: in the end there was no way to compete with the larger farms and herds to the west, or the speed and economy of long-distance transport, or their children's dreams of other lives. The tide has rinsed the marshes countless times since those large harvests, and almost a century later I have found only one voice that fully remembered that world. John Fogg's written account is hidden away in the special collection of a coastal town's library: *They poled enough hay for the stack and I raked the scatterings. When they got the bottom started I layed the stack. When pitching they had to watch how the stack was being layed out over the staddle. The one on the stack had no way of telling how it was looking. They needed some more hay to make a good top. While they were out getting more, I had then a chance to look around and what a sight to see, all those doing the same as we were, stacking as far as the eye could see in all directions.*

Then, even a boy could build a haystack so solid that a weighted, tarred rope alone would protect it from autumn's storms and winds, so solid the memory of it might give an old man a foothold in a soft and noisy life where summer bathers—insubstantial in the shimmer of heat, smelling of sunscreen—face the open water, radios going. *That was the best part of the stacking, standing on the very top of the stack,* he remembers. And it seems he stands for a whole grounded world waiting for liftoff. It's sweltering in August. His face is flushed from work. Salt hay scratches at his eyes and throat. The sun goes in and out of high, fair clouds, and everything around him—the homes and home fields, the distinct salt sea, the marsh and its harvesters—is lit, in its turn, by the brilliance.

WHEN I READ John Fogg's account of salt haying, it was the scope as much as the work itself that drew me in—the world entire: *all those doing the same as we were.* A farm does need to be part of a world, not so different from the way an owl or even the common deer needs contiguous habitat—greenways, they are called—for survival. Now that agriculture is marginal here, a farmer might find himself traveling far to talk. "I go down to the gun show in Worcester, just to be around the fellows. They come from all over. A lot from the West—their places go on for days . . . You wouldn't believe it, they have tourists on the cattle drives now, taking the cows up to the summer pastures," one of my neighbors told me during an evening at the Grange. He went on: "Oh, how your father would love to hear that."

The Grange Hall still stands in the heart of our town, foursquare, muted and staid against the lit signs of the tanning salon beside it and the convenience store across the way. It's often dark except during the Monday evening meetings, the fall bake sale and yard sale, and the annual roast beef dinner. Most of the few dozen men and women who remain members of our local chapter have known one another all their lives, and have belonged since they were young. Up on the second story of the hall I've heard you can still read names carved into the rafters above the dance floor, which was said to be the finest in the lower valley for the way it would give under their stomping, kicking boots in the days when they flirted and tossed their partners—the Yankees whose ancestors had come over on the second sailing of the *Mayflower* mingling with the children of Poles, French Canadians, Greeks, Armenians, and Lebanese whose families had moved out of the cities to take over farms at the end of the nineteenth century.

There haven't been dances for decades. The largest gatherings are for the fall dinners, when we all contentedly eat thin-sliced beef, whipped squash, and green beans, and wash down the apple and lemon meringue pies with hot coffee. Henry Clough, who still keeps a straggling orchard that crests the top of one of the hills, is too blind to drive anymore, so he comes with Agimah, the foreman of his Jamaican apple-picking crew, and Agimah eats silently in our midst, a whole head above us, spine straight, not looking to his left or right, helping himself to seconds and thirds, while at his shoulders men and women cough and clear their throats and talk on about the old days. Most don't farm full time anymore (the younger farmers who do don't have time for the Grange). Their hay no longer feeds dairy herds, but is a fragmented commodity shipped out across southern New England to area racetracks and horse stables. Bales of the lesser grades are used by the construction industry as a buffer between work projects and wetlands, or for lawn decorations in the fall, or garden mulch. When I see trucks laden on the highways I imagine a field condensed and moving. The bales teeter, barely contained by wires and wooden cribs, and dry grass swirls through the air, kept aloft for a time by the turbulence and speed of oil tankers and long-haul trucks and cars going by. I can imagine the last wisps of it floating over the coastal hills, the slopes and curves of which swirl and pool with countless kinds of glacial soils—muck and gravel and clay and loam, soils which the Algonquin with their wooden and stone tools had no choice but to follow and which the first colonists tried to remake in the image of the world they'd come from, the swirl that exists still, even now that prosperity has been parsed from cultivation.

The land where we live, monetarily, is worth more now than ever before, while the Grange as a political force has become as marginal as the hay crop. The National Grange had been organized in the second part of the nineteenth century with the intent of consolidating the power of farmers so as to counter an increasing exploitation by the railroads and middlemen. It had a social aim as well: against the urban world, the farming life was seen as backward, and the Grange founders hoped to help farm families beat back that growing perception. They were determined to shore up the place of the farmer, both to dignify it again—revitalizing Jefferson's dream of the cultivator of the earth as the ideal citizen—and to make the farmer fit to meet the modern world. *Attend to every duty promptly,* directs a nineteenth-century pamphlet, *and keep constantly before the minds of the members, the important fact that the great and grand object and the crowning glory of our organization is to "EDUCATE AND ELEVATE THE AMERICAN FARMER."* But it was no longer a farmer's world alone, and the place they maintained would always be self-conscious, would always exist in relation to another kind of life. I'll always remember that my father would wash up and change his clothes before going to the bank, and he would always take the car, not the truck, to do his business.

Though its political aims have long since faded, the rituals of the Grange, which were tied to the structure of the English manor house, remain intact. Officers continue to carry titles commensurate with that world: Overseer, Master, Steward, Assistant Steward . . . When there is a new inductee for membership, the Assistant Steward leads the blindfolded candidate around and around the hall floor, which stands in for an English field: *To reach the Master's office we must cross an enclosed field, which we enter by crossing this stile.* [*Leads the candidate over the steps.*] *This portion of the field is being drained. Before us is a deep ditch, across which is a narrow bridge. Step with care.* [*Leads the candidate over the bridge.*] *We now enter the woods where the Laborers are at work chopping, preparatory to clearing the field.* [*The sound of chopping is imitated by four of the officers clapping their hands with regularity.*] *We here come to a narrow path in which there are obstructions. Step slowly and with care.* [*Leads the candidate over the obstructions.*] *As we leave the woods we find before us a freshly plowed field, which we must cross. The ground is soft and mellow, step carefully.* [*Leads the candidate over plowed ground.*] Once, twice, three times, four turns around the hall, and with each turn a world comes more fully into order. First a new member is granted status as a Laborer, then Cultivator, Harvester, Husbandman. Or Maid, Shepherdess, Gleaner, Matron. There are now more women than men at the meetings. Daughters of farmers, widows of farmers.

"Your father was a member for years," Mrs. R___ said when she asked me to join the Grange in the months after his death. Her request reached the part of me that has always been uneasy with home, careful not to completely define myself as *here.* After all these years back, I am far more comfortable as an *I* in this place than as part of its *we.* It's an awkwardness that appears even when I sleep:

I've dreamt of walking uphill in a dry field of waist-high hay. The way the field slopes uphill, I think it must be the one I used to pass as a child every weekday on the school bus, and every Sunday on the way to church. It would encompass my entire view for the moment we drove past: green in spring, flecked with insects and birds; the shadows and wind across the long grass in June, and again in late summer just before it was mowed a second time. Dried stubble. A field of snow. I think I was drawn to it even before I was aware it might be beautiful, the way it rose up towards the tall pines just beyond its crest where trees met sky. In my dream I hear it whispering as I wade through it. Then I come to a place where the field has been covered—tautly, neatly—with blankets and quilts. Some neighbors of mine appear and say they covered it because rain was coming and they wouldn't be able to get the harvest in before the storm. Their son, who used to take care of the haying, is no longer here, though I think he must still be everywhere in their world: even I think of him more often now that he has died, and always when I pass their house and the barn is wide open, and packed to the rafters with bales of hay. As I walk further up the field, I worry that I've made a mess of their blankets, which become more haphazard anyway, the more I dream, until the last parcel of the field is covered with bits of cloth of all kinds—calico and ticking, frayed corded bedspreads, rough canvas and old pillowcases . . .

I couldn't refuse Mrs. R___'s request to join, however awkward I felt, and I have been surprised by how comfortable I am when I occasionally attend Grange meetings and take my seat in a circle facing the others, my back to the stacks of folded tables along the north wall, or the jumbled goods along the south wall—toys and knickknacks, old baskets and racks of dresses and coats half covered with cloths and pushed into the corners to wait for the next yard sale. The expanse of maple floor between us shines with a half sheen: rich, old, worn into the present day. *Let the labors of the day begin,* the Gatekeeper announces. I touch my heart in greeting, and open my palm towards the center of the circle. Along with faith and charity, I pledge fidelity. I'm advised of the work of the season, just like everyone else: *Time to lift the mulch from your strawberries, time to put in the peas . . .*

When the piano starts up my voice strays a little behind the faithful as I sing "Simple Gifts" with a cluster of people who know almost nothing of my larger life but who knew me when I was young, and who've known my family always, whose quandaries are also my family's quandaries. It's a relief, a kind of simple seeing, far from my own ambitions and desires, since nothing is expected of me other than that my old child-self stand as representative of my family. What fidelity I have to the farm is taken for granted, a proper right, however much it can seem a mystery in the larger world.

Part of my ease, I know—and part of the assurance of the place—has to do with the fact that everyone remembers my father in all the stages of his life, and they remember him to me often; it is through him that I'm included in the conversation after the formal part of the meeting breaks up, after everyone has set down their instruction manuals and laid aside their velvet sashes. The voices that had so stiffly recited the ceremony relax as we all become neighbors again, sitting down for coffee and cake, dwelling in the past. I hover at the table, a little awkward in the freer moments, without the ritual to give me my place. "Sit down, sit down—have some cake. How's your mother?" Mr. R___ will say to me. The stories start: "I remember it was the first time I ever saw electric lights, and I just kept turning the switches on and off, I couldn't believe it. Everyone thought I was simple . . ." Inevitably: "Your father, Christ, he was over eighty, and picking corn three times as fast as those boys . . ." Somewhere in the future may come a time when my memory of him will be confined to my own thoughts, but as long as the anecdotes are repeated and stories I've never heard continue to unfold, memory feels as if it is more white smoke than ash.

At the Grange, on those evenings, if you were to ask about the future, you would be met with a quietness, feel a privacy close in. There's not a farmer there who doesn't know, now that land prices have shot up, that the stone soil they've spent their lives plowing and planting, haying, weeding, has an untold worth, too much worth for more acreage to be devoted to farming. In farm families, when the last of the oldest generation dies, even if the land has been farmed for nearly four hundred years, when it is appraised for inheritance purposes it must be appraised for its "highest and best use," so it is divided into theoretical house lots. The appraiser establishes frontage distances and roadways, and imagines dwellings at fixed intervals apart across back fields and lower fields. Since it is often well drained and clear, farmland makes attractive land for development. Lately in our town, prime one-acre house lots along back roads have sold for more than $150,000. The agricultural value placed on the same land wouldn't be more than $3,000.

If no extraordinary provisions are made beforehand, such an assessment often raises the value of a farm far beyond what the inheritors will be able to pay in taxes. One of the very few ways for the family to carry the farm through to another generation is to sell the development rights to the land. This procedure constrains the future use of the land, restricting it to farming alone, and this restriction is written in the deed so that every future owner of the land must abide by it. No one ever discusses what will happen to the land when there are no more farmers. With the worth of the land devalued, much of the tax burden is lifted. Most families who choose to follow such a route must make use of a combination of development rights bought by the state and bought by the town to secure their land. They may have to go before a town meeting to ask for public funds. Even in the most generous case, farmers won't receive the full development value of the land, since the state and town cannot afford to pay them what a developer would pay them. Such a course of action not only opens up the private family world to the scrutiny and debate of the public, it forecloses future choice. It can never do justice to all a family's wishes—each child has his or her own interest and ideas.

It's not surprising to me that few farmers avail themselves of the option to sell their development rights: it goes against a grain deep in most of them, who have viewed the land itself as their only legacy, integral to their lives and to their sense of independence, which has always gone along with working the land. They cannot imagine it may be different after their time. *We live in a dream world,* writes the contemporary political commentator George Monbiot. *With a small, rational part of the brain, we recognise that our existence is governed by material realities, and that, as those realities change, so will our lives. But underlying this awareness is the deep semiconsciousness that absorbs the moment in which we live, then generalises it, projecting our future lives as repeated instances of the present. This, not the superficial world of our reason, is our true reality . . . The future has been laid out before us, but the deep eye with which we place ourselves on Earth will not see it.*

In wider conversations about land preservation, the idea of the farm can be swept up with the more general idea of saving "open space," which might include any land not yet built on—swamps and woods, brook beds. Sometimes the fields themselves have not for a long time been seen as private but rather, in the diminished state of farming, have become associated with the character of the town, with the idea of a place, and their visibility is deemed their most attractive asset. The public conversation often turns to talk, not of crops and what can be best grown on a given field, or the quality of the soil and the incline of the hills, but of "viewsheds"—the way the farm looks to those going by on the road. The workscape has become a landscape pulled out of reason and proportion by a community's desires and memories.

Our own farm lies on both sides of one of the main roads into town, and the selectmen and town manager are eager to have it preserved. "We have always been a farming town," they say, "and what better view to give those entering town than real working fields." It will take a consensus of the wider family to preserve it. The original farm of the 1901 deed is owned by my father's remaining brothers. The newer fields my father had acquired in his life are owned by my mother, and the farm as a modern operation is viable only if all the acreage and buildings on both properties are used in concert. We were never prepared for this time—maybe no one ever is. The farm we were raised on had at times seemed to be the possession of all of the extended family. At other times it felt to me as if it was possessed by my father and brother alone. Sometimes I believe their quarreling was one of the ways the farm remained theirs, that it kept the rest of us from true involvement, kept anyone else from venturing a say. "I would like to have tried it," my sister says about farming here, all these years later, though she has built her own cleared place a hundred miles away.

Even in the days just before my father died you could see the boundaries they had set together. My father was in the hospital by then, in and out of coherence. My brother was late in coming. Where was he? When he finally walked through the door, I saw how nothing in the world could have touched the light across my father's face. I suppose if he had any last things he wanted to say, he would have wanted to say them to my brother. As it was, his unsteady voice was full of wonder and gratitude: "You came."

I count ourselves lucky that Dave has stayed on through the years and has kept the cultivation in a steady state. We have had the luxury of time to consider what to do with the farm, though we are all still unsure whether these days mark a beginning or an end, or remain an interim. Since no one in the family currently farms the land, the idea of its future feels diffuse, there's nothing specific to rally round, or to oppose. When my brothers and sister come back to visit, we have tentative discussions about how we might go about preserving the farm for the long future. We each have our own idea of what its future should be, just as we each have our own memories of it. To be responsible for so much land and yet to be separated from the work of the land itself is its own weight: the farm in our minds is growing more distant and abstracted. We are a loose confederation of siblings, no longer tied to place in fact, only in memory. We try to imagine what Dad would have done. We talk about what our mother might need for the future. How do you know what the future needs? It may be that the only way to preserve the farm would be to turn it into a kind of agrotourist destination with hayrides, apple picking, popcorn, petting zoos. Then it may no longer be understood as necessary to any kind of daily life, but may stand only as a symbol of necessity. Preservation, then, would be the most self-conscious of acts.

Sometimes I imagine this place will become a somewhere as clear and clean as the salt marshes. Removed from the human press of time, it will empty itself of memory, tracing itself back before our incursions on the land, and all the decisions, all the weight of human desires that were once imposed on it: the accumulation of judgments, the calculations, the care and the exploitations, attempts, histories, and abandonments. It will become close to an abstraction—"open space"—absent work, absent that exhaustion at dusk after a long day in the open air. Absent, too, will be the way a house feels when you come in from working in a storm, or in the cold, when the warmth rushes your face and makes you weak, tells you you don't need strength or endurance anymore—the relief of that—and the indoor quiet amid the bright domestic colors leaves you feeling a little deaf, even when you come in from the silence of a rimy October morning. "Open space" ready to be filled with a hundred dreaming gazes.

If it is open, at least it isn't abandoned or empty, the way places on the prairie now stand empty only a scant century after almost all the big bluestem was plowed under and planted in Old World grains and corn, after the frontier moved on and what was left behind came to be known as the heartland where you can trace by air the grid that laid out a world into great lots. *First it was a dirt road, narrow between two hedges, with a car crawling along it dragging a tail of dust,* noted a pilot flying over the Midwest. *Then the road turned off, but the line went straight ahead, now as a barbed wire fence through a large pasture, with a thin footpath trod out on each side by each neighbor as he went, week after week, year after year, to inspect his fence. Then the fence stopped, but now there was corn on one side of the line and something green on the other.*

Many midwestern counties are losing their population, and what rural jobs exist are not often agricultural: agriculture long ago married industry, and is now dependent on subsidies. It is a story that repeatedly appears in the *New York Times*: *Big farmers used their government checks to expand their acreage, buying small neighboring farms, and increased their production, which pushes down the world market price. They are still profitable, because government subsidies make up the difference . . . In Nebraska, nearly 70 percent of all farmers rely on government largesse to stay in business. Yet the biggest economic collapse is happening in counties most tied to agriculture—in spite of the subsidies.* Those bigger farms require more and larger machinery, which demands more production to make the enterprise profitable. There is a liability in the expanse of the land, in endless possibility. More production demands more acreage for farming, and farmers find themselves competing for whatever becomes available: *In the complicated equation that is modern agriculture,. the mere size of one's farm can mean the difference between prosperity and failure. The competition to snap up whatever becomes available—the "land battle," one farmer here called it—can be fierce. Increasingly expensive tractors and combines and other mounting overhead costs . . . have led some farmers, whose families once planted just a few hundred acres, to farm as much land as possible . . . "You have to spread these costs over more acres . . . Every farmer feels the need to grow. That's just the way the business is."*

Meanwhile the grid has moved skyward: we are more familiar with images, not of men guiding self-scouring plows, but of men walking on steel beams seventy stories above street level, while the same story plays itself out again and again as populations all over the world leave the countryside for cities, making much the same sojourn as the New England farm girls had in the early nineteenth century, and as my grandparents had in later decades. Now there are almost twice as many people in New York City as there are on all the farms in America. *We are a cross-section of the entire world . . . a densely crowded ethnic hodgepodge, and the potential for chaos is enormous,* Paul Auster writes from the city. And: *We are the true heartland.* The word feels no more fitting for New York City than it does for the Midwest in a country where human experience refuses to be fixed in place, where every place absorbs change and shifts beneath our feet, and in our imagination, country of eternal immigrants now with our inner resistances in tow, now with our inner resistances breached.

In the prayers of the Grange the afterlife is called *the Paradise not made with hands.* It is this I keep remembering long after I've forgotten our secret password and secret handshake, when to stand up, sit down, after the rest of the rote and ritual fall away. The phrase itself suggests both an allegiance to the love of the labor and the dream of release from the love of the labor. I wonder what this place will be without these farmers. What does it mean to want their life to remain, to count on them for a sense of gravity, even as I could never abide the same demands for myself? I keep thinking of this one evening when after cake and coffee we all go out into the night. The clocks have been turned back, it has been dark for hours, the air is dense as sable. My neighbors are hunched into their bulky coats and they call to each other—"Cold . . ." "Maybe rain . . ." "Goodnight . . ." " 'Night . . ."—as they walk towards their cars by ones and twos. It seems to me that the dark mythologizes them. I feel myself mythologize them, my own separate life mythologizes them. They remind me of Brueghel's drawing *The Beekeepers*, with its three methodical, slow-moving men clothed in protective frocks, isolated each to each. One is carrying a hive away, one is lifting the lid off another hive, the central figure is moving towards or away from a task. They aren't looking at one another or moving outside the work as they walk among their hives concentrating and alone. Their stout bodies are hardly animate, hardly even clay—when I first saw the drawing I thought it was unfinished, that the faces had not yet been drawn, and they reminded me of ancient, severely pruned apple trees with limbs cut off mid-twist, no longer reaching skyward in taut and graceful contortions, but querying forward, until slowly I understood it was really the beekeepers' gauze masks I was seeing, covering the stiff extended hoods of their robes. Their facelessness makes them appear a little furtive as they go about their tasks, though in truth they are no match for the heavy but agile boy—free of protective clothing—who, behind their backs, is sneaking up a tree, advancing towards a wild hive. *He who knows the nest knows it*, says the proverb in the lower left corner of the drawing. And *He who steals it has it.*

Yet even as I wonder about the old life of farming, a new, less obvious life is taking hold. There are always a few men and women, new to farming, who cultivate a handful of acres here and there, or come back to old places after having lived away. Their holdings may be no larger than the first few acres set atop the village at Patuxet, and they are sometimes dismissed as "hobby farmers." The markets they serve may seem rarified: goat cheese and antique apples, for instance. That they have established a new world of their own choosing may be a part of the dismissal: they can appear freer on the land, cultivating a past free from the past, without the confinements of a personal history. We always seem to want farmers to be flinty and difficult, born to it somehow. I caught myself at it when I learned that José Bové, the French activist farmer who'd dismantled a McDonald's and has served time in jail in his stand against agribusiness and genetically modified crops, had come to farming after years of being a college activist. I'd felt a little disappointed that his passion hadn't risen up out of the old life.

My neighbor who has come back to her family's place after a lifetime of teaching in other states, wanting to meet a little way the idea of a self-sufficiency, to raise a few chickens, hay, have a garden of her own, says, "We need to keep the land open, there might come a day when we'll need it again." Her words may at first seem fanciful, or wishful in their implication of necessity, but she is perilously correct. Massachusetts grows only 15 percent of its own food now, and however full and gleaming the supermarkets appear, there is only three days' supply of food on the grocery shelves. You need all kinds of men and women working the land to preserve agriculture, to do the work of feeding a thousand, or five, or one. Yes, you save a small patch of land, and the hope is that one here, one there, discontinuous as it may be, small as it may be, will somehow keep a ceiling lifted, and bring pressure to bear from below as the world moves on. It isn't inconceivable at all that we may need to fall back on ourselves and our labors, and the land underneath our feet, which carries the mistakes and knowledge of centuries, that we may need the land to mean once again what it had meant. *We heard a distant tapping on the road*, Edwin Muir wrote in his poem "The Horses," as he imagined recovery after catastrophe:

> *A deepening drumming; it stopped, went on again*
> *And at the corner changed to hollow thunder.*
> *We saw the heads*
> *Like a wild wave charging and were afraid.*
> *We had sold our horses in our fathers' time*
> *To buy new tractors. Now they were strange to us*
> *As fabulous steeds set on an ancient shield*
> *Or illustrations in a book of knights.*
> *We did not dare go near them. Yet they waited,*
> *Stubborn and shy, as if they had been sent*
> *By an old command to find our whereabouts*
> *And that long-lost archaic companionship.*
> *In the first moment we had never a thought*
> *That they were creatures to be owned and used.*
> *Among them were some half-a-dozen colts*
> *Dropped in some wilderness of the broken world,*
> *Yet new as if they had come from their own Eden.*

Since then they have pulled our ploughs and borne our loads,
But that free servitude still can pierce our hearts.
Our life is changed; their coming our beginning.

THE BOUNDARIES OF our farm have stayed the same for as long as I've known them: what has been woods has remained woods; fields, fields. But as the world surrounding it has increased in density, as most other farms have gone out and the valley population has grown and spread into suburbs, the possibilities of the farm have grown. The small farm stand I worked in as a child had become a store by the time I returned—my brother saw to that—and now, along with what is grown on the farm, Dave sells milk, pies, bread, and juice. He travels to the Chelsea wholesale market to bring back lemons, limes, potatoes, and onions. He is blurring the line between here and there. His customers are busy people and would like to stop once for everything they need.

"It's something to see," Dave had said of the Chelsea Market, and he asked me if I might like to go in one morning. "Sure." We left well before sunup, maybe 4 a.m., though already the interstate into Boston was scattered with early drivers heading for work, and more and more joined us as the city skyline grew closer. It looked as it often does to me, approaching it from the north—a kind of cartoon of a city hugging the curve of the earth. You see the twentieth century—the financial district, the Prudential, the Hancock—long before you see the seventeenth, eighteenth, nineteenth centuries: Beacon Hill, the State House, the Customs House tower with its eagles of Cape Ann granite. Since it was summer the towns along the way were hidden behind maples, oaks, and white pine, and the city seemed to rise out of green. It appeared as if we were only an artery flowing towards skyscrapers.

We exited the highway before the descent into the city, just as traffic was slowing a bit, and entered into flat country full of rivers and creeks wending their way towards Boston Harbor. I could sense right away how it must once have been marshland—the lay of the land was so low and level. On the land beyond the warehousing and houses and industry, I knew that phragmites would have taken hold. It's a taller, coarser grass than salt hay—it's wild with height, as high as the big bluestem of the old prairie: when you are in it, you can only see straight up. In other places in the world phragmites had been used for thatch and for paper. Here it is considered an invasive species, a threat to the remaining salt marshes since it grows voraciously in places where the flow of tides has been restricted by development, where the salinity is lower. Phragmites proliferates by rhizomes, the way big bluestem once had, and like bluestem comes back stronger if it is burned. Its plumes of purple go gray in fruiting time. They say it's going to outlast everything.

In a small truck like Dave's you feel overwhelmed on the road to Chelsea long before you reach the market. You are swamped in a phalanx of long-haul trucks gearing down, trundling along the narrow streets among dark sleeping houses. All night long they've been hauling their goods against the flow of the old migrations, making noise like the sea at a distance, hauling their cargoes across international borders and state lines, through weigh stations and tollbooths, across bridges, through tunnels, carrying the world and all its seasons. The sound of their gears is heir to the whistle of the train Thoreau had heard from his cabin *sounding like the scream of a hawk sailing over some farmer's yard, informing me that many restless city merchants are arriving within the circle of the town, or adventurous country traders from the other side . . . Here come your groceries, country; your rations, countrymen! Nor is there any man so independent on his farm that he can say them nay. And here's your pay for them! screams the countryman's whistle . . . I watch the passage of the morning cars with the same feeling that I do the rising of the sun, which is hardly more regular. Their train of clouds stretching far behind and rising higher and higher, going to heaven while the cars are going to Boston . . . There is no stopping to read the riot act, no firing over the heads of the mob, in this case. We have constructed a fate, an* Atropos, *that never turns aside. (Let that be the name of your engine.)*

Once we were among the market buildings—long, low warehouses that could just as well have been housing air conditioners, washing machines, or latex paint—I could see men built like stevedores zip pallets stacked with crates across the cement floors. I could hear voices amid the bleating of the backup warnings. "Got to move product." Everything seemed to be called "product": clementines, frost-sensitive avocados, winter-hardy cabbages, Chinese eggplant and Italian eggplant, tomatoes sized and sorted and priced for every market: grape and cherry, greenhouse slicers grown in a soilless mix, field-ripened from the southern hemisphere, tomatoes still on their vines, wrapped in paper, or wrapped in plastic. There's little reason anymore to be bound by soil or climate, latitudes and ocean currents, or the till the ice ages left behind. Everyone can have peaches, rhubarb, and strawberries in midwinter, though the price and frequency and quality still depend on soils somewhere, and on sun and rain. On any given day in the Chelsea Market the pineapples may be at their peak and smell sweet, but the melons may still be green, the grapefruit of fair quality, and the quantity of plums insufficient to quote. The air never clears of the myriad scents of the tropic and the northern, all latitudes, every scent converging along with every season. Most food now travels more than 1,300 miles before it settles on someone's dinner plate—20 percent of the traffic on the roads serves the food system—and before it settles on a plate it must converge in a terminal market like Chelsea where the same varieties of apples may be offered from China, Washington state, New York, and maybe even New Hampshire, and they will be far more uniform than the ones Thoreau found on his walks more than a hundred and fifty years ago: *In 1836 there were in the garden of the London Horticultural Society more than fourteen hundred distinct sorts. But here are species which they have not . . . There is, first of all, the Wood Apple (*Malus sylvatica*); the Blue-Jay Apple; the Apple which grows in Dells in the Woods (*sylvestrivallis*), also in Hollows in Pastures* (campestrivallis); *the Apple that grows in an old Cellar-Hole (*Malus cellaris*); the Meadow Apple; the Partridge Apple; the Truant's Apple (*cessatoris*), which no boy will ever go by without knocking off some, however* late *it may* *be; the Saunterer's Apple,—you must lose yourself before you can find the way to that; the Beauty of the Air (*decus aëris*); December-Eating; the Frozen-Thawed (*gelato-soluta*), good only in that state . . .*

I believed the entire world was converging there on that built-up marsh, and would be bartered for, and then dispersed to other corners of the earth. Amid the noise, the rush, the large trucks, the pallets stacked high with crates, I felt how small our own enterprise was, and how defensive was my feeling for it. I wanted to believe the smaller world would persist as something more than a last illusion, as a stay against the confusion, and I thought I could see neighborhood shopkeepers going from merchant to merchant each on their own worn path—the same as a deer through a woods—to favored vendors. The Chinese proprietors sought out their own, and would carry lemongrass and Asian greens back to their neighborhoods; the Italians, eggplant and peppers and squash; the Hispanics, yucca, calabaza, sugarcane.

As we walked through, Dave talked mildly about buying in some early corn: "Everybody is already asking for it." It would be another month before his own would be ripe, but the stores had long been stocked with sugar-enhanced and supersweet varieties—sweeter than the local ever had been. For decades there had never been a crop more tied to locale than corn, available in our part of the world from mid-July to the frost. It had evolved from Indian meal to sweet corn, and sweetness had become its most fleeting, desired characteristic: its sugars turned to starch the minute it was picked from the stalk. Sayings grew up around it. "You've got to have the water boiling before you pick it." "It's no good the second day." Local farm stands depended on that urgency, they were the only ones who could assure a daily freshness, but in the last several decades science has enabled sweetness to keep, a sweetness that swamps the old cornier and mild sweetness, that can make corn taste like candy, really, even days after it's been picked and shipped. In doing so it has removed some of the allure from the local.

Long since those eight sketched fields of it shown on Samuel de Champlain's map of Port St. Louis, corn has become the predominant crop on American soil. It now has more uses than the Wampanoag could have dreamed with their steamed cakes and parched maize in their hunting pouches. We don't even think about most of the corn we consume. High-fructose corn syrup—so much cheaper than sugar—flavors sodas and sweets, and the cattle who once thrived on a diet of salt hay, and then tame grasses, now live on a diet of corn, which enables beef cattle to gain weight more quickly and to be raised year-round. The beef is marbled and fatty, which is what we have come to prefer even though corn plays havoc with their health, making antibiotics a necessity.

Feedlots have shrunk the need for large areas of pasture and hay, and the herds are concentrated down into a muddy, trampled pen where cattle stand in piles of their own manure, and in the hot dry evenings plumes of manure dust travel on the wind. Lagoons of manure leak and sometimes burst, or send ammonia and methane into the air. Zinc and copper, which had been added to the feed, build up in soils because of the overapplication of manure onto cropland. In her book on new infectious diseases, *Secret Agents*, Madeline Drexler writes: *The site of modern meat production is akin to a walled medieval city, where waste is tossed out the window, sewage runs down the street, and feed and drinking water are routinely contaminated by fecal material. Each day, a feedlot steer deposits 50 pounds of manure, as the animals crowd atop dark mountains composed of their own feces.*

What does a field mean now, a field that had its beginning in the vast expansions of those first colonial settlements into space for their cattle, the dream of the English field, its order and correctness and idea of husbandry, the space that spurred the war with Philip, that distinguished a world from the wild, that spread even to the salt marshes? As the early planes take off from Logan and the sun rises over Deer Island, over the bones of Christian Indians who died there during the winter of Philip's War, the produce prices in the market are dropping amid the smells of concrete, waxed cardboard, plastic, diesel, of vegetable rot and of ripeness: cold keeps the produce fresh only for so long—once it is on the open market, it begins to spoil, and tomorrow more will be coming from Florida, Mexico, Australia, New Zealand, and California. They say "the produce starts in California and comes across the country. After Boston, there's only Boston Harbor." The Haymarket vendors are moving in to take advantage of the morning surplus, and representatives from the food banks scour the place for donations. Somewhere beyond us the phragmites stirs in the first breezes of the morning.

6
Wilderness

IN RECENT YEARS a few of the local farmers, those who've been determined to continue, have sold their family holdings in the area and moved north to Vermont, New Hampshire, or Maine, following the old migrations up the rivers, inching towards the limits of wide cultivation, adapting to different soils, to colder valleys with shorter growing seasons, nearer the steeper lands that, before white settlement, had been largely the province of hunter-gatherer tribes. The sale of their old farms buys two, three times the acreage in the north and also, strangely enough, places them in a stronger farming community—or at least in a less suburban one, where many yards still contain vegetable gardens fenced off from deer. Though the move may appear to be sudden—after the death of a child, say—always I think the decision must have been long in coming. Often it leaves rumor whispering behind—"How much did you say he sold the land for?"—that persists even after the old fields and orchards have filled in with horseshoe streets, and other dreams have taken hold. For myself, I still think of the north as its own country, one that takes me out of the world I know, and I go there when I simply need to walk clear of it all.

It's said New Hampshire's White Mountains were named by sailors who saw the distant winter peaks from their ships, elemental and inhospitable as the gray-green Atlantic they were sailing on. By now all the summits of those ranges have been climbed, blazed, recorded, and owned, written about, legislated, and regulated, though to me they can feel as if they're still purely imagined, the way they were before maps, before trails cut through them. The name, anyway, is as true now as then: *white*, meaning fully luminous, the color of salt and milk, of frantic seas and the albatross. Even on a clear, warm June day when they are gray-blue rock against a summer sky, the summits of the Franconia Range—Lafayette, Lincoln, Little Haystack—seem nearly inaccessible and beyond human bearing—as far as a sailor's glimpse of them. *Those sublimer towers*, Melville called them, *whence, in peculiar moods, comes that gigantic ghostliness over the soul at the bare mention of that name . . .*

Thoreau knew what it meant to climb such peaks. Maine's Mount Katahdin caused him to write: *Some part of the beholder, even some vital part, seems to escape through the loose grating of his ribs as he ascends. He is more lone than you can imagine. There is less of substantial thought and fair understanding in him, than in the plains where men inhabit.*

His reason is dispersed and shadowy, more thin and subtile like the air. Vast, Titanic, inhuman Nature has got him at disadvantage, caught him alone, and pilfers him of some of his divine faculty. She does not smile on him as in the plains. She seems to say sternly, why came ye here before your time? This ground is not prepared for you. Is it not enough that I smile in the valleys? I have never made this soil for thy feet, this air for thy breathing, these rocks for thy neighbors. I cannot pity or fondle thee here, but forever relentlessly drive thee hence to where I am *kind.*

As I stand at the trailhead at the base of Mount Lafayette on this June day I do feel free and clear, but also a little overwhelmed at the thought of the climb—elevation 5,260 feet—for the top really does look like another country, and the clusters of hikers I see picking their way across an exposed ridge look like burdened pilgrims heading for another life. But the trail in front of me, an old bridle path, is wide and obvious from nearly two hundred years of use. As early as the 1820s large hiking parties ascended Lafayette—first known as the Great Haystack—on foot and horseback: rusticators, adventurers, tourists, and botanists who'd gained the trailhead by way of wagon, then steam rail. They climbed through the Industrial Revolution—dressed in dun-colored woolens and leather boots—and a handful of wars; they slept in the grand hotels in the Notch and took their ease at the top in a stone summit house; they climbed through the haze of timberland fires burning on the flanks of the mountains, then through the atomic age, and the age of information, until it is our own strange selves walking in their footsteps, arrayed as we are in reflective and fluorescent microfibers, big in the shoulders, never out of reach of voices, looking like bright daubs of paint against the subtle colors of the landscape. So many have ascended that experienced hikers sometimes derisively call this path the "Bridle Trough."

Even in its well-worn state it feels entirely challenging to me. As a child I'd wandered freely through our mild, cut-over woods around the farm, but I'd climbed only a few modest peaks when I was young, and since I'd been part of a larger group—from school, say, or 4-H camp—I was always insulated from the larger challenges of the trail. Now I'm always cautious when I climb on my own, but I like the alertness it takes, the way the need to feel a bit apprehensive also makes you feel a bit new, the way I do now as I face the trail, which begins by rising up on gradual switchbacks sheltered in the dappled light of birches. It's easy going for the first mile, on a path of moss and soft dirt that is still tied to a kind of valley-world forest with its familiar, moist smells of duff and acid earth. Then a set of stone steps carries me towards the ridge, and the birches drop away. The trail narrows and steepens as it begins its climb through straight, true red spruce and hemlock. I get the first hint of the tonic scent I have been waiting for: spicy, sharp, northern, the scent of a world that cannot possibly be cultivated. Where red spruce grows, the frost-free days are too brief, and the growing season too short to be of much use for farming. The woods become less and less characteristic as I go. Soon the spruce begins to grow stunted, no taller than human height, and toughened by the cold and exposure of the higher elevation. The ridgeline trail narrows, the smells grow even sharper and less intimate, the soil on the path is gone. It's no longer a thoughtless ascent as I clamber over rocks and work up narrow chutes of tumbled stone, and sometimes rely on the staunch trail-side trees to pull myself up. Some have been grabbed by so many hands that their bark is polished smooth.

On every side of me the season is retreating: flowers that had already bloomed and shriveled in the birch woods are in full bloom among the spruce: drooping yellow heads of the bluebead lilies nod towards the earth, and the bunchberries are a creamy whorl of petals. I grow cold when I stop for more than a minute. Further on, the bluebeads and bunchberries are still buds, and red spruce gives way to balsam and black spruce, which in turn grow lower and lower—shoulder height, then waist height, growing as they can, sculpted and half killed by wind, rime, ice, and snow. It's not that they were made for the place, it's that little else can survive. Tight, uncountable growth rings mark incremental yearly progressions. They may be older than any trees that tower over the valley trail, maybe they'd been old even back in the heyday of the grand hotels in the Notch, when the logging railroads followed the East Branch of the Pemigewasset River into the prime forests. Now two young girls, their backpacks sagging under the weight of hats and windbreakers, chocolate, water, and raisins, pause among a ragged clump of them. The ridge drops off steeply from where they are gazing out at the fir-covered, rock-scarred slope of Mount Lincoln. I hear one, in a clear, pure voice, exclaim: "Don't you just want to fly? Don't you wish you were a bird?"

If you climb high enough, you enter a world with its own vocabulary, one specific to geologists and climbers: boreal, alpine, strange. Two hours from the valley floor, the shallow intractable eye of a nearby bog is called a *tarn.* And *col,* meaning neck, little saddleback, is a dip in the ascent, a sheltered respite where wood sorrel spangles the duff. What trees there are lie nearly prostrate, no more than knee high, with sparse flecks of green among mostly bare and bleached tops the color of driftwood, great rags of it, flagging stiffly away from the prevailing winds. It's called *krummholz,* and everything weak about it has been pruned away by scouring, steady gales and winter ice that beards the branches. Thoreau, when he climbed Mount Katahdin, walked on top of such trees, and they supported his full weight. On a perfectly clear June day, the sun strong, the warblers darting in and out of cover, krummholz alone makes the harshness of the place fathomable.

Several trails converge near the col, and there is a wooden shelter where hikers can get fresh water, splash the salt off their faces, and sleep overnight in summer. Benches, wood floors, pine tables, cookstove, much of it was hiked in more than half a century ago. In early afternoon the caretaker stands in the long cool shadows of the kitchen punching down bread dough and shaping long loaves for a second rise. An old man just down from the summit rinses out his shirt and hangs it over a railing to dry. On summer nights the hut comes to life at the end of the day as hikers who'd arrived in ones, twos, fours, sevens congregate for a noisy and jovial dinner. Afterward there's chatting, and games among children. Some walk off to see the sun go down over the western mountains, the moon come up in the east. The night distorts the distances so that the ridges ahead—the way to go—seem unfathomable, and the lit town to the northeast feels as far off as the summer stars. In that remoteness you can feel strange among so many people, maybe a little resentful that so many have come the same way. You might want a clear quiet.

On a night I spent there years ago I lay in bed after the lights were put out and imagined how we all disappeared to that world, to the million stars, to the dark humps of scarred and toughened land. Were we dreaming of becoming kin to the people Crèvecœur had imagined after agriculture—*a still simpler people divested of every thing beside hope, food, and raiment of the woods: abandoning the large framed house, to dwell under the wigwham?* I tossed fitfully in my bunk in a room where twenty others slept, each one curled under three gray wool blankets. In the night someone opened the door for air, a little later someone else closed it against the cold. Below us: stunted spruce running down into the valley; above, the soilless, exposed world beyond treeline. And we slept, souls on the edge of *something*. Someone talked on and on in his sleep, another sighed and turned towards the wooden wall. Snores deep and grainy, snores light and aspiring, puffs of breath, simple even breathing, stops and starts, and—once in a while—startling gaps of no breath at all made a kind of night music of the exhausted.

Past the hut and the col, the spruce—even in its wind-stunted state—soon disappears entirely, and I lift myself above treeline where cairns mark the way over rocks and boulders and tumbled stones. Beyond agriculture, and then beyond timber, there is nothing to do but put one foot in front of the other and look out on the vast stretches of earth, at the patches of grasses and alpine flowers pushed to this extreme by the advance of the spruce forest when the last of the ice melted away. Here they can thrive without competition, as they once thrived in the ice ages, and do now in the Arctic, small and hunkered to bear the winds, the cold, the short growing season. Meadow grasses, and flowers, worts, mosses. Lichen on everything, on the last of the blown twigs, on rocks and scoured earth. On such land Thoreau sounded as bewildered as a spaceman tumbling out beyond Pluto: *This was that Earth of which we have heard, made out of Chaos and Old Night. Here was no man's garden, but the unhandselled globe . . . What is it to be admitted to a museum, to see a myriad of particular things, compared with being shown some star's surface, some hard matter in its home! I stand in awe of my body, this matter to which I am bound has become so strange to me. I fear not spirits, ghosts, of which I am one,—that my body might,—but I fear bodies, I tremble to meet them. What is this Titan that has possession of me? Talk of mysteries! Think of our life in nature,—daily to be shown matter, to come in contact with it,—rocks, trees, wind on our cheeks! the* solid *earth! the* actual *world! the* common sense! Contact! Contact! Who *are we?* where *are we?*

How different it must have been for him, who had to cut his own way up the mountain, who'd climbed above treeline out of an agrarian society in a time when the quality of soil marked your life and determined your chances, when almost everyone lived according to the seasons, and a summit was the limit of the extreme, yet attainable, world, back before speed and flight, and the modern city rising up in glass and steel, before small talk and golf on the moon, and the photograph of our solitary blue planet, floating within the permutations of its clouds in the black *no* of space.

To ascend still feels rare to me, especially in the last moments of the climb, when I am hunched down, clambering over the strewn stones, placing every foot and hand, reading the hard surfaces from cairn to cairn until I lift myself up to the highest point, and am at last free and upright and seeing everything of a vast, ever-flowing green world, all the living, and fallen, and decaying merged in a trackless expanse of green and green and green as my labored breath and heart slow and settle bit by bit in the absolved, exhausted ease of attainment. The top of Mount Lafayette is as high as I have ever climbed on earth.

Even here dozens are with me, milling around, resting, drinking, eating. I'm sure many have been higher—for some this is an almost usual day, for others this is a singular aspiration, and together we form a mountaintop society. Whispers and murmurs, talk rising and falling. "Do you have some water?" "Sure is beautiful." "Are you getting cold?" "We should get going before we stiffen up." A young girl applying moleskin to her heel reminds me of the alabaster statue in Rome of a boy pulling a thorn from his foot. Someone is crouched in the last remnants of the old summit house—all but the stone foundation has blown away—her face towards the sun. For now it *is* kind on the summit, the wind softly, pleasantly, buffets my ears, and nothing, no trees or edifices of any kind, stops it above me. To the west, and tiny, tiny down below, are the highway through the gap, the Cannon tram, white spires, clock towers, white clapboards on white trim, white, white, the settlers thought best to stand for order in the wilderness. Valley people contented with the valleys. My father never climbed a mountain so far as I know, nor my mother either.

Almost as striking as the vast views is the path that links summit to summit along the Franconia Range—just scrabble, the gravel of old ice ages—defined by a low scree wall: Lafayette slung to Lincoln, Lincoln to Little Haystack, Little Haystack to the peak beyond. If you follow it south you will reach Georgia, if you follow it north you'll reach Thoreau's Katahdin. The way to go on the ridge has been made as narrow—and full of possibility—as a walk beneath the barrel vaults of a monastery. The reason for the path is practical: we are too many to wander freely. Inheritors and intensifiers of the rusticators' dreams, we walk in such numbers one after another that a litany of alpine flowers have disappeared from the popular peaks. DO NOT STRAY FROM THE PATH, the signs have been saying along the way. CARRY OUT WHAT YOU CARRY IN. NO CAMPING. NO FIRES. NO SLEEPING ABOVE TREELINE.

All the footfalls have taken away some of the apprehension of the wild, since they largely free you from maps and compasses and the need to scale old snags to gain a long view. They've dissipated, too, the intense concentration and keenness that comes with relying on yourself to keep your place at all times, conscious of every footfall, and of what might be coming towards you: *Though in many of its aspects this visible world seems formed in love, the invisible spheres were formed in fright.* In good weather even a path the width of a harrow line through dense stands and flagrant undergrowth, one kept clear by only casual passings, one that might be lost in dark or rain or snow, is an ancestral trace not so distant from the arrow in the bone and the light of cooking fires. Then you can fully believe you only need keep clear what has been made, and go on. But in the instance of a cloudburst racing across from the west, you suddenly understand that the ease is all chance. We walk in danger of sudden storms, and fronts moving in, fog and cold and rain, and voices veering away like birds caught on a wind.

The eastward view from the top of Lafayette has no villages in it, no spires or white paint, just mountains beyond the mountains until they meet the soft and hazy far sky. This is the Pemigewasset Wilderness, the heart of the White Mountain National Forest, officially designated and protected. You are not gazing all the way back before design and ambition. It is only distance, the overview, that makes it appear as some kind of primordial beauty. All that unbroken forest has been inextricably mixed up with history, a part of our own complicated past, with its exploitations, its ideas of the romantic and the sublime. Had I stood in my place among the ageless lichen a century ago, I'd have seen huge clear-cuts where stands of trees had been felled, trimmed, and dragged out. The cuts, some reaching up the side of Mount Lafayette, would have been littered with slash and deadfall. The East Branch and Lincoln Railroad stretched into the timberlands by following the Pemigewasset River, and there were logging camps throughout, painted red, and built so they could be taken down and transported by train to be set up again farther in, near new cutting operations along Cedar Brook, along the flank of Mount Guyot, at North Fork junction . . .

It took far less than a century to clear those tracts of land, and for a while it was a world entirely its own. French Canadian loggers, Nova Scotians, and New Englanders lived the winter in the woods, sawing and striking, felling tree after tree. On a winter day snow squalls could obscure the entire Franconia Range, and still they worked on, not seeing beyond the world of their saws and chains, and the surrounding snow-draped trunks of spruce. Some were trying to make money for the sake of a little farm in kinder country, others looked forward to a marriage, or worked to send money back home. For some this was all, a life made of following the work. And in the hours that weren't work, in the long northern night, a hundred men slept two to a bunk, snoring, spitting on the floor, singing French songs in their sleep. Frost on the windows as they slept on in the cold. Meat and doughnuts, brown bread and beans, salt cod on Fridays. Sundays one of the men, sitting in a rocker cut from a barrel, might read a paper to the others. Someone in a corner might whittle an ax helve. Emerging from the woods to return to their farms at sap season must have been like emerging from a fogbound sea to realize they'd lived a parallel life, completely separate from village and farm. Remote, isolate, partial.

It is desperately clean work that is going on in the East Branch Wilderness. There was little talk—simply concentrated effort and energy, and throughout it all a perfectly apparent genius of direction, wrote one visitor to the logging camps. For however particular their individual dreams were, their collective efforts cleared the valleys, cleared half the east flank of Lafayette, and other peaks on the Franconia Range. Slash lay in great black heaps on the snow, then was covered by new snow, then shone with meltwater. It dried in the sere exposed heat of summer until a spark from the railroad lighted on the refuse, or lightning struck, say, on the east side of Owl's Head, as it did in August 1907, and the ensuing fire swept Mount Bond and burned over the easterly slope of Mount Garfield, the slope of Mount Guyot, and the northeasterly flank of Mount Lafayette.

The smoke, driven before the northwest wind, obscured farther mountains. It went beyond Lake Winnipesaukee and lay over the valleys to the ocean. Another current carried a long streak of smoke towards Vermont. The glow reflected on the clouds and caught the attention of summer visitors watching from their porches, for tourism was on the rise on the west flank of the range even as the timber prospects on the east side were playing out. Though large parties of men worked at fighting the fire, only rain and the dying of the winds ended it. A country, then, of charred stumps and dried streambeds. Though there had long been tensions between the logging and tourist industries in the mountains, when the smoke cleared it was obvious that tourism and logging could not coexist on the Franconia Range, at least not on the scale desired by both industries.

Now in the Pemigewasset Wilderness, the railroad bridges have dropped into abysses. The main hiking trail into those woods is the old railbed that follows the East Branch of the Pemigewasset. You slip onto it off a major highway, and walk on and around the last of the decaying wooden sleepers. Iron spikes shine out of the trail dust and softening wood. The old logging camps along the way have been burned to keep hikers from sleeping in them, or have been salvaged—some say you can still spot the distinctive red boards in the nearby towns—or have collapsed. But there are stone foundations here and there, and mossy, lichen-covered bridge abutments. The river that runs alongside has its source on the flank of Mount Lafayette. Its bed is of tumbled stone and boulders, and when the spring meltwater subsides, there are places shallow enough to cross on the rocks, and pools deep enough to swim in. I don't know why, but even five minutes in that water will leave your old skin feeling softer than a child's.

Almost in answer to all the ambition and excess determination of the timber barons comes an ambition and determination not only to obliterate the marks of those logging days but to create something absolutely pristine. The Wilderness Act of 1964 decrees that the land it protects shall be *an area where the earth and its community of life are untrammeled by man, where man himself is a visitor who does not remain . . . retaining its primeval character and influence, without permanent improvements or human habitations, which is protected and managed so as to preserve its natural conditions* . . . One of the insistences of a designated wilderness area is that no preexisting human structures are to be maintained, that the past should be encouraged to disappear. They say if you walk far enough into the Pemigewasset Wilderness you'll see the last railroad trestle still standing. I imagine it shrouded in woods and silence, the rock-strewn river running under it: care and time and effort abandoned just as in my own valley world. It stands odd, harmless, representative, and argued about vehemently. Some insist it is a nonconforming artificial structure in the wilderness that must be removed; others, an important historical artifact that by law must be preserved, suggesting everything of the long quarrel between remembrance and forgetting, of how much more than the swirl of snow, or the smoke and heat of fire, or the rate of decay separates us from the loggers who once worked there. It can seem that more than a century separates us. We hikers are now so many that we constitute our own genius of direction as the known world and the felt world, the maps of science and the maps of belief, become more distinct from one another.

And now the desire is not only to forget but to feel new in the wilderness. The extreme is considered more and more to be a testing place, matching the will against elevation, stone, and cold to make a hurt that keeps you from absence. The paths are not only maintained but made difficult. The Appalachian Trail has departed the valleys and towns and roadways—where habitation and wilderness might mingle—and has become more a trail of mountain ridges, full of desolate challenges. If you want to complete the Appalachian Trail today you'll have to wade through rough waters and crawl through icy caves. At the end of his journey, Earl Shaffer, a seventy-nine-year-old hiker who, in 1998, walked the trail for the third time in his life, couldn't disguise his bitter disappointment at the way the route had changed over time: *In 1965 the trail was perfect*, he said, *but they were not satisfied. They make all these changes. They seem to be obsessed with the idea that you have to make it as rough as possible . . . It's an almost impossible trip. Who wants to go out and wade across a 100-foot icy river barefoot?*

Who *are we?* where *are we?* Thoreau had asked, and it seems to me the extreme did not define him. It undefined him.

When it's time to climb down I do so in a blur of weariness, picking my way over the tumbled rocks on the path. I have to hold myself back all the five miles and favor my left knee. Halfway down I start to hear the trucks passing through the notch on the highway. As the air grows more dense and the trees regain their stately forms, I find myself half dreaming of the summit still, and also half desiring home. The trail grows softer underfoot as soil begins again to pad the trail. Everything is humming, including my nerve endings, and hums still as I drive down the highway through the gap and into the foothills where the Pemigewasset River flows into the higher lakes of the region, where you can hear the loons calling off the lakes—a call out of a time older than almost any other living thing. Loons thrived in prehistoric times, and they once thrived in the kinder valley worlds to the south, though they've been driven north by loss of habitat where their call sounds rare and arresting: yodels, tremolos, wails, hoots, and laughter at all hours across the water. They can set the coyotes howling. Even when they are coaxed back to more southerly waters, they don't return to the smaller ponds, which are crowded now with camps and homes. They return to the man-made Quabbin Reservoir, and call their weird calls over the drowned towns there.

Soon after the water leaves the north lakes it becomes the Merrimack River, and once I glimpse it—the interstate crosses and recrosses the Merrimack—I begin to feel I am burning back into my own atmosphere, happily tired, past ripening strawberry fields and peas with their laden pods hanging heavy on the trellised vines. I can see greenhouses with their sides peeled back and pink tomatoes within. The first mowing cut and baled across the fields. The water that is so soft to the skin in the East Branch is unenterable less than a hundred miles farther south, having been severely compromised by fertilizers, pesticides, and salts coming off fields and parking lots and neighborhoods and businesses. The metals and dyes of closed factories are still buried in its mud. I see the river slowed and widened, shining in the last sun of the day. It cannot support salmon, nor can the little rivers running into it. Now in the valleys of the Merrimack and its tributaries, frequently the town hall doors—hundreds of years old, some of them, the green paint peeling—the doors we all walk through to pay taxes or to purchase hunting licenses or building permits, to register to vote, to look up the wetlands maps and zoning maps—have warnings tacked to them concerning the Beaver Brooks and Black Brooks, the Long Ponds and Silver Lakes that find their way into the Merrimack. A fish in a red circle with a slash through it: PREGNANT WOMEN AND CHILDREN SHOULD NOT EAT FISH TAKEN FROM THE FOLLOWING LAKES AND STREAMS . . . And still porch lights go on one by one. The color of the pines approaches their shadows, the blue sky deepens into dusk, and the sun sets in a red blaze behind a stand of trees, the color glinting off the side of a white barn, and then the peak of the barn.

Everything calls me back in the long June evening, and yet I can't get used to not seeing the shapes of the mountains. By the time I arrive home it's nearly dark, and I stand at my back door, holding open the screen, gazing, believing it should be enough to take in the orchard and its first fruits, and the white pines beyond. It's almost always been enough these last years, and calming in and of itself, this world that *is* characteristic. But tonight I feel there should be more behind the immeasurable night woods, some steep, rising, and challenging land. I keep listening for silence, keep desiring more, the *more* that makes unstable my allegiance to all I know. I keep imagining the wind-stunted red spruce, keep seeing them in dreams, seeing them for days wherever I look—and the ridge, the long walk in front of me on top of the world, the gaps, rock slides, and screes, the near trees ragged and ravaged by the elements, and beyond, as far as I can see, the vast blue assuaging Pemigewasset Wilderness.

7
Ghost Countries

THERE ARE TIMES when, as night approaches, I sit on the stone step at the back of my house and watch the contours of the land grow more severe in the coming dark: the orchard hillside steepens as the swale disappears into the depths, and the pines tower in their shadows as if they were rising up to an older stature. After a while I can't tell where the tended acres end and the woods begin, and it feels as if the wild is crowding the farm back to the size of the first settlers' places, back before cattle arrived in this country, before tame grasses were cultivated and wide expanses were enclosed, when to every person was given only one acre of land. Later, when I lie down to sleep, I can imagine an unseen world emerging from the pines: foxes hunting among the apple trees, deer rooting out old squash and nosing the soil in the freshly turned beds. I sometimes hear the coyotes' erratic baying over their kill, or the owls calling to one another—softly at first, with each hoot distinct, then the calls quicken as they strengthen, growing closer and closer together until they overlap to make their concerted music. *The death of my father left me sad and depressed for a couple of months,* wrote Thomas Merton in the time before he began to search out his belief. *But that eventually wore away. And when it did, I found myself completely stripped of everything that impeded the movement of my own will to do as it pleased.* What is it that makes the wild feel so close? The foxes? The owls, calling from the bare boughs? Is it the deaths of so many that seem to have come so quickly, or the few who remain and are all that impede the movement of my own will?

After so many deaths and decisions leave you feeling light on the land, something has to move in. These subsided years on the farm are the time in which I've come to know my mother best. Our talk over meals isn't deflected by details about the farm or the work of settling my father's estate. Time isn't fraught with trips to the lawyers and the large decisions about what needs to be done. The conversations usually involve family, what everyone is doing, plans for visits, the holidays. My brother, who for so long ran the farm along with my father, is out west now, in Arizona, doing some carpentry, a little farming for people on the side, taking care of an old apple orchard that a retired couple took ownership of, helping with a vineyard, planting a garden. When we mention him the conversation is mild and incidental.

It's late for many things, and largely our relationship occupies present moments. I laugh a little now at myself, and the idea I had for my mother after my father's death—one I now see I devised for myself. I had imagined independence for her, that she would manage her own finances, travel some, at least take small trips throughout New England. Perhaps I wanted not just to ease my own day-to-day life but to ease some future memory of her. If she were in the world, going to the Grange, involved in the community, there would be a larger repository for memory: all the talk, talk, talk in a room like the Grange Hall where stories could grow in time.

My mother has always been quiet and inward, and she is even more so now that she is growing forgetful, but she is dreamy these days, too. It's as if she is freeing herself from almost everything, though she still remembers my father in a militant way, and she will never willingly leave "the house your father built for me." Yet as she has it, it is growing in over him. The exterior appears largely the same, modestly set back on its green lawn, but the place that accommodated us all easily is impractical for her alone, and now seems like a chambered shell. The rooms above her are clean and empty, and she lives in the center of the first floor amid a clutter of papers, of work she means to do, things she means to read, her lists, the TV on loud, her excitable dog by her side. She lets everything impractical reign. I aim to give in to it, saying to myself there is a great freedom in letting everything be, but when I enter I have to resist turning down her TV and putting up the shades during the day so as to make it more like the place I once knew.

I imagine when her memory goes a whole unknown world will disappear; she is the last in her own family's generation with any kind of memory: her parents are gone, and her two older sisters are well into their eighties, and remember little. When the three of them are together my mother is by far the most talkative. Most of the time among them is occupied with a contented silence, with their gazing out at the world. "They were always so close," she says of her sisters, the Irish twins—just eleven months apart. "I was so much younger, they just left me behind. And now . . ." Now one Irish twin will certainly recognize the other. We ask: "Do you know who this is?" "Yes, yes . . . my sister . . ." But between them, they barely patch together a coherent past, and both are forgetting my father. "Who did our sister marry?" they might ask each other. "What was his name?"

A highway runs through the first place they lived in the city of Lawrence, and urban renewal has razed the last. Almost all the stores they shopped at are gone—a few bakeries remain, a few markets tucked back in the side streets. You rarely hear Italian. For all the ambitions of the industrialists, that world did not endure as long as the world of the first white settlers had, or the world of the Algonquin before them. The jobs hardly lasted out my grandfather's working life, and little more than a hundred years after Lawrence was founded, the textile jobs moved south for the cheaper labor. Then the jobs that had gone south moved overseas, and the testimony of young Asian women who came out of their villages to work in the factories chimes with those of the young women who worked in the Merrimack Valley factories almost two centuries ago. *People have asked me, how can I tolerate this?*

Most of the people I work with come from up-country and use their pay to help the people back home. They're eighteen or nineteen, and they still have the energy to work the hours, a young Thai woman says. *All we can think about is that we have to catch the bus. We have to work. We get home at 2 or 3 a.m. and wash our clothes and try to get some rest . . . My parents had twelve children. I was the youngest. Since I wasn't as good at school as my sister, and we couldn't all go, I let her be the one to go to school. I sacrificed since I wasn't as clever. I'd rather send money home . . . If we don't do the work, other people would . . . So it's like this, they lower prices—and still we sew . . . During the holidays we work all night, go home to shower, and come right back . . . If they catch you yawning they fine you 500 baht.* The factories can move away quickly, in the night, when the labor gets too expensive: *. . . then suddenly the factory closed down. I can't even understand why it closed. It can't be that there was no work. We were busy up until the very last day.*

Of everything contained in the original dream of Lawrence, Charles Storrow's dam has proved to be the most sturdy. It has required almost no repairs. The wooden fishways were occasionally replaced until a concrete one was built in 1917. In July 1915 a freshet carried away a portion of another dam on the Merrimack, and so they decided to secure the Lawrence dam crest with steel bars. In 1945 small cracks in several of the crest stones were repaired. During the 1936 flood, when the Merrimack rose over its bridges and washed tenements miles downstream, it overflowed the crest by nearly fourteen feet, and you could only distinguish the dam by a ripple in the flow of water.

When the waters subsided it was as intact as ever. If I tell people, especially local people, that the Lawrence dam was once the largest in the Western world, they have trouble believing me.

In the world high above the roofs, the world of steeples and bell towers and smokestacks, it can seem as if the nineteenth century still exists. The hunkered power of Gothic and Romanesque Revival churches had seemed such a natural match for the immigrant workers and the squat earth-tied cities, churches built partly of local brick and granite and oak, partly of marble from Italy and Yugoslavia, and deep-grained woods from Cuba and Brazil. But the drafty, cavernous interiors have proved impossible to heat, and even block and brick need upkeep. The parishes began to shrink after the work went to the Carolinas, and as children and grandchildren of the immigrants moved to the suburbs. I have no doubt that one day soon the local and the foreign, the fine-wrought and the rough-dressed, the tracery, woodwork, marbles, and granites with all their varying qualities, weights, and hefts, will lie for a moment smashed together in heaps of rubble before being packed into dumpsters and hauled away.

At street level, decay is evening out the old distinctions. There had been, when my mother was growing up, a marked difference between life on the hills of the city and that in the flats. The grand, ornate Victorians on the rises have been cut up into apartments now. Some are even abandoned; their windows are boarded up. They are slouching towards the tenements, and the tenement life is creeping up to meet them. New immigrants from the

Dominican Republic, Puerto Rico, Cambodia are nesting in the old world. One high summer day as my mother and I were driving down Water Street the kids in the neighborhood had all spilled out into the street, their sandals flopping on the pavement as they ran beneath open windows full of music and chatter, past their parents fanning themselves on the stoops. My mother turned to me and said, "I don't know how people live here—the houses are so close together."

"But when you grew up here, your neighborhood was just as crowded."

"I know."

As for the granite quarries that helped to build the foundations of such cities, you can find vestiges of them if you follow the shore road along Cape Ann from Gloucester towards Rockport. You'll see all kinds of hammered stone along the way, in the curbing and foundations and walls and posts. You'll pass by many of the derelict granite pits, though you might not detect them with all the scrubby growth of seaside trees, and the houses tight together now. Eventually you'll arrive at Halibut Point State Park, site of the former Babson Farm Quarry, which has been filling with water for nearly a century, constantly, from underground springs and rain. If the water is a little salty, the salt is from sea spray and not from the sea itself. Mineral streaks stain the rock at the rim, plant life sprouts from the seams.

It's not that the rock ever ran out. The advent of asphalt, concrete, and steel—cheaper, faster, easier to transport—meant an end to the great demand for granite. Once the quarrying stopped, the pumps stopped. Water began to creep over the bottom ledges of the pits, then over the derelict ladders canted against them. Water rose over the quarrymen's exhaustion and ambition and anger at being exploited, over the 1879 strike for a ten-hour day and the 1899 strike for a nine-hour day. It rose also over their pride in the work: all the innovations that would never be taken further, the precision and technique that would never be handed down, and after a while, if the men looked back at where they'd once labored, they'd gaze into an absolute calm that reflected salt-stunted trees and a high, fair, twentieth-century sky. In their old age they led their horses to drink at the filled pits. They wet down their wagon wheels. In the palpable silence they could hear the surf again, and they let themselves believe that if the sea tosses up our losses, then the quarries keep them hidden. For decades after its closing in 1930 the Rockport Granite Company, the last owners of Babson Farm, had to keep the names of all their workers on file, and the dates each one worked, to determine the benefits due when silicosis—stonecutter's consumption—got the best of them. It almost always got the best of the later generation. Once the more efficient pneumatic drills replaced chisels, hammers, and quarrymen's spoons, dust flew everywhere, though mostly it hovered in front of them—its own silty gray twilight, level with their breath.

On Cape Ann, every once in a while, piles of chipped and discarded stone are hauled away to be used for drainage, or a quarry is reopened, the water pumped out, and a certain amount of granite for a particular job is cut from it, safely and efficiently, with state-of-the-art saws and drills. The workers flood water over the surfaces as they cut the stone to keep down the dust. They wear masks. One man works his own quarry still and cuts stone for the upscale home trade—the greens and sapstone pinks have a distinctive decorative grace—but the stone is rarely used for its strength, since there's steel for strength. Maybe only an old cutter can really feel the difference between the nostalgia for the surface and the certain massiveness of the block: *When they use stone now, it's different*, one of them says. *The stone is only one or one-and-a-quarter inches thick and they use it as facing. It's practically like hanging it in a frame. Our work was another way of doing something at another time.*

At Babson Farm men and women stroll along while schoolchildren play hide-and-seek or tag in a stunted oak woods littered with chunks of quarried stone. Metal clasps protrude from the rock, some of the edges are gouged with chisel marks. More than half the time in the quarries the granite split awry at a knot or along an unpredicted seam, and there were inaccurate cuts and slips in the hewing and hammering and dressing, so there is enough discarded stone at Halibut Point to form a mountainous pile—called a grout pile—which has been filled to create a heightened outlook. If you stand upon it on a low, cloudy day when the fog has socked in the coast, your eye narrows to what is beneath you, and all you can see is those hundreds of feet, it must be, of cast-off, half-squared stone, the discard of old use tumbled into a hard-edged beauty that now insists on itself like some socialist realist monument, with its sharpness and shadows and shades of gray. PLEASE DO NOT CLIMB ON THE SIDES OF THE GROUT PILE, a sign says. THE STONES ARE UNSTABLE AND VERY DANGEROUS. They look both sturdy and precarious, a force unfolding, and frozen, and forever falling down and down to the gray-green waters of the Gulf of Maine and the sea-washed stone along the tideline, to granite stained with salt life, with barnacles, fringed wrack and moss, which thrive on unpredictable surf and relentless tides—*the pounding of the surf is part of the normal life . . . the sea is not its enemy.* The stone smells of iron and brine, and the rise and run of the tides has worn it smooth as if to return it to its old molten state, to granite's first meaning: "plutonic," after the god of the netherworld. *We are so easily baffled by appearances,* writes the poet Hugh MacDiarmid:

> *And do not realise that these stones are one with the stars.*
> *It makes no difference to them whether they are high or low,*
> *Mountain peak or ocean floor, palace or pigsty.*
> *There are plenty of ruined buildings in the world but no ruined stones.*

FEBRUARY, MARCH, SOMETIMES into April, after the snow melts back, I spend a couple of hours every afternoon pruning the apple trees on the slope in back of my house. It's my last real tie to the working life of the farm: three weeks' worth of work, five or six or seven trees a day—work apart from necessity now, since the orchard is just barely commercially viable: apples are produced more profitably in Washington state and China, and in most of New England the money is elsewhere. "One more bad year, and I'm selling," you hear the growers say. For now Dave takes care of the spraying and harvesting of the trees, but the wholesale price he can get for apples no longer warrants the expenses, and it won't be long before he, too, stops thinking of production. Even so, the orchard has stayed central to my idea of this place, and pruning is a little allegiance to what was, no more than that. You can tell by the way, in the last few years, I've gotten less businesslike about the work, taking my time as I look through the intricate branches into a milky blue spring sky. Every once in a while I let a branch or two at the top go—one that I know I ought to prune—just because I like the way it spirals away freely. Even so, as always, if I spend a long afternoon in the orchard I see a tangle in front of me when I close my eyes to sleep.

What the trees want is to send their many shoots upward as much as they can to gather all of the full sun. What I want—by lopping and sawing and weighting down limbs with water jugs if need be—is to keep the branches sturdy and horizontal, which encourages large fruit on every tier of the tree. So I make my way through the rows, sawing away the crowded centers, snipping off the water sprouts, puzzling out ideal shapes in my head, all the while talking to myself: "This one . . . that . . . not that one . . . how about—this . . . yes . . . as I deliberate over competing branches, and lop whatever is growing towards the trunk or down or crossing another branch. After more than a quarter of a century the struggle has thickened into the shapes I see now: measured row after row of squat gray trunks and sturdy, spare branches with all the supple growth at the far edges insisting on its way in spite of the efforts of previous years—a beauty all its own. In the weeks after I've finished, however much it heartens me to see the buds gradually swelling and the forms of the trees softening, I always feel a little regret to think that soon the weird struggle—so apparent in their stripped-down life—will be hidden.

Last winter the first heavy snow fell before the ground froze, and snow kept falling—almost two feet in just one night—all through December, January, February. "An old-fashioned winter," everyone kept saying, as town after town ran down their snow-removal coffers. We didn't see bare ground until March, and it was nearly April before I could get into the orchard to prune. Even then the snow in the shadows was over my knees: a winter hard on wildlife, and harder on the orchard. In my first walk through I saw that many of the lower branch tips had been bitten off by hungry deer—I'd seen them grazing the buds under the light of the full moons—and worse: starving mice had tunneled under the snow and gnawed away swaths of bark—six or eight inches—partially, or entirely, around the bases of some of the trees. They'd even nudged down through the soft ground and eaten bark beneath the soil line. I panicked when I saw it—the bark is the living part of the tree, and those that had been completely girdled would die. I quickly made my way through the rows to gain the extent of the damage: out of a little more than a hundred, numerous trees were partially damaged, and twenty, twenty-one, twenty-two were fully girdled.

Such a large sense of failure, though I knew there was little I could have done. With the winter ground unfrozen, not even the wire guards wrapped around some of the trees kept the mice from gnawing at the bark. The depth of that feeling of failure—and of responsibility that accompanied it—was a double sting because I'd started to believe I was freeing myself of the place and the past, but then and there I felt rooted to the ground every bit as much as I had immediately after my father's death years before. During his last days, when I was the daughter he'd wanted, taking care of the bills, tending to the house and farm needs, I remember also wishing—though never saying so—that he would in his last moments relent from the practical concerns of his life, let go of his control of everything. Maybe it was the part of me that is so much like him—the responsible one—that I dreamt of unmaking.

For all the conflicted feelings I had as I stood there, I was fairly sure the only thing to do was to let the trees stand and see which ones died off as the season progressed. My one hope was that they'd contain enough stored energy to put out a last crop of apples. My neighbor, who sometimes runs his dogs in the orchard, had larger hopes, though, and I happily believed him when he said it would be worth it to at least try to apply bridge grafts to the half dozen partially damaged trunks. He came by again a few days later—I remember just a little green was showing in the buds—to give it a try. Without his hounds running ahead of him with their bounding, scattered energy, he is a compact and deliberate man, patient, a hunter who brought the same intense concentration of the hunt to the task of grafting over the mice-bitten bark. He bent before each tree and surveyed the damage, peering in close through his wire-rimmed glasses, then backing away a bit and peering over the rims. He cut measures of slim new growth from branch tips, trimmed each end at a slant with a razor, and slipped one end under the bark at the upper edge of the wound and the other at the lower edge. Every three or so inches over each girdle he cobbled together these little conduits in hopes they'd reconnect the life of the roots to the tree above the wounds. He nailed them in place with brads, then daubed the grafts with tar to keep them from drying out.

I followed him as he worked, handing him a brad or the tub of tar when he needed it. Our talk was sporadic and casual: the trees, the winter, the deer, the birds, how the orchard would be giving way to time soon. "When I hunt up north," he said, "some of the most beautiful places I come across are the old orchards in the woods. The hillsides up there are full of them. There's always more light where they stand, and the soil is always a little richer, even after all the years. The wildlife love it." He was animated by the remembrance of it, and it lifted my spirits to hear him talk—it was something to imagine: all the human struggle let loose, with just a ghost of the old prunings to mark a place, while every year the Macs and Macouns and Northern Spies soften and drop and disintegrate to earth in a time beyond our own division of things into the beautiful, the vexing, the useful, when whatever we had cursed—scab and curculio—and whatever we had admired and cursed at the same time—such as deer—would thrive as they could. I thought of Thoreau's wild apples: *fruit of old trees that have been dying ever since I was a boy and are not yet dead, frequented only by the woodpecker and the squirrel . . .*

He loved the way apple trees were scattered through the countryside, remnants from a time when *vast straggling cider orchards were planted, when men both ate and drank apples, when the pomace-heap was the only nursery, and trees cost nothing but the trouble of setting them out. Men could afford then to stick a tree by every wall-side and let it take its chance.* Everything I know of orchards, and think beautiful—the measured rows of certain varieties trimmed without fail each year—Thoreau saw as marking the passing of what he loved. *Now that they have grafted trees, and pay a price for them,* he wrote, *they collect them into a plat by their houses, and fence them in,—and the end of it all will be that we shall be compelled to look for our apples in a barrel.* And after these orchards pass? Some of the newer ones in the region are planted in dwarf trees, kept no higher than five feet and set close together and strung on trellises like grape vines across the worn old hills. Even so, they are only a slight gesture. There's no real successor to the unreturning days.

Sometimes when it comes back, the wilderness that had been struggled against for so long really does seem assuaging. Maybe not at first, not when a hayfield goes uncut or the cattle aren't turned out to pasture, and the grass stalks stiffen and shatter. Then it can feel chastising, the way it's so easily released from its old order, as burrs and tickseed are carried in by sparrows, and fluff blows in on the wind. The following spring a scattering of primrose and vetch nose up among the clover and timothy, and as the months go on hawkweed and chicory force their way. Oxeye daisies, black-eyed Susans, butter-and-eggs—who knows anymore which are native and which came across the Atlantic in the ballast of ships? By late August, as the birds are flocking and the sun is beginning to wane, goldenrod surges and asters begin their late bloom while the umbels of the Queen Anne's lace wither and draw in. Milkweed pods open and their milkseed flies away. The old hayfield is now a tangle to walk through—like wading through an incoming tide—a confusion of clotted, leaning life where everything is both living and dying, and there is no order except the one imposed by the demand for striving. After a few years' time, when pine seedlings begin to cast shadows and the last of the hay dies back, the grass seeds become buried in an eternity of falling needles and leaves. Life crowds up high in the treetops, the lower branches of the pines grow brittle and bare in the scant light. Footfalls grow softer and more scarce within. Grown trees fall and molder to earth. Then it is almost forgiving in the way it lets you forget—even if it is not the great forest of the frontier, it asserts itself enough so that those without living memory are incredulous that there ever was a field.

My Aunt Bertha, when she had grown old, when the stone boundaries of her childhood had become labyrinths through the woods, walked among full-grown pines in the same place where, just decades before, cows had grazed. If it was December, in the afternoon—a brief and brilliant sun piercing its way through the trees—she would gather creeping partridgeberry and club moss to make a Christmas terrarium for the farmhouse kitchen. The stinging smell of bruised evergreen would come to life as she pulled up moss out of the cold duff. All over the forest floor they run, those mosses, a vestige of the immense forests of three hundred million years ago, forests that had once grown far taller than the pines we call *ours.* Massive stands of them had spored patiently and inefficiently, and when they couldn't compete with seed-bearing trees they evolved to persist close to the ground, shaded, nearly hidden. Over time the dead giants were buried where they'd fallen, and down millions of airless years they compressed into the coal beds we have brought to light. Now they're burning and drifting on the northwest wind, and falling in the rain.

A few weeks after my neighbor applied the bridge grafts to the apple trees, the blossoms on all of them—healthy and damaged alike—emerged pink, then whitened into full bloom. New leaves formed and hid the deer-bitten tips of the branches, and grass grew high enough to obscure both the grafts and the girdles, which had darkened with the weeks, anyway. Without the weird contortions of the branches to capture my attention, my eyes wandered to the air above the trees, to the flights of birds: goldfinches dipping in and out of the rows in wild undulations, the kingbird keeping itself aloft with rapid wingbeats as it searched the grass for food, the sparrow hawk hovering and crying, cardinals skirting the windbreak, the great swoops of tree swallows taking all the joy in flight, one undeterred heron passing over the orchard, and then over the woods. When a hawk drifted above the trees, the sparrows would flutter from the grass and hide in the branches, waiting silently. When the coyote trotted through, peremptory, they'd make a racket of warning.

Most were claiming territory and foraging for pine straw, old grass, twigs, and feathers to build their nests. Some of the smaller ones were vying for the half dozen birdhouses my neighbor had been putting around the perimeter of the orchard in the last several years. We always hoped for bluebirds, and we've had half a dozen pairs over time, though they are docile by nature, and everything that can will take the opportunity. Later in the season wrens fill the boxes with false nests full of sticks if nothing else has claimed them, or a pair of tree swallows skirl overhead, seeming to toss themselves on the wind to create a boundary around a box. They line their nest with feathers, and won't scare. If you lift the lid and peer in you'll see the female holding tight, tense: a bright eye deep in.

Even when bluebirds nest successfully sometimes their eggs are eaten by predators, or the chicks are attacked by house sparrows. I once saw a whole nest of nearly fledged chicks pecked to death—"by a crazy bachelor," my neighbor said. After such a failing you can see what defense makes of life: there's a large emptiness where the male bluebird once stood guarding the box—nothing, no kingbird teasing, no skirmishes with the swallows, just quiet. That absence is what you train your eye on for a time—though there are countless other nests still thriving, suspended tenuously from a branch, or saddled in a crook, or packed in a hollow bole—until one evening in a muggy dusk you hear the last sound of the day coming from the woods beyond: the fluting notes, clear and piercing, of a wood thrush—only one—calling from somewhere low in the pines. When I hear his singing I feel it coming straight towards me, and what with his solitariness, and me listening alone as the increasing dark extinguishes the boundary between the orchard and woods, it seems to suggest everything of *the inexpressible privacy of a life,—how silent and unambitious it is.*

There are some inheritances I know I'll never be able to shake. Some dream of order that won't stand for death in life. Deadwood, which can look fine and defiant in the wild, ruins the idea when it's within bounds. Cultivation *is* a possession, an allegiance intertwined with necessity. In August, after the thrush grew silent, and I could smell the Astrakhans on a light wind, the foliage on the girdled trees—even on the ones my neighbor had grafted—began to yellow. I couldn't imagine letting the dying remain through until spring, when they'd be skeletal amid healthy blossoming trees, so I marked them all with surveyor's tape before they lost their leaves entirely, and had them cut down. It took two men part of the morning to make a pile of brush to be chipped come March, and a smaller pile of wood to be cut into stove-length pieces. Strange: the trees that remained were still so insistent that, from a distance, I hardly noticed the skips in the rows where the damaged ones had been cut down.

Then came a winter that could not be more different from the one before. We were nearly snowless in January, and were already having rain in February. "I don't know, this is so weird," people said. I rarely saw the deer in the orchard, and the mice had their pick of other, easier food. Still, the light was spare as ever those late winter days, and by the time I settled down to work the sky was just beginning to lighten in the east. Quiet, the air above the orchard. The crowns of the trees, dense with last season's growth, seemed to be an impossible, wild tangle of work that lay ahead of me. A hawk sentried at the edge of the woods—a wary lookout all winter, finding the small life—was there when I looked up from my desk, and later when I walked out among the trees his high *keeeeeeer* claimed the ground, making me think to myself: *Not yours.*

Afterword

A Last Look Back

I'D BEGUN TO sketch out what would become *Here and Nowhere Else* years before I returned to our farm. I was trying to capture my childhood by concentrating on the things of that world, which remained, after more than a decade away, palpable to me: the scent of the softening Macs in the apple cellar, the heft of Blue Hubbards, the light in the old milk room—and its mineral must. The white pines, the particular roll of our low hills, the last strains of my grandparents' Lebanon in the stuffed grape leaves and eggplant we ate for dinner. That I was living in Greater Boston at the time, and writing at a table where I looked out on the rooftops of 1920s houses, only magnified the distance between here and there, now and then. I was in my early thirties, and still uncertain as to the direction of my life, but I felt sure about the world I was describing.

In time, I joked to friends that I wrote my way home. Though not entirely without ambivalence. We'd been raised to move on. The stability that had meant so much to my grandparents when they bought the farm at the turn of the twentieth century gave me the freedom to wander: Even during my teenage years I expected my future was elsewhere, and going away to college was the means of going forward. By the age of eighteen, I couldn't wait to leave. But during the late 1980s, while I was writing at that table in an apartment on the outskirts of the city, however fixed the world of my childhood felt in my imagination, on the farm hearts were failing, and kidneys, and memories. The land surrounding ours was being transformed from running brooks, old woods, and neighboring farms into house lots and industrial complexes.

What did I think I would accomplish with my return? I understand now, at a distance of decades, that it didn't mean seeing the farm into the future. All that had made the world of my childhood feel so secure, especially the inextricable relations among us, was also a source of its vulnerability. My father didn't own all of the land he farmed, and there were always the shared and sometimes competing interests of the extended family, many of whom were aging, to accommodate.

But even if I'd been free to do so, I had no desire to make the farm my entire life: My return had more than one purpose. I imagined working on the farm in its season, and then writing during the late fall and winter. I even told myself that my hopes for a writing life were at the heart of my decision to return. So I was surprised at how quickly the farm—as it was then—took on a profound resonance for me. Some long summer evenings I'd stand in my back doorway and look out over land that had lasted longer in our family than my grandparents' native tongue, and I could not imagine anything other than being surrounded by the fields and orchards, then the woods.

As the older generation thinned, the households grew quieter and quieter. With each death our world grew increasingly fragile. Time was coming for us and, more than ever, I felt a part of the place.

I BELIEVED I'D written what I could when I finished the first book, the one that begins with *We*. Yet, after a time, I realized the world I'd conjured was still changing in fact and in my mind. I also began to see how much our farm was connected to history, and to the larger landscape of the country, so I started researching the deeper past of the farm and its roots in the valley. Another book. And then a third. I thought of them as describing concentric circles, each successive one reaching farther beyond the boundaries of our place, both in geography and in time.

During the years I worked on what became the trilogy, I moved from typewriter, to computer, to—most significantly—computer-also-connected-to-the-world; from mailing manuscripts to my editor that were held together with elastics and snugged in large envelopes, to sending, with the press of a key, a document that had no more material life than bits of light.

❦

MY RETURN LASTED for more than fifteen years. In the end, the only voice I had was there on the page. However free I'd been to imagine the world I lived in, the vagaries of old wills and new trusts left the farm's future in the hands of others in my extended family. But I already knew that, with *Clearing Land*, I was writing a farewell even before I discovered the last words of that book: *Not yours.*

Now, twenty years later, after other books on other subjects, after settling into another life, in another state, in another house—which sits on less than half an acre in the middle of a town—I sometimes stand at my back door at evening and cannot imagine anything other than looking out on the calm surround of maples and conifers and old gardens I keep fussing with. Sometimes the world I was so intent on describing feels as profoundly separate from me as my grandparents' old country was from them. Sometimes, helplessly, I feel an aching thinness when I realize how far I've traveled from that life.

My first three books have their place on my laden bookshelves. When I take one down I see that, however much has been lost, I've managed to describe the past, at least as I saw it. I also realize the story of our farm gave my hopes for writing an intent, and what my return eventually meant for the place can't be separated from what it meant for my writing life.

Even though I now live just a few hours' drive north of the Merrimack Valley—not so far away that there aren't echoes of its landscape in stands of white pines, in old orchards bearing Macs and Cortlands and Northern Spies—I almost never return. I know that back there is a gas station where my father always planted a stand of corn. And another old field—I still see it green with winter rye in late November, and the berries on the black alder bushes at its edge glowing in an early dusk—has three oversized houses on it. More are planned for the dry woods beyond. I wouldn't begin to know how to write about the place it has become.

—Jane Brox
Brunswick, Maine
2023

NOTES

HERE AND NOWHERE ELSE

Some of the names of family members and other figures in this story have been changed.

I am indebted to many books for understanding and inspiration. Among them are Betty Flanders Thomson's *The Changing Face of New England* (New York: Macmillan, 1958), William Cronon's *Changes in the Land: Indians, Colonists, and the Ecology of New England* (New York: Hill and Wang, 1983), Paul Hudon's *The Valley and Its Peoples* (Windsor Publications, 1982), *Indian New England Before the Mayflower* (Hanover: University Press of New England, 1980) and *A Long Deep Furrow: Three Centuries of Farming in New England* (Hanover: University Press of New England, 1982), both by Howard S. Russell, and *A Natural History of Trees* (Boston: Houghton Mifflin, 1991), by Donald Culross Peattie.

FIVE THOUSAND DAYS LIKE THIS ONE

135 Epigraph, part I: *The Selected Letters of Emily Dickinson*, Thomas H. Johnson, editor (Cambridge: The Belknap Press, 1958), 141.

151 The information on soils in chapter 2 and throughout the book have been derived from *Soil Survey of Middlesex*

County, Massachusetts, Series 1924, United States Department of Agriculture and the Massachusetts Department of Agriculture.

154 *The quality of mercy*: Shakespeare, *The Merchant of Venice*, act IV, scene 1, lines 190–92.

160 "Marching Through Georgia," by Henry C. Work.

The statistics in chapter 3 and throughout on farming in the nineteenth century have been derived from *Soil Survey of Middlesex County, Massachusetts, Series 1924*. The information on milk inspections here and in chapter 4 has been derived from *The Report of the Lawrence Survey* (Lawrence, MA: Trustees of the White Fund, 1912). The survey was commissioned by the trustees of the White Fund to look into living conditions and the state of public health in the city of Lawrence. See "Part II: Public Health," by Frank Sanborn, 147–216. The quotes in this chapter concerning the conditions of milk farms are from the captions to photographs within these pages. A map of the Lawrence milk supply is published in the report.

162 *As we rode along*: Silas Coburn, *History of Dracut* (Lowell, MA: The Courier-Citizen Press, 1922), from an account by George Mozley, 180–81.

164 *Born in Ireland*: *The Lawrence Telegram*, Lawrence, MA, November 11, 1918.

189 Epigraph, part II: Henry David Thoreau, *A Week on the Concord and Merrimack Rivers* (Orleans, MA: Parnassus Imprints, 1987), 191.

192 *It was already*: Thoreau, *A Week*, 100.

193 *All along the river*: William Wood, *New England's Prospect* (Amherst: University of Massachusetts Press, 1977), 64.

194 *being the first*: John Pendergast, *The Bend in the River* (Tyngsboro, MA: Merrimack River Press, 1991), from a Contact Period map reproduction, 78.

195 *The living were in no wise*: Pendergast, *The Bend in the River*, 43.

unmapped, unmarked: Howard S. Russell, *Indian New England Before the Mayflower* (Hanover, NH: University Press of New England, 1980), 201.

196 *it was thought by some*: Thoreau, *A Week*, 263.

197 *haste to get past the village*: Thoreau, *A Week*, 306.

At the time of our voyage: Thoreau, *A Week*, 306–307.

200 *Our hands are soldiers' property*: Excerpt from a letter of Dorothy Dudley of Cambridge, Massachusetts, in Catherine Fennelly, *Textiles in New England 1790–1840* (Meriden, CT: Meriden Gravure Co., 1961), 3.

201 For the section on the spinning wheel collection at the Museum of American Textile History, I am indebted to Joan Whittaker Cummer's *A Book of Spinning Wheels* (Portsmouth, NH: Peter E. Randall, 1993).

202 *They say a humming wheel rises*: See Jane C. Nylander, *Our Own Snug Fireside* (New Haven: Yale University Press, 1994), 175. Francis Underwood describes the work of spinning: "Then, while the hum of the wheel rises to a sound like the echo of wind in a storm, backwards she steps, one, two, three . . ."

203 *I have got the most of my wool spun*: From a letter of Malenda Edwards, in Thomas Dublin, ed., *Farm to Factory* (New York: Columbia University Press, 1981), 86–87.

206 *Since our voyage*: Thoreau, *A Week*, 264.

207 *The English came*: Coburn, *History of Dracut*, 38–39.

208 *At first, the sight of so many bands*: Benita Eisler, ed., *The Lowell Offering: Writings by New England Mill Women (1840–1845)* (York: Harper and Row, 1977), 180.

209 *There are girls here for every reason*: Eisler. ed., *The Lowell Offering*, 60–61. This excerpt is from "The Letters of Susan,"

composed by Harriet Farley for publication in *The Lowell Offering*. "Mother-in-law" refers to a person we would call a stepmother today.

210 *One would swear*: Charles Dickens, *American Notes for General Circulation* (Boston: Ticknor and Fields, 1867), 76.

211 *You ask if the work*: Eisler, ed., *The Lowell Offering*, from "The Letters of Susan," 53.

The Overseers: Eisler, ed., *The Lowell Offering*, from the Regulations for the Appleton Company, 25.

I have sometimes stood: Eisler, ed., *The Lowell Offering*, from "The Letters of Susan," 57.

212 *Chairs, chairs*: Eisler, ed. *The Lowell Offering*, from "The Letters of Susan," 73.

It is now: Eisler, ed. *The Lowell Offering*, from "The Letters of Susan," 55–56.

213 *I also wish you could see a prairie*: *The American Agriculturist*, vol. 1, no. 1, April 1842, 14–15. The excerpt is from a letter to the editor entitled "Something about Western Prairies," by Solon Robinson.

The children of New England: Quoted in Eisler, ed., *The Lowell Offering*, 11.

214 *Lowell Hall was always*: Eisler, ed., *The Lowell Offering*, 32–33. From the comments of Professor A. P. Peabody, who lectured every winter for the Lowell Lyceum.

215 *I had closed my book*: Eisler, ed., *The Lowell Offering*, 209.

217 *At the time I was just about fourteen*: Transcribed from the tape "The Lawrence Textile Strike: Viewpoints on American Labor," produced by Myles Jackson (New York: Random House, 1971).

218 *And it was hot*: Mary H. Blewitt, *The Last Generation: Work and Life in the Textile Mills of Lowell, Massachusetts 1910–1960* (Amherst, MA: University of Massachusetts Press, 1990), 111.

219 *They had advertised*: Tamara K. Hareven and Randolph Langenbach, *Amoskeag: Life and Work in an American Factory-City* (Hanover, NH: University Press of New England, 1978), 221.

220 *The spinning room*: Hareven and Langenbach, *Amoskeag*, 127, 129.

221 *Everybody got only so much black cloth*: Hareven and Langenbach, *Amoskeag*, 216.

222 *There were so many black frames*: Hareven and Langenbach, *Amoskeag*, 189.

My work was handlooms: Hareven and Langenbach, *Amoskeag*, 206.

We tied each end with a knot: Blewitt, *The Last Generation*, 77–78.

They never wanted to see a loom stop: Blewitt, *The Last Generation*, 87.

And it was gloomy: Blewitt, *The Last Generation*, 111.

225 *You didn't dare . . . They'd come around to your house*: Transcribed from "The Lawrence Textile Strike: Viewpoints on American Labor."

226 The testimonies of strikers are drawn from *Report on Strike of Textile Workers in Lawrence, Mass. in 1912*, prepared by Chas. P. Neill, Government Printing Office, Washington, D.C., Document no. 870. These excerpts are taken from the testimony of Samuel Lipson, 32–41; Charles Dhooghe, 157–59; Auguste Wante, 147–49; and John Boldelar, 153–55.

229 The discussion of "The Internationale" is transcribed from "The Lawrence Textile Strike: Viewpoints on American Labor."

230 *the eloquence of an Italian*: Transcribed from "The Lawrence Textile Strike: Viewpoints on American Labor."

231 *This human being*: Excerpt from Joseph Ettor's speech in front of the Lawrence City Council.

I sent my child away: From the testimony of Samuel Lipson, *Report on Strike of Textile Workers*, 44.

238 "Influenza 1918" owes much to Alfred Crosby's *America's Forgotten Pandemic: The Influenza of 1918* (Cambridge: Cambridge University Press, 1989). I am indebted to the Oral History Collection of the Lawrence History Center (formerly the Immigrant City Archives), in Lawrence, MA, where I was able to consult the Records of the Board of Health, January 1918 to April 1931, and the Board of Health Influenza Journal, 1918 to 1920. In addition, the Archives houses recordings of oral histories. Listening to these voices helped in my understanding of the atmosphere of the time. The recordings made of the recollections of Daniel Murphy and Sister Jeanne D'Arc were particularly helpful. Also useful were copies of the *Lawrence Telegram* and the *Lawrence Sun American*, September to November 1918.

248 The title and epigraph of chapter 12, "The Pure Element of Time," are taken from the opening chapter of Vladimir Nabokov's *Speak, Memory* (New York: Vintage, 1989).

I'd be walking to Lowell High: Blewitt, *The Last Generation*, 307.

249 *You go up and down the aisle*: Hareven and Langenbach, *Amoskeag*, 381.

251 *When I talk about my work*: Hareven and Langenbach, *Amoskeag*, 201–202.

256 *Below the Lawrence falls*: Transcribed from the recorded interview with Ernie Russell conducted by the Oral History Collection of the Lawrence History Center (formerly the Immigrant City Archives) in Lawrence, MA. Tape #303, Counter Nos. 111–19.

257 The section on the Ayer clock owes a debt to Octavio Paz's poem "I Speak of the City," in *The Collected Poems of Octavio Paz 1957–1987*, Eliot Weinberger, ed. (New York: New Directions, 1991), 510–17. The facts of the clock restoration

were drawn from Jeanne Schinto, *Huddle Fever* (New York: Knopf, 1995). See 279–84.

260 *It's been tough*: The *Boston Globe*, December 13, 1995, 37.

261 Epigraph, part III, from "At the Fishhouses" by Elizabeth Bishop, *The Complete Poems 1927–1979* (New York: Farrar, Straus and Giroux, 1979), 66.

268 *As we glided over*: Thoreau, *A Week on the Concord and Merrimack Rivers*, 93–94.

272 *How can something be good*: "Path of the Pipeline's Progress," *Boston Sunday Globe*, December 7, 1997, B16–B17.

273 *we all think*: From *Twenty-Four Conversations with Borges: Interviews by Roberto Alifano, 1981–1983* (Housatonic, MA: Lascaux Publishers, 1984), 12.

282 *It is strange*: From Anton Chekhov's short story "Gusev."

291 Current farming statistics in Middlesex County, Massachusetts, in chapter 16, "White Clover," and in succeeding chapters, have been derived from a 1992 agricultural census conducted by the United States Department of Agriculture. Information on seed survivors such as white clover was gleaned from *A Sierra Club Naturalist's Guide, Southern New England* by Neil Jorgensen (San Francisco: Sierra Club Books, 1978). See 211–13.

299 *None of the farmer's sons*: From *The Journal of Henry David Thoreau*, Bradford Torrey and Francis H. Allen, eds. (Boston: Houghton Mifflin, 1906), vol. 3, 237–38.

309 *I fear that he*: From Thoreau's journal entry for November 16, 1850.

313 *Though we have*: Thoreau, *Walden* and *Civil Disobedience* (New York: Viking Penguin, 1983), 100.

To what end, pray: Thoreau, *Walden*, 100–101.

321 *The open and sunny interval*: Thoreau, *A Week*, 266.

322 *The frontiers are not*: Thoreau, *A Week*, 379–80.

CLEARING LAND

Throughout *Clearing Land* I depended for agricultural information on Howard S. Russell's *A Long, Deep Furrow: Three Centuries of Farming in New England* (Hanover, NH: University Press of New England, 1982) and his *Indian New England Before the Mayflower* (Hanover, NH: University Press of New England, 1980) as well as William Cronon's *Changes in the Land: Indians, Colonists, and the Ecology of New England* (New York: Hill and Wang, 1983) and *Nature's Metropolis* (New York: Norton, 1991). *Larding the Lean Earth: Soil and Society in Nineteenth-Century America* (New York: Hill and Wang, 2002) by Steven Stoll and *New England Forests Through Time: Insights from the Harvard Forest Dioramas* (Cambridge, MA: Harvard University Press, 2000) by David R. Foster and John F. O'Keefe were also helpful. Jill Lepore's *The Name of War: King Philip's War and the Origins of American Identity* (New York: Vintage, 1999) was essential for my understanding of possession in both the colonial and Algonquin worlds. O. J. Reichman's *Konza Prairie: A Tallgrass Natural History* (Lawrence: University Press of Kansas, 1987) and Lauren Brown's books, *Grasses: An Identification Guide* (Boston: Houghton Mifflin, 1979) and *Grasslands: The Audubon Society Nature Guides* (New York: Alfred A. Knopf, 1985), illuminated the prairie and salt marsh habitats. Andro Linklater's *Measuring America: How an Untamed Wilderness Shaped the United States and Fulfilled the Promise of Democracy* (New York: Walker, 2002) helped me to understand the settlement of the prairie. John Fogg's *Recollections of a Salt Marsh Farmer*, edited by Eric Small (Seabrook, NH: Historical Society of Seabrook, 1983), made the salt hay harvest come alive.

I am grateful to Nathaniel Philbrick's *Away Offshore: Nantucket Island and Its People, 1602–1890* (Nantucket: Mill Hill Press, 1994) and Obed Macy's *History of Nantucket* (Boston: Hilliard, Gray, 1835) for historical information on Nantucket. Barbara Blau Chamblerlain's *These Fragile Outposts: A Geological Look at Cape Cod, Marthas Vineyard and Nantucket* (Yarmouth Port, MA: Parnassus Imprints,

1981) and *From Cape Cod to the Bay of Fundy: An Environmental Atlas of the Gulf of Maine* (Cambridge, MA: MIT Press, 1995), edited by Philip Conkling, were helpful for information on the geology of the island. The website of the Nantucket Conservation Commission (www.nantucketconservation.com) provided me with information on the natural habitat of the Moors, as did *The Nature of Massachusetts* (Reading, MA: Addison-Wesley, 1996) by Christopher Leahy, John Hanson Mitchell, and Thomas Conuel. The Nantucket Historical Association was helpful with general research on the island.

Barbara H. Erkkila's *Hammers on Stone: A History of Cape Ann Granite* (Gloucester, MA: Peter Smith, 1987) helped with my understanding of nineteenth-century quarries. I'm also indebted to the Cape Ann Historical Association and the museum at Halibut Point State Park, Rockport, Massachusetts.

"Wilderness" owes much to *Into the Mountains* by Maggie Stier and Ron McAdow (Boston: Appalachian Mountain Club, 1995), *Logging Railroads of the White Mountains* by C. Francis Belcher (Boston: Appalachian Mountain Club, 1980), *Field Guide to the New England Alpine Summits* by Nancy G. Slack and Allison W. Bell (Boston: Appalachian Mountain Club, 1995), and *The Appalachian Mountain Club Guide to the White Mountains* (Boston: Appalachian Mountain Club, 1992). For the descriptions of the logging camps I relied on "A Logging Camp c. 1900" at the website greatnorthwoods.org.

331 *Our signals from the past*: George Kubler, *The Shape of Time: Remarks on the History of Things* (New Haven: Yale University Press, 1962), 17–18.

INHERITANCE

333 *Horseman, pass by*: William Butler Yeats, "Under Ben Bulben," in *The Collected Poems of W. B. Yeats* (New York: Macmillan, 1956), 341–44.

334 *However it is*: Robert Frost, "In Hardwood Groves," in *The Collected Poems of Robert Frost* (New York: Holt, Rinehart, Winston, 1964), 37.

351 *Cultivators of the earth*: Thomas Jefferson in a 1785 letter from Paris to John Jay, in *Thomas Jefferson: Writings*, ed. Merrill D. Peterson (New York: Library of America, 1984), 818.

That's what I despise: John D. Fogg, *Recollections of a Salt Marsh Farmer*, ed. Eric Small (Seabrook, NH: Historical Society of Seabrook, 1983), 69.

AGRICULTURAL TIME

355 *They sounded the harbor*: William Bradford, *Of Plymouth Plantation, 1620–1647* (New York: Modern Library, 1981), 79–80.

That neither he nor any of his should injure: ibid., 88–89.

356 *how far these people were*: ibid., 92.

of 100 and odd persons: ibid., 85.

357 *stood them in great stead, showing them*: ibid., 94–95.

The soil is for general: William Wood, *New England's Prospect* (Amherst: University of Massachusetts Press, 1977), 33–35.

We found after five or six years: quoted in Howard S. Russell, *Indian New England Before the Mayflower* (Hanover, NH: University Press of New England, 1980), 119.

359 *The Indians are not able to make use*: quoted in Jill Lepore, *The Name of War: King Philip's War and the Origins of American Identity* (New York: Vintage, 1999), 76.

The ploughmen ought to be men of intelligence: quoted in Georges Duby, *Rural Economy and Country Life in the Medieval West*, trans. Cynthia Postan (Philadelphia: University of Pennsylvania Press, 1998), 387.

360 *a class of genuine full-time agriculturists*: Plato, *Timaeus and Critias*, trans. Desmond Lee (New York: Penguin, 1977), 134.

the rich, soft soil: ibid., 134.

I have wondered: René Dubos, "A Family of Landscapes," in *The Norton Anthology of Nature Writing*, eds. Robert Finch and John Elder (New York: Norton, 2002), 456.

361 *Every one of our commune*: quoted in Duby, 407–8.

362 *A man may travel many days*: quoted in Russell, 204.

363 *And to every person was given*: Bradford, 160–61.

they began now highly to prize corn: ibid., 160.

For now as their stocks increased: Bradford quoted in James Deetz and Patricia Scott Deetz, *The Times of Their Lives: Life, Love, and Death in the Plymouth Colony* (New York: Anchor Books, 2001), 79.

364 *There is so much hay ground in the country*: Wood, 34.

366 *Our beasts grow lousy*: quoted in John Stilgoe, *Common Landscape of America, 1580–1845* (New Haven: Yale University Press, 1982), 182.

[They] took scythes out: John Teal and Mildred Teal, *Life and Death of the Salt Marsh* (New York: Ballantine, 1969), 24–25.

367 *Rise free from care before dawn*: Henry David Thoreau, *Walden and Civil Disobedience* (New York: Penguin, 1983), 254–55.

368 *As for the Natives*: Roger Williams quoted in Andro Linklater, *Measuring America: How an Untamed Wilderness Shaped the United States and Fulfilled the Promise of Democracy* (New York: Walker, 2002), 28.

You have driven us out of our own Countrie: quoted in Lepore, 95.

369 *Though English man hath provoked us to anger & wrath*: quoted in ibid., 69.

You know, and we know: quoted in ibid., 95.

Colonial writers understood: ibid., 74.

370 *In Narraganset, not one House*: quoted in ibid., 71.

371 *meat to the people*: quoted in ibid., 174.

took off the Jaw: quoted in ibid.

372 *The scene is truly savage*: quoted in Stilgoe, 173.

374 *He, who would wish to see America*: J. Hector St. John de Crèvecœur, *Letters from an American Farmer* (Oxford: Oxford University Press, 1997). 47, 51–52, 54.

and thus the path is opened: ibid., 54, 15, 27.

375 *I wish for a change*: ibid., 187, 196–99.

376 *Thus shall we metamorphose ourselves*: ibid., 211.

Those who labour in the earth: Jefferson, *Notes on the State of Virginia, Query XIX,* in *Writings*, 290.

You are becoming farmers: Jefferson, "To the Chiefs of the Cherokee Nation," in *Writings*, 561.

377 *I have been pleased to find*: Jefferson in a letter to Lafayette, April 11, 1787, in *Thomas Jefferson: The Farm and Garden Books*, ed. Robert C. Baron (Golden, CO: Fulcrum, 1987), 179.

378 *May 4. the blue ridge*: Jefferson, *The Farm and Garden Books*, 67.

380 *I think our governments will remain*: Jefferson in a letter to James Madison, December 20, 1787, in *Writings*, 918.

There are but two means: Jefferson quoted in Linklater, 213.

381 *Of prospect I have a rich profusion*: Jefferson in a letter to William Hamilton, July 1806, in *The Farm and Garden Books*, 191.

382 *The last 12 miles*: quoted in Lauren Brown, *Grasslands: The Audubon Society Nature Guides* (New York: Alfred A. Knopf, 1985), 38.

The children of the American Revolution: quoted in William Least Heat-Moon, *PrairyErth: A Deep Map* (Boston: Houghton Mifflin, 1991), 11.

When I saw a settler's child: quoted in Brown, 30.

383 *for you see no one*: Herman Melville, *Moby-Dick* (New York: Penguin, 1992.), 379–80.

a perch of poor soil: Linklater, 16.

Northward lys the lott: quoted in ibid., 41.

384 *At the end of 22 yards*: ibid., 77.

6 Chains, 60 links: quoted in Joseph W. Ernst, *With Compass and Chain: Federal Land Surveyors in the Old Northwest, 1785–1816* (New York: Arno Press, 1979), 43.

385 "*When you see how easy it is*": quoted in Linklater, 259.

A uniform, invariable shape: ibid., 174.

386 *peopled the Western States*: Jefferson quoted in Wendell Berry, *The Unsettling of America: Culture and Agriculture* (San Francisco: Sierra Club Books, 1997), 144.

388 *We must now place*: Jefferson, *Writings*, 1371.

390 *Emigrants too from the Mediterranean*: Jefferson, *The Farm and Garden Books*, 177.

397 *The unharming sharks*: Melville, 625.

THE NEW CITY

406 *I get along very well*: quoted in *Farm to Factory: Women's Letters, 1830–1860*, ed. Thomas Dublin (New York: Columbia University Press, 1993), 127.

I arrived here safe and sound: "Letters from Susan: letter first," in *The Lowell Offering: Writings by New England Mill Women (1840–45)*, ed. Benita Eisler (New York: Harper and Row, 1980), 46, 49.

407 *It is* very *hard indeed*: quoted in Dublin, 129.

408 *They contemn the calling*: quoted in ibid., 35.

At the time of our voyage: Henry David Thoreau, A *Week on the Concord and Merrimack Rivers* (Orleans, MA: Parnassus Imprints, 1987), 306–7.

410 *For two centuries the river*: J. W. Meader, *The Merrimack River: Its Source and Its Tributaries* (Boston: B. B. Russell, 1869), 289.

411 *In all the region*: Maurice B. Dorgan, *History of Lawrence, Massachusetts* (privately printed, 1924), 39, 41.

412 *The site had advantages*: Peter M. Molloy, *Nineteenth-Century Hydropower: Design and Construction of Lawrence Dam, 1845–1848, Winterthur Portfolio*, vol. 15, no. 4, Winter 1980 (Henry Francis du Pont Winterthur Museum, 1980), 319.

413 *far and near*: J. S. Amherst and C. Adams, *Final Report on the Geology of Massachusetts* (Massachusetts Geological Survey, 1841), 270.

no person shall dig: Arthur W. Brayley, *History of the Granite Industry of New England* (Boston: National Assoc. of Granite Industries, 1913), 13.

I see him there: Robert Frost, "The Mending Wall," in *The Complete Poems*, 48.

414 *As it is from the surface*: Crèvecœur, 16.

416 *You can look through those*: quoted in Marion Knox, "The Glorious Day of Granite," in *Stone Slabs and Iron Men: The Deer Isle Granite Industry* (Stonington, ME: Deer Isle Granite Museum, 1997), 12, 10.

417 *You have to take advantage*: quoted in ibid., 12.

People couldn't imagine: Barbara H. Erkkila, *Hammers on Stone: A History of Cape Ann Granite* (Gloucester, MA: Peter Smith, 1987), 3.

418 *My life and health*: quoted in Dublin, 126.

419 *It is not a home but a tool-box*: *The Report of the Lawrence Survey, 1911* (Andover, MA: White Fund, privately printed), 109.

I used to hear my mother call: quoted in Ardis Cameron, *Radicals of the Worst Sort: Laboring Women in Lawrence, Mas-*

sachusetts, 1860–1912 (Urbana: University of Illinois Press, 1993), 91.

You know how we workers: quoted in ibid., 93.

ISLAND

424 *And I have sent them a shell*: Obed Macy, *History of Nantucket* (Boston: Hilliard, Gray, 1835), 263.

427 *Then it grew a crown*: Henry Moore and John Hedgecoe, *Henry Moore: My Ideas, Inspiration and Life as an Artist* (London: Collins & Brown, 1986), 157.

The landscape is so bleak: ibid., 156.

429 *The open talk that boomed Nantucket*: Florence Bennett Anderson, *A Grandfather for Benjamin Franklin* (Boston: Meador Publishing, 1940), 121.

Liberty in America: D. H. Lawrence, *Studies in Classic American Literature* (New York: Penguin Books, 1977), 13.

431 *The humanization of the Greek wilderness*: Dubos, 454–56.

When first settled by the English: Macy, 9–10.

432 *At the time of the settlement*: ibid., 10.

The soil will not produce: ibid., 112.

433 *a mere hillock*: Melville, 69.

The island being owned: Macy, 22.

434 *Who would have imagined*: Crèvecœur, 86.

They found it so universally barren: ibid., 92.

435 *When they come within our harbors*: quoted in Deetz and Deetz, 248.

In the year 1690: quoted in Macy, 33.

436 *these sea hermits*: Melville, 70–71.

437 *In that gale*: ibid., 116.

and sometimes more than a dozen wrecks: Henry David Thoreau, *Cape Cod* (Princeton: Princeton University Press, 1993), 125.

438 *Another . . . showed me, growing*: ibid., 90.

something green growing: ibid., 130.

When I remarked to an old wrecker: ibid., 126.

The traveler stands: J. H. Merryman, *The United States Life-saving Service–1880* (Grand Junction, CO: Vista Books, 1989), 33.

441 *one could not*: Anton Chekhov, "Easter Eve," in *The Bishop and Other Stories*, trans. Constance Garnett (New York: Ecco Press, 1985), 49.

445 *I don't want pictures*: quoted in John Golding, *Paths to the Absolute: Mondrian, Malevich, Kandinsky, Pollock, Newman, Rothko, and Still* (Princeton: Princeton University Press, 2000), 46.

446 *[T]here has to be*: Adrienne Rich, "When We Dead Awaken: Writing as Re-Vision," in *On Lies, Secrets, and Silence: Selected Prose, 1966–1978* (New York: Norton, 1979), 43.

I am here alone: May Sarton, *Journal of a Solitude* (New York: Norton, 1992), 11.

448 *the left wing of the day*: Melville, 462.

450 *And sometimes, during northeast storms*: Fred Bosworth, *Last of the Curlews* (Washington, DC: Counterpoint Press, 1995), 54–55.

451 *We have built roads*: Peter Dunwiddie quoted in "Uncommon Ground" by Jill Evarts, *Cape Cod Life*, November 1998, 48.

453 *I was always aware that I was treading*: Rachel Carson, *The Edge of the Sea* (Boston: Mariner Books, 1998), 140.

455 *away off shore, more lonely*: Melville, 69.

GRANGE

457 *Have salt in yourselves*: Mark 9:50.

458 *They used to tell you*: John D. Fogg, *Recollections of a Salt Marsh Farmer*, ed. Eric Small (Seabrook, NH: Historical Society of Seabrook, 1983), 40.

459 *On the south side of the path*: ibid., 29.

I would like to tell you: ibid., 33.

460 *One afternoon he was going*: ibid., 65.

461 *They poled enough hay*: ibid., 33.

That was the best part of the stacking: ibid., 33.

464 *Attend to every duty promptly*: *Private Instructions to Officers and Members of the Patrons of Husbandry* (pamphlet, 1889), 20.

465 *To reach the Master's office*: ibid., 10.

470 *We live in a dream world*: George Monbiot, "With Eyes Wide Shut," *Manchester Guardian* (online), August 12, 2003.

473 *First it was a dirt road*: as quoted in Stilgoe, 87.

474 *Big farmers used their government checks*: Elizabeth Becker, "A New Villain in Free Trade: The Farmer on the Dole," *New York Times* (online), August 25, 2002.

In Nebraska: Timothy Egan, "Pastoral Poverty: The Seeds of Decline," *New York Times*, December 8, 2002, 3.

In the complicated equation: Monica Davey, "A Farmer Kills Another and Town Asks, How Did It Come to This?," *New York Times*, October 23, 2003, A20.

475 *We are a cross-section of the entire world*: Paul Auster, "The City and the Country," *New York Times* (online), September 11, 2002.

478 *We heard a distant tapping*: Edwin Muir, "The Horses," in *Selected Poems of Edwin Muir*, ed. T. S. Eliot (London: Faber and Faber, 1965), 86.

481 *sounding like the scream of a hawk*: Thoreau, *Walden*, 161–63.

482 *In 1836 there were in the garden*: Henry David Thoreau, "Wild Apples," in *The Natural History Essays* (Salt Lake City: Gibbs Smith, 1980), 204.

485 *The site of modern meat production*: Madeline Drexler, *Secret Agents: The Menace of Emerging Infections* (Washington, DC: Joseph Henry Press, 2002), 86.

WILDERNESS

487 *Those sublimer towers*: Melville, 209.

Some part of the beholder: Henry David Thoreau, *The Maine Woods* (Princeton: Princeton University Press, 1972), 64.

493 *This was that Earth*: ibid., 70–71.

496 *Though in many of its aspects*: Melville, 211.

498 *It is desperately clean work*: quoted in C. Francis Belcher, *Logging Railroads of the White Mountains* (Boston: Appalachian Mountain Club, 1980), 132.

499 *an area where the earth*: Wilderness Act, Public Law 88–577, Eighty-eighth Congress, S-4, September 3, 1964.

501 *In 1965, the trail was perfect*: quoted in "Challenges Taking Root on Appalachian Trail," *Boston Globe*, November 2, 1988, Metro section, B-3.

GHOST COUNTRIES

505 *The death of my father*: Thomas Merton, *The Seven Storey Mountain* (New York: Harcourt Brace Jovanovich, 1976), 85.

507 *People have asked me*: The quotations from women workers in Thailand are from the *Bill Moyers NOW* (PBS television series) program on globalization, September 5, 2003.

512 *When they use stone now*: Deer Isle, Maine, stoneworker Bob McGuffie as quoted in Knox, 13.

513 *the pounding of the surf*: Rachel Carson, *The Edge of the Sea* (Boston: Houghton Mifflin, Mariner Books, 1998), 55.

We are so easily baffled: Hugh MacDiarmid, "On a Raised Beach," in *Stony Limits and Scots Unbound and Other Poems* (Edinburgh: Castle Wynd Printers, 1956), 46.

518 *fruit of old trees*: Thoreau, "Wild Apples," 197.

vast straggling: ibid., 209.

Now that they have grafted trees: ibid.

522 *the inexpressible privacy*: Thoreau, "The Natural History of Massachusetts," in *The Natural History Essays*, 4.

PERMISSIONS

FIVE THOUSAND DAYS LIKE THIS ONE

Excerpts from *The Last Generation: Work and Life in the Textile Mills of Lowell, Massachusetts, 1910–1960*, by Mary H. Blewett (Amherst: University of Massachusetts Press, 1990), are reprinted by permission of University of Massachusetts Press. Copyright © 1990 by the University of Massachusetts Press.

Excerpts from *Amoskeag*, by Tamara K. Hareven and Randoph Langenbach. Copyright © 1978 by Tamara Hareven.

Excerpt from the taped interview of Ernie Russell from the Oral History Collection of the Lawrence History Center (formerly the Immigrant City Archives) is reprinted by permission of the Lawrence History Center.

Excerpt from "At the Fishhouses" from *POEMS* by Elizabeth Bishop. Copyright © 2011 by The Alice H. Methfessel Trust. Publisher's Note and compilation copyright © 2011 by Farrar, Straus and Giroux. Reprinted by permission of Farrar, Straus and Giroux. All Rights Reserved.

CLEARING LAND

ACKNOWLEDGMENTS

HERE AND NOWHERE ELSE

Acknowledgment is due to the following publications where some of these pieces first appeared: *Fiction*, *The Georgia Review*, *The Gettysburg Review*, *Merrimack: A Poetry Anthology*, *New England Review*, *The Ohio Review*, *Salamander*, and *Taos Review*.

I'd like to thank the National Endowment for the Arts and the Massachusetts Cultural Council for literature grants, which gave me time to complete this book. Thanks also to the MacDowell Colony, Yaddo, the Millay Colony, VCCA, the Ucross Foundation, and the Banff Centre for providing quiet places to work.

Over the years many friends and teachers have given thoughtful attention to parts of this book, and to them I owe my deepest gratitude. Thanks in particular to Linda Hess, Kathy Aponick, E. F. Weisslitz, Paul Marion, and Maggie Brox for their encouragement, and to Martha Rhodes and Alan Brown for their attention to an earlier version of this manuscript. A debt of gratitude to Michael Ryan for first suggesting a form for these pieces, and to Thomas Lux and Heather McHugh for their guidance. Finally, I'd like to thank Deanne Urmy at Beacon Press for her clear-eyed intelligence and her unstinting support.

FIVE THOUSAND DAYS LIKE THIS ONE

Acknowledgment is due to the publications where some of these essays appeared: *The Georgia Review* originally published "Afterwards," "Baldwins," and "Influenza 1918," which was later reprinted in *Best American Essays*, 1996. *The Georgia Review* also published the first serializations of "Twilight of the Apple Growers" and (under the title "The Wilderness North of the Merrimack") an adaptation of "The Quality of Mercy" and "By Said Last Named Land." "Last Look" was originally published in a slightly different form as "The New Pruner" in *Orion*.

I'd like to thank the staff at the Immigrant City Archives of the Lawrence History Center Oral History Collection, in Lawrence, Massachusetts, in particular Ken Skulski and Mary Armitage, for their invaluable guidance. Thanks also to Sarah Blake and Kathy Aponick for their close attention to the manuscript in its final form, to Elizabeth Brown for her help with musical research and her attention to detail, and to Blake Tewksbury for the loan of his family's Farmer's Diary.

Gratitude, always, to the MacDowell Colony for providing me with a quiet place to work, and to Stanley Lindberg at *The Georgia Review* for his continued support over the years. Special thanks and gratitude to Deanne Urmy at Beacon Press for her clear-eyed faith in these pages.

CLEARING LAND

I'm grateful to the MacDowell Colony for providing a place to work during the years it took to complete this book. Gratitude, also, to Blue Mountain Center and Wellspring House. Many thanks to my friends for their support, in particular to Sarah Blake for her insight, and to Elizabeth Brown for an early reading of "Squam" and her patient attention to detail. Thanks to Deanne Urmy, always, for her continued faith, to Cynthia Cannell for her unstinting efforts on behalf of my work, and to Becky Saletan at North Point Press, whose intuition and precision have guided these pages all along.

IN THE MERRIMACK VALLEY

It's been an unexpected delight to see *Here and Nowhere Else*, *Five Thousand Days Like This One*, and *Clearing Land* collected together in one volume. I'm deeply grateful to Joshua Bodwell at Godine for his abiding interest in my work, and his enthusiasm for *In the Merrimack Valley*. I am fortunate to count Suzanne Berne as a dear friend, and I thank her wholeheartedly for her foreword. I couldn't have asked for a more thoughtful, graceful, encompassing meditation on these pages.

A NOTE ABOUT THE AUTHORS

Jane Brox's *In the Merrimack Valley: A Farm Trilogy* gathers together her first three books about her family's farm: *Here and Nowhere Else*, which won the L.L. Winship/PEN New England Award; *Five Thousand Days Like This One*, which was a finalist for the National Book Critics Circle Award; and *Clearing Land: Legacies of the American Farm*. She is also the author of *Silence: A Social History of One of the Least Understood Elements of our Lives* and *Brilliant: The Evolution of Artificial Light*. Her essays have appeared in many journals and have been included in *Best American Essays* and other anthologies. She has been awarded grants from the Guggenheim Foundation and the National Endowment for the Arts. She lives in Brunswick, Maine.

Suzanne Berne is the author of five novels, including the Orange Prize–winning *A Crime in the Neighborhood*. She has written frequently for the *New York Times* and the *Washington Post*, and published essays and articles in numerous magazines. For many years she taught creative writing, first at Harvard University and then at Boston College. Berne lives outside of Boston.

A NOTE ON THE TYPE

In the Merrimack Valley has been set in Caslon. This modern version is based on the early-eighteenth-century roman designs of the British printer William Caslon I, whose typefaces were so popular that they were employed for the first setting of the Declaration of Independence, in 1776. Eric Gill's humanist typeface Gill Sans, from 1928, has been used for display.

Book Design & Composition by Tamsyn Leigh Bodwell

GODINE NONPAREIL

Celebrating the joy of discovery with books bound to be classics.

Godine's Nonpareil paperback series features essential works by great authors—from stand-alone books of nonfiction and fiction to collections of essays, stories, interviews, and letters—introduced by celebrated contemporary voices who have deep connections to the featured authors and their trove of work.

ANN BEATTIE More to Say: Essays & Appreciations
Selected and Introduced by the author

HENRY BESTON Herbs and the Earth
Introduction by Roger B. Swain and Afterword by Bill McKibben

JANE BROX In The Merrimack Valley: A Farm Trilogy
Introduction by Suzanne Berne and Afterword by the author

GUY DAVENPORT The Geography of the Imagination: Forty Essays
Introduction by John Jeremiah Sullivan

ANDRE DUBUS The Lieutenant: A Novel
Afterword by Andre Dubus III

MAVIS GALLANT Paris Notebooks: Essays & Reviews
Foreword by Hermione Lee

LAURIE LEE Cider with Rosie
Introduction by Simon Winchester

WILLIAM MAXWELL The Writer as Illusionist: Uncollected & Unpublished Work
Selected and Introduced by Alec Wilkinson

JAMES ALAN MCPHERSON On Becoming an American Writer: Nonfiction & Essays
Selected and Introduced by Anthony Walton

BHARATI MUKHERJEE Darkness: Stories
Introduction by the author and Afterword by Clark Blaise

ROBERT OLMSTEAD Stay Here with Me: A Memoir
Introduction by Brock Clarke

ADELE CROCKETT ROBERTSON The Orchard: A Memoir
Foreword by Betsy Robertson Cramer and Afterword by Jane Brox

ALISON ROSE Better Than Sane: Tales from a Dangling Girl
Introduction by Porochista Khakpour

ALEC WILKINSON Midnights: A Year with the Wellfleet Police
Foreword by William Maxwell and Afterword by the author

ALEC WILKINSON Moonshine: A Life in Pursuit of White Liquor
Introduction by Padgett Powell and Afterword by the author

GEOFFREY WOLFF Providence: A Novel
Afterword by Ian Maxtone-Graham

MONICA WOOD Any Bitter Thing: A Novel
Introduction by Cathie Pelletier